SEEKING HIGHER GROUND

THE CRITICAL BLACK STUDIES SERIES
INSTITUTE FOR RESEARCH IN AFRICAN-AMERICAN STUDIES
COLUMBIA UNIVERSITY

Edited by Manning Marable

The Critical Black Studies Series features readers and anthologies examining challenging topics within the contemporary black experience—in the United States, the Caribbean, and Africa, as well as across the African Diaspora. All readers include scholarly articles originally published in the acclaimed quarterly interdisciplinary journal *Souls*, published by the Institute for Research in African-American Studies at Columbia University. Under the general editorial supervision of Manning Marable, the readers in the series are designed both for college and university course adoption, as well as for general readers and researchers. The Critical Black Studies Series seeks to provoke intellectual debate and exchange over the most critical issues confronting the political, socioeconomic, and cultural reality of black life in the United States and beyond.

Titles in this series published by Palgrave Macmillan:

Racializing Justice, Disenfranchising Lives: The Racism, Criminal Justice, and Law Reader
Edited by Manning Marable, Ian Steinberg, and Keesha Middlemass

Seeking Higher Ground: The Hurricane Katrina Crisis,
Race, and Public Policy Reader
Edited by Manning Marable and Kristen Clarke

Transnational Blackness: African Americans Navigating the Global Color Line
Edited by Manning Marable and Vanessa Agard-Jones

The Islam and Black America Reader
Edited by Manning Marable and Hisham Aidi

The New Black History: The African-American Experience since 1945 Reader
Edited by Manning Marable and Peniel Joseph

Beyond Race: New Social Movements in the African Diaspora
Edited by Leith Mullings

The Black Women, Gender, and Sexuality Reader
Edited by Manning Marable

Black Intellectuals: The Race, Ideology, and Power Reader
Edited by Manning Marable

SEEKING HIGHER GROUND
THE HURRICANE KATRINA CRISIS, RACE, AND PUBLIC POLICY READER

Edited by Manning Marable and Kristen Clarke

palgrave
macmillan

First published in 2008 by
PALGRAVE MACMILLAN™
175 Fifth Avenue, New York, N.Y. 10010 and
Houndmills, Basingstoke, Hampshire, England RG21 6XS.
Companies and representatives throughout the world.

PALGRAVE MACMILLAN is the global academic imprint of the Palgrave Macmillan division of St. Martin's Press, LLC and of Palgrave Macmillan Ltd. Macmillan® is a registered trademark in the United States, United Kingdom and other countries. Palgrave is a registered trademark in the European Union and other countries.

ISBN-13: 978-1-4039-8396-1 (hardcover)
ISBN-10: 1-4039-8396-8 (hardcover)
ISBN-13: 978-1-4039-7779-3 (paperback)
ISBN-10: 1-4039-7779-8 (paperback)

Library of Congress Cataloging-in-Publication Data is available from the Library of Congress.

A catalogue record of the book is available from the British Library.

This volume includes articles which originally appeared in *Souls*, published by the Institute for Research in African-American Studies, Columbia University.

Design by Scribe, Inc.

First edition: January 2008

10 9 8 7 6 5 4 3 2 1

Printed in the United States of America.

CONTENTS

ILLUSTRATIONS

SEEKING HIGHER GROUND: RACE, PUBLIC POLICY AND THE HURRICANE KATRINA CRISIS

MANNING MARABLE

UNQUESTIONABLY, THE SEPTEMBER 2005 HURRICANE KATRINA was the largest natural disaster in U.S. history. Yet, contrary to the assertions of President George W. Bush that no one could have "anticipated the breach of [New Orleans's] levees" and the massive flooding and destruction of one of America's historic cities in the wake of a major hurricane, the catastrophe we have witnessed was widely predicted for decades.[1] A 2002 special report of the New Orleans *Times-Picayune*, for example, warned, "It's only a matter of time before South Louisiana takes a direct hit from a major hurricane. . . .Levees, our best protection from flooding, may turn against us." The *Times-Picayune* predicted that such a disaster might "decimate the region" from flooding and that in New Orleans, "100,000 will be left to face the fury."[2] That same year, in a *New York Times* editorial, writer Adam Cohen predicted coldly, "If the Big One hits, New Orleans could disappear." A direct major hurricane strike, Cohen estimated, would certainly force Lake Pontchartrain's waters "over levees and into the city. . .there could be 100,000 deaths." Thousands "could be stranded on roofs, surrounded by a witches' brew of contaminated water."[3]

A natural disaster for New Orleans was statistically inevitable. But what made the New Orleans tragedy an "unnatural disaster" was the Federal government's gross incompetence and indifference in preparing the necessary measures to preserve the lives and property of hundreds of thousands of its citizens. The Federal Emergency Management Agency (FEMA), established in 1979, has been plagued for years with financial mismanagement, administrative incompetence, and cronyism.

The litany of FEMA's bureaucratic blunders has been amply documented: its insistence that vital supplies of food, water, and medical aid were impossible to deliver to thousands of people stranded at New Orleans's downtown Morial Convention Center, though entertainers and reporters easily reached the site; its inability to rescue thousands of residents marooned on the roofs and in flooded houses for days; the failure to seek deployment of active duty troops in large numbers until three days after Hurricane Katrina struck the Gulf Coast region. But the incompetence goes deeper than that. FEMA Director Michael Brown actually instructed fire departments in Louisiana, Mississippi, and Alabama not to send emergency vehicles or personnel into devastated areas unless local or state officials communicated specific requests for them—at a time when most towns and cities lacked working telephones, fax machines, and internet access. Florida's proposal to send 500 airboats to assist rescue efforts was blocked by FEMA. Thousands of urgently needed generators, communications equipment, and trailers and freight cars of food went undelivered for weeks. Meanwhile, hundreds of dead bodies floated in New Orleans's streets and rotted in desolated houses. Millions of desperate Americans who attempted to phone FEMA's 800 telephone number for assistance heard recorded messages that all lines were busy or were disconnected.[4]

Even before Katrina struck, it was obvious that the overwhelming majority of New Orleans residents who would be trapped inside the city to face the deluge would be poor and working-class African Americans, who comprised nearly 70 percent of the city's population. As the levees collapsed and the city's Ninth Ward flooded, tens of thousands of evacuees were herded into the Superdome and Convention Center, where they were forced to endure days without toilets and running water, food, electricity, and medical help. Hundreds of black evacuees seeking escape on a bridge across the Mississippi River were confronted and forcibly pushed back into the city. One paramedic witnessing the incident stated, "I believe it was racism. It was callousness, it was cruelty."[5]

As the media began to document this unprecedented tragedy, the vast majority of New Orleans's victims were "the faces at the bottom of America's well—the poor, black and disabled," as reporters Monica Haynes and Erv Dyer of the *Pittsburgh Post-Gazette* observed. "The indelible television images of mostly black people living in subhuman conditions for nearly a week have prompted some to ask whether race played a role in how quickly or how not-so-quickly federal and state agencies responded in [Katrina's] aftermath."[6]

However, much of the media coverage cruelly manipulated racist stereotypes in its coverage. In one well-publicized example, the Associated Press released two photographs of New Orleans residents, wading through chest-deep water, carrying food obtained from a grocery store. The whites were described as carrying "bread and soda from a local grocery store" that they found; the black man pictured was characterized as having "loot[ed] a grocery store."[7] A London Financial Times reporter, on September 5, 2005, declared New Orleans had become "a city of rape" and "a war zone," with thousands subjected to "looting" and "arson."[8] Administrators in Homeland Security and FEMA justified their lack of emergency aid by claiming that they had not anticipated that "people would loot gun stores. . .and shoot at police, rescue officials and helicopters." The flood of racialized images of a terrorized, crime-engulfed city prompted hundreds of white ambulance drivers and emergency personnel to refuse to enter the New Orleans disaster zone. Television reports locally and nationally quickly proliferated false exposés about "babies in the Convention Center who got their throats cut" and "armed hordes" highjacking ambulances and trucks. Baton Rouge's mayor Kip Holden imposed a strict curfew on its facility that held evacuees, warning of possible violence by "New Orleans thugs."[9] That none of these sensationalized stories were true hardly mattered: as Matt Welch of *Reason* magazine noted, the "deadly bigotry" of the media probably helped to "kill Katrina victims."[10]

The terrible destruction of thousands of homes and businesses, and relocation of over one million New Orleans and Gulf area residents, was perceived as a golden opportunity by corporate and conservative political elites who had long desired to "remake" the historic city. Even before the corpses of black victims had been cleared from New Orleans's flooded streets, corporations closely associated with George W. Bush's administration secured noncompetitive, multibillion dollar reconstruction contracts. Brown and Root, a subsidiary of Halliburton, for example, was awarded the contract to reconstruct Louisiana and Mississippi naval bases. Bechtel was authorized to provide short-term housing for several hundred thousand displaced evacuees. Shaw, the Louisiana engineering corporation, received lucrative contracts for rebuilding throughout the area. Bush waived provisions of the Davis-Bacon Act, allowing corporations to hire workers below the minimum wage. After Congress authorized over $100 billion for the region's reconstruction, Halliburton's stock price surged on Wall Street.[11] Local corporate subcontractors and developers who directly profited from federal subsidies set into motion plans for what local African Americans feared could quickly become a gentrification removal of thousands of black households from devastated urban neighborhoods.

Behind the plans to "rebuild" New Orleans may be racially inspired objectives by Republicans to reduce the size of the city's all-black voting precincts. About 60 percent of New Orleans's electorate is African American, which normally turns out at 50 percent in local elections. All-white affluent neighborhoods have turnout rates exceeding 70 percent. In the 1994 mayoral race, only 6 percent of the city's white voters supported successful black candidate Marc Morial.[12]

African American political analyst Earl Ofari Hutchinson speculated that "the loss of thousands of black votes" could easily "crack the thirty years of black, and Democratic dominance of City Hall in New Orleans." The seat of black Democrat William Jefferson, who represents the city in Congress, could be in jeopardy. Even more seriously, Hutchinson observed, the massive African American vote in New Orleans in 2000 and 2004 "enabled Democrats to bag many top state and local offices, but just narrowly. A shift of a few thousand votes could tip those offices back to Republicans."[13]

Nationally, most African American leaders, public officials, and intellectuals were overwhelmed and outraged by the flood of racist stereotypes in the media, as well as their government's appalling inaction in failing to rescue thousands of black and poor people. They observed that the most devastated sections of the city were nearly all black and mostly poor. Local blacks had been largely ignored in preparations for evacuating the city.[14] Beverly Wright, the director of Xavier University's Deep South Center for Environmental Justice, expressed the general sentiment of most African Americans by declaring, "I am very angry, and I really, really believe that [the crisis] is driven by race. . . .When you look at who is left behind, it is very disturbing to me."[15] Wright's viewpoint was echoed by many black intellectuals. For example, Harvard professor Lani Guinier observed that in American society "poor black people are the throw-away people. And we pathologize them in order to justify our disregard."[16] Some reporters assigned to the Katrina crisis soon began to reflect these mounting criticisms. *Detroit Free Press* columnist Desiree Cooper drew parallels between the economic devastation of New Orleans and that of Detroit, noting, "The poverty rate in both cities rivals that of Third World nations. So as I watched the hurricane coverage with racism and poverty, creating the perfect storm, I couldn't help but think: If Detroit were underwater, no one would bother to rescue us either."[17]

By mid September, 2005, 60 percent of African Americans surveyed in a national poll believed that "the federal government's delay in helping the victims in New Orleans was because the victims were black." By contrast, only 12 percent of white Americans agreed.[18] In response, the Bush Administration unleashed its black apologists to deny any racial intent of its policies and actions. Secretary of State Condoleezza Rice insisted, "Nobody, especially the President, would have left people unattended."[19] Black conservative ideologue John McWhorter, a senior fellow at the Manhattan Institute, ridiculed the accusations of racism as "nasty, circular, [and] unprovable. . . .It's not a matter of somebody in Washington deciding we don't need to rush [to New Orleans] because they're all poor jungle bunnies anyway."[20]

African Americans were stunned and perplexed by white America's general apathy and denial about the racial implications of the Katrina catastrophe. On a nationally televised fundraiser for the hurricane's victims, rap artist Kanye West sparked controversy by denouncing "the way America is set up to help the poor, the black people, the less well off as slow as possible."[21] Blacks were especially infuriated with the descriptions of poor black evacuees as "refugees" by officials and the media. Black Congresswoman Diane Watson protested vigorously, "'Refugee' calls up to mind people that come here from different lands and have to be taken care of. . . .These are American citizens."[22] But the racial stigmatization of New Orleans's outcasts forced many African Americans to ponder whether their government and white institutions had become incapable of expressing true compassion for the suffering of their people. Prominent Princeton professor Cornel West, at a Columbia University forum sponsored by the Institute for Research in African-American Studies, pondered whether "black suffering is required for the preservation of white America."[23]

West's provocative query ought to be explored seriously. The U.S. government and America's entire political economy were constructed on a racial foundation. Blacks were excluded by race from civic participation and voting for several hundred years; they were segregated into residential ghettoes, denied credit and capital by banks, and relegated to the worst jobs for generations. Over time, popular cultural and social attitudes about black subordination and white superiority were aggressively reinforced by the weight of discriminatory law and public policy. Psychologically, is the specter of black suffering and death in some manner reaffirming the traditional racial hierarchy, the practices of black exclusion and marginalization?

Even before Katrina's racial debate had receded from the media, the question of racial insensitivity was posed again by former Reagan Education Secretary William Bennett's remarks in a national radio broadcast. In early October 2005, Bennett announced to his radio audience: "I do know that it's true that if you wanted to reduce crime, you could—if that were your sole purpose—you could abort every black baby in this country, and your crime rate would go down." Perhaps covering his racial gaffe, Bennett immediately added, "That would be an impossible, ridiculous and morally reprehensible thing to do, but your crime rate would go down." *New York Times* columnist Bob Herbert interpreted Bennett's remarks as the central aspect of the Republican Party's "bigotry, racially divisive tactics and outright anti-black policies. That someone who's been a stalwart of that outfit might muse publicly about the potential benefits of exterminating blacks is not surprising to me at all. . .Bill Bennett's twisted fantasies are a malignant outgrowth of our polarized past."[24] Bennett's repugnant statements, combined with most white Americans' blind refusal to recognize a racial tragedy in New Orleans, illustrate how deeply rooted racial injustice remains in America.

Has the public spectacle of black suffering and anguish evolved into what might be defined as a "civic ritual," reconfirming the racial hierarchy with blackness permanently relegated to a subordinate status? During the summer of 2005, the U.S. Senate seemed to confirm Cornel West's hypothesis as it was forced to confront the civic ritual of lynching. Between 1882 and 1927, over 3,500 blacks were lynched in the United States, about 95 percent in the South. An unknown number of additional African Americans were killed, especially in rural and remote areas where we have few means to reconstruct these crimes.

In Marion, Indiana, on August 7, 1930, a massive white mob stormed the jail in the local county courthouse, seizing two incarcerated African American teenagers, Thomas Shipp and Abram Smith, who had been accused of raping a white woman. Within less than an hour, a festive gathering of several thousand white women and men armed with baseball bats, crowbars, and guns beat and then lynched the two black boys. A photograph of the Marian lynching was reproduced in my book co-authored with Leith Mullings, *Freedom*. It depicts smiling young adults, a pregnant woman, teenage girls, and a middle-aged man, pointing proudly to one of the dangling corpses.[25]

A third young African American, a sixteen-year-old shoeshine boy named James Cameron, was also seized and beaten by the mob that night. Several men lifted Cameron up, and a noose was slipped around his neck. Just at that moment, a local white man in the crowd pushed forward and declared that young Cameron was innocent. Years later, on June 13, 2005, speaking at a U.S. Senate new conference, ninety-one-year-old James Cameron recalled: "They took the rope off my neck, those hands that had been so rough and ready to kill or had already killed, they took the rope off my neck, and they allowed me to start walking and stagger back to jail, which was just a half-block away."[26] Cameron, the only known survivor of an attempted lynching, had come to the Capitol as part of an effort to obtain a formal apology from the Senate for its historic refusal to pass federal legislation outlawing lynching. For decades, Southern senators had filibustered legislative attempts to ratify anti-lynching legislation, denouncing such bills as an unnecessary interference with states' rights. Prompted by the emotional testimony of Cameron and the family members and descendants of lynching victims, the Senate finally issued an apology for lynching—the first time in U.S. history that Congress has acknowledged and expressed regret for historical crimes against African Americans—in a formal resolution. What was most significant, perhaps, was that only eighty-five of the one hundred U.S. senators had co-sponsored the resolution when it came up for a voice vote. The fifteen senators who did not initially co-sponsor the bill were Republicans. Belatedly, seven senators subsequently signed an oversized copy of the senate's anti-lynching resolution that was to be publicly displayed. The eight senators who still refused to concede an apology were Lamar Alexander (R-Tennessee); Thad Cochran (R-Mississippi), John Cornyn (R-Texas), Michael Enzi (R-Wyoming), Judd Gregg (R-New Hampshire), Trent Lott (R-Mississippi), John Sununu (R-New Hampshire), and Craig Thomas (R-Wyoming).[27]

Why the steadfast refusal to acknowledge the forensic evidence and the obvious human pain and suffering inflicted not only on the victims of racist violence but upon their descendants? Because, in a racist society—by this I mean a society deeply stratified with "whiteness" defined at the top and "blackness" occupying the bottom rungs—the obliteration of the black past is absolutely essential to the preservation of white hegemony, or domination. Since "race" itself is a fraudulent concept, devoid of scientific reality, "racism" can only be rationalized and justified through the suppression of black accounts or evidence that challenges society's understanding about itself and its own past. Racism is perpetuated and reinforced by the "historical logic of whiteness" that repeatedly presents whites as the primary (and frequently sole) actors in the important decisions that have influenced the course of human events. This kind of history deliberately excludes blacks and other racialized groups from having the capacity to become actors in shaping major social outcomes.

In this process of falsification, two elements are crucial: the suppression of evidence of black resistance and the obscuring of any records of white crimes and exploitation committed against blacks as an oppressed group. In this manner, white Americans can more easily absolve themselves of the historical responsibility for the actions of their great-grandparents, grandparents, and parents—and of themselves. Thus the destructive consequences of modern structural racism that can be easily measured by social scientists within contemporary U.S. society today, as well as the human suffering we have witnessed in New Orleans—can be said to have absolutely nothing to do with "racism." Denial of responsibility for racism permits the racial chasm in America to grow wider with each passing year.

When the "unnatural disaster" of the New Orleans tragedy of race and class is examined in the context of American structural racism, the denial by many whites of the reality of black suffering becomes clear. It parallels the denial of the Turkish government of the massive genocide of the Armenian population committed by the former Ottoman empire in 1915–1916. It mirrors the repulsive anti-Semitism of those who to this day deny the horrific reality of the Holocaust during World War Two. Until the denial of suffering ceases, there is no possibility of constructing meaningful, corrective measures for addressing the racial chasm that continues to fracture the foundations of democratic life and a truly civil society in America.

NOTES

1. Ted Steinberg, "A Natural Disaster, and a Human Tragedy," *Chronicle of Higher Education* 52, no. 5 (September 23, 2005), 811–12.

2. "Washing Away," New Orleans *Times-Picayune*, June 23–27, 2002.

3. Adam Cohen, "If the Big One Hits, New Orleans Could Disappear," *The New York Times*, August 11, 2002. Also see Jon Nordheimer, "Nothing's Easy for New Orleans Flood Control," idem., April 30, 2002.

4. Editorial, "Truly Clueless at FEMA," *Boston Herald*, September 8, 2005; Editorial, "Political Appointments, Loss of Focus, Crippled Disaster Relief Agency," *USA Today*, September 8, 2005; Tina Susman, "FEMA: Effort Mired in Bureaucratic Hash," *Newsday*, September 11, 2005; Jonathan S. Landay, Alison Young, and Shannon McCaffrey, "Was FEMA's Brown the Fall Guy?" *Seattle Times*, September 14, 2005; Angie C. Marek, Edward T. Pound, Danielle Knight, Julian E. Barnes, Judd Slivka and Kevin Whitelaw, "A Crisis Agency in Crisis," *U.S. News and World Report*, September 19, 2005; Editorial, "FEMA: Just a Money Pit?" *Hartford Courant*, September 23, 2005.

5. Andrew Buncombe, "'Racist' Police Blocked Bridge and Forced Evacuees Back at Gunpoint," *Independent* [London], September 11, 2005.

6. Monica Haynes and Erv Dyer, "Black Faces Are Indelible Image of Katrina," *Independent*, September 4, 2005.

7. Aaron Kinney, "'Looting' or 'Finding'?" *Salon*, Serptember 1, 2005, http://dir.salon .com/story/news/feature/2005/09/01/photo_controversy/index.html.

8. Guy Dinmore, "City of Rape, Rumour and Recrimination," *Financial Times* [London], September 5, 2005.

9. David Caruso, "Disaster Official at NY Symposium: Planners Didn't Anticipate Gun Problem after Katrina," *Newsday*, September 12, 2005.

10. Matt Welch, "The Deadly BigoTry of Low Expectations? Did the Rumor Mill Help Kill Katrina Victims?" *Reason Online*, http://www.reason.com/links/links090605.shtml.

11. Katherine Griffiths, "Firms Linked with Bush Get Katrina Clean-Up Work," *The Independent* [London], September 17, 2005; Scott Van Voorhis, "Katrina Boon to Builders," *Boston Herald*, September 6, 2005.

12. Coleman Warner, "Primary Turnout Makes Black Vote Crucial in Runoff," New Orleans *Times-Picayune*, February 7, 1994.

13. Earl Ofari Hutchinson, "Katrina Wallops Black Voters," available from hutchinson report@aol.com.

14. Jonathan Curiel, "Disaster Aid Raises Race Issue; Critics Say Poor Blacks Not Considered in Planning for Emergencies, Evacuations," *San Francisco Chronicle*, September 3, 2005.

15. Alex Tzon, "Katrina's Aftermath: Images of the Victims Spark a Racial Debate; Some Say Authorities' Response Time Is Affected by the Victims' Skin Color," *Los Angeles Times*, September 3, 2005.

16. Lynne Duke and Teresa Wiltz, "A Nation's Castaways: Katrina Blew In, and Tossed Up Reminders of a Tattered Racial Legacy," *Washington Post*, September 4, 2005.

17. Desiree Cooper, "Outrage, Caring Mix in Katrina Response," *Detroit Free Press*, September 15, 2005.

18. CNN, *USA Today*, and the Gallup poll released September 13, 2005, cited in Ibid. Other opinion polls confirmed that most black Americans believed that racism was behind the federal government's inaction to aid Katrina's victims. A Pew Institute poll, for example, indicated that 66 percent of blacks surveyed "felt the government would have reacted faster if the stranded victims had been mainly white than black." See Alex Massie, "Racial Tensions Simmer as Blacks Bear Brunt of Slow Official Response," http://www.Scotsman.com/?id=1920892005.

19. Elisabeth Bumiller, "Gulf Coast Isn't the Only Thing Left in Tatters; Bush's Status with Blacks Takes a Hit," *The New York Times*, September 12, 2005.

20. Duke, "A Nation's Castaways."

21. Kanye West, quoted on *The O'Reilly Factor*, Fox News Network, September 8, 2005.

22. Robert E. Pierre and Paul Farhi, "'Refugee': A Word of Trouble," *Washington Post*, September 7, 2005.

23. Cornel West, "When Affirmative Action Was White" (remarks, Institute for Research in African-American Studies, Columbia University, New York City, October 1, 2005).

24. Bob Herbert, "Impossible, Ridiculous, Repugnant," *The New York Times*, October 6, 2005.

25. See Manning Marable and Leith Mullings, *Freedom: A Photographic History of the African American Struggle* (London: Phaidon, 2002), 132.

26. Sheryl Gay Stolberg, "Senate Issues Apology Over Failure on Anti-Lynching Law," *The New York Times*, June 14, 2005.

27. "Eight U.S. Senators Decline to Cosponsor Resolution Apologizing for Failure to Enact Anti-Lynching Legislation," *Journal of Blacks in Higher Education Weekly Bulletin*, June 30, 2005; Avis Thomas-Lester, "Repairing Senate's Record on Lynching," *Washington Post*, June 11, 2005.

PART I

POLITICS AND PLACE

THE NEW ORLEANS MAYORAL ELECTION: THE VOTING RIGHTS ACT AND THE POLITICS OF RETURN AND REBUILD

RONALD WALTERS

INTRODUCTION

The first regular post-Katrina election for the city of New Orleans, originally set for February 2, 2006, was washed out following the damage wrought by Hurricane Katrina. An important post-Katrina case study is to analyze the extent to which the hurricanes, which affected the physical condition of so many African Americans, also impacted their fundamental citizenship rights. In particular, it is important to assess whether those rights, especially the right to vote, would be protected given the scope of general issues that encompass the rehabilitation process. Given the state of disorganization, despair, and urgency for some of the displaced New Orleans residents that I encountered in visits to the region, voting in the election was a lower priority relative to other rebuilding issues such as finding housing, obtaining economic subsistence, finding and burying relatives, and other critical issues. Nevertheless, what was at stake in the election was the potential displacement of black political power, not only in the city of New Orleans, but as it affected the status of black presence in the state legislature and in the Congress of the United States as well.

Also at stake in the election was a test of democratic theory—that is to say, whether the selection of the leadership for the city by the electorate would represent a choice for a particular set of desired public policy. Theorist Robert Dahl's arguments have particular relevance to the situation in New Orleans; he has suggested that "[a] good deal of Democratic theory leads us to expect more from national elections than they can possibly provide. . .[i.e., that they] reveal the 'will' of the majority on a set of issues."[1] Rather, he argues, most often the normal result of voting by majorities is that they express the underlying consensus on policy but that the particulars of policy are most often the work of minorities.[2] Thus, what "underlying consensus" was expressed in this case? In post-Katrina New Orleans, the concern was not merely the process by which the votes would be cast, but the presumption among those who advocated for the right to vote that participating in the election would have an empowering effect, as I described in *Freedom Is Not Enough*.[3] In that sense, as the mayoral election addressed the monumental crisis caused by Katrina, the election would be the instrument through which the citizens of New Orleans could make manifest their views about their right to return and rebuild. The election carried with it, then, decisions about who would lead and what would be the priorities in the agenda of their leadership.

Pre-Katrina census data indicate that the city of New Orleans had 462,269 residents prior to the storm. The stakes of this election were dependent upon the nearly impossible task of mobilizing all voters in a city where 70 percent of the population was black and over 200,000 residents remained displaced. The result was that the large number of persons who remain displaced, particularly those in various parts of Louisiana and among forty-four other states, left the city with well under 50 percent of its black population to participate in the election. Thus, issues concerning the status of the black vote immediately drew the specter of voting rights activists given the prohibition in the Voting Rights Act (VRA) that the black vote should not be diluted. Their attention was focused on the extent to which the Louisiana State Legislature and the city would, in pursuing the election, help preserve the ability of blacks to vote, and thus ensure that the voting system would provide fair access to the ballot. By extension, the problem of addressing barriers to vote in New Orleans was set in the context of the actions and attitude of the Bush administration with respect to the manner in which the Justice Department implemented provisions of the Voting Rights Act that were in the process of being reauthorized. This analysis will proceed from the perspective of the black community's understanding of the issues surrounding these elections and the dilemmas involved.

ELECTORAL DISENFRANCHISEMENT

In the uncertainty that characterized whether the post-Katrina municipal election could be conducted in New Orleans in light of extensive hurricane damage, then-Governor Kathleen Blanco announced that the election date would be moved from February to April 22. Soon thereafter, the state legislature met in an emergency session in February to consider the parameters under which the election could be conducted, given the extraordinary circumstances that existed in New Orleans. It passed Act 40, which attempted to establish a new set of procedures for voting in the context of the crisis, but that expanded access only slightly.[4] The leading local election official in New Orleans, Katherine Butler, indicated that of the normal 427 election precincts only 300 were operable and that her usual staff of 3,000 election officials had shrunk by more than half. Moreover, the infrastructure of the city was eroded such that it made transportation, administration, and other vital functions of election administration particularly onerous.[5]

Meanwhile, it became clear both to local officials and outside observers that holding an election under circumstances in which the population had been so widely dispersed presented a unique challenge, and thus a debate developed over whether the election should be delayed further or held as soon as possible. Those who felt that it should be delayed reasoned that it would take some time to determine where voters had been displaced and to reorganize the election system to permit a satisfactory process. On the other hand, those who felt that it should be held sooner believed that the delay would promote conditions under which displaced residents would adjust to their local circumstances outside of the state and become registered voters and permanent residents in those areas. In any case, there was considerable angst and concern in the national civil rights community that black voters would be disenfranchised, as indicated by Bruce Gordon, President of the NAACP, who said: "We are afraid that many African-American voters will be disenfranchised due to unclear directives, misinformation, and acts of omission on the part of those officials charged with ensuring equal access to the polls."[6]

Therefore, what came to be regarded as a "two-track" strategy developed. The first track focused on preparations for the election while attempting to remove the barriers in the process—this plan, established by then-Secretary of State Al Ater's office, concerned absentee voting, first-time voting, validating voter rolls, presenting accurate information to voters, et cetera. This track was led by the NAACP, the NAACP Legal Defense Fund (LDF), National Bar Association, the National Coalition of Black Civic Participation (NCBCP), the Advancement Project, the Lawyers Committee for Civil Rights Under Law, the Congressional Black Caucus (CBC) and the Louisiana Black Legislative Caucus (LBLC). Beginning with a meeting between the NAACP head, Bruce Gordon, and U.S. Attorney General Alberto Gonzalez, these groups pushed to ease the rules by which displaced voters could participate in the election.

Other local and national leaders such as the Reverend Jesse Jackson Sr., the Reverend Al Sharpton, Louisiana State Senator Cleo Fields, and others were concerned however, that the election itself was being held under apparent illegal circumstances.[7] They pointed to the formidable barriers to participating in the election process, including changes in the voting procedures that had not been precleared by the Justice Department in accordance with the requirements of the Voting Rights Act given Louisiana's status as one of the Act's covered states. In this respect, an important mid March meeting was initiated by the Reverend Jackson and Senator Fields with the U.S. Assistant Attorney General and representatives of the Voting Rights Section of the Civil Rights Division to voice concerns about the lack of Justice Department preclearance and the existence of formidable barriers facing voters in the upcoming election. Preclearance had not yet occurred, in violation of the Act's requirements, despite the fact that the Louisiana state legislature had enacted, and the Secretary of State had implemented, a set of changes to voting procedure in

New Orleans. Some of these changes included the merger of voting precincts in the predominantly black sections of the city where the damage was greatest. The response from Justice officials to these issues was a noncommittal and cautious promise to review all the facts before making a finding.

A summary analysis of the barriers that impacted the rights of displaced voters follows:

Demand for Satellite Voting outside Louisiana

As indicated above, the majority of the dispersed residents were relocated to several nearby states, such as Mississippi, Alabama, Texas and Arkansas. However, the State of Louisiana established only ten satellite voting sites, all located inside the state. The prevailing view among black leaders was that satellite voting sites that would capture the votes of a sizeable number of displaced citizens could be mounted in other states. The historical precedent for this demand was that for the first democratic election in Poland, election sites were set up in Chicago, Illinois, and for the December 2005 Iraqi election, out-of-country voting was also allowed to take place within the United States. With respect to the most recent example, the Independent Election Commission of Iraq conducted voting at sites in six cities and Washington, DC. The conditions under which the voting was conducted were even more liberal than in some U.S. states, allowing for same-day voter registration, permitting two among seventeen kinds of identification cards to be used, and permitting those Iraqi citizens held in detention to vote.[8]

Voting Precincts That Are of Questionable Validity

As a result of the hurricanes, many voting precincts were damaged or destroyed, and thus, election officials merged and relocated some voting stations to reflect this situation. Yet, the NAACP regional office in New Orleans also physically reviewed each of the sites selected by the election officials and confirmed that several sites were in unattractive and unusable locations. A review of merged voting precincts by this writer revealed that 60 percent of those moved or merged were in areas largely inhabited by blacks.[9]

Moreover, changes in the voting precincts were not decided upon nor publicized to voters in a manner and in sufficient time that would allow those voters to identify their assigned polling location in advance of the election. Neither was polling place access made easy, given a recent decision by a district court judge that required individuals to observe a state law that required individuals who were monitoring the elections to maintain a distance of 600 feet from the polling station.

The Unavailability of Voter Rolls

The Federal Emergency Management Agency (FEMA) had at its disposal a list of evacuees that it serviced in various states—a list that also was useful in that it reflected the location of both actual and potential voters. The list had to be cross-referenced or merged with the pre-Katrina voter registration list in order to establish the whereabouts of registered voters, but it was a potentially valuable resource for identifying the universe of eligible voters in this election. However, this list was given by FEMA to the Louisiana Secretary of State Ater, who refused to provide it to city officials. This meant that candidates for various offices were unable to use the list to contact potential voters to campaign, thus denying some potential voters important information about those who were running for various offices. The maladministration and mishandling of the displaced resident list is not only a change in voting procedure that required clearance by the Justice Department, it is also a violation of the National Voter Registration Act, which aimed to make voter participation easier.[10]

The Difficult Procedures of Absentee Voting

Act 40 was passed within a context that sought to extend to displaced voters the kind of benefits that overseas and military voters received, including liberal access to the absentee ballot. However, the process required individuals who would vote by absentee ballot to submit an application that could be obtained from various sources, including the state's Web site. Thus displaced individuals had to apply for the ballot, receive the ballot (which is twenty-five pages long), then mail it back to the Secretary of State or the New Orleans Registrar's Office, whose mailing address had been temporarily moved to Baton Rouge. Given the abnormalities with the mail system in the state, the city of New Orleans announced in March that it would construct "cluster boxes" for the receipt of mail for those areas that experienced considerable storm damage. Nonetheless, persisting problems with mail service may have resulted in some displaced not receiving their ballots in a timely manner. As a result of these problems, absentee ballot request forms were distributed by officials and local community organizations to displaced persons and organizations serving them outside the city and the state.

The Requirement for First-Time Voters to Appear in Person

Louisiana election law requires individuals voting for the first time to vote in person, but allowed those who registered in person to cast their vote by absentee ballot if they submitted an affidavit notarized two witnesses with an understanding that they waived their right to a secret ballot. The problem with arriving in New Orleans in person to register, however, was that under the current financial circumstances of Katrina evacuees, this requirement amounted to a modern day poll tax.

The Placement of Satellite Voting Station inside the State

Satellite voting sites were located within ten cities around the state, but many evacuees who resided in trailers were located in trailer parks outside of the cities, where public transportation was generally not available. Thus, although a displaced resident may have been located within the state, they still may not have had efficient access to polling stations without adequate transportation networks. Therefore, local as well as national organizations such as the National Coalition on Black Civil Participation, the Coalition of Black Trade Unions, and the National Association for the

Advancement of Colored People focused on providing transportation to such sites. Why the state could not mobilize extensive transportation networks is not known.

Given that black residents of the city of New Orleans were found statistically to have experienced the brunt of the hurricane because they were less likely to own private transportation, the stationing of such in-state satellite sites in a context where officials fail to provide transportation should be regarded as a practice that violates Section 2 of the Voting Rights Act, which prohibits actions that dilute the black vote.

THE JUSTICE DEPARTMENT PRE-CLEARS THE ELECTION

Despite the extensive problems outlined above, some of which clearly violate the Voting Rights Act, the Justice Department precleared the election in New Orleans on March 16.[11] As described, the state election plan called for sending mass mailings to evacuees to inform them of voting procedures, such as procedures for requesting absentee ballots and the establishment of satellite voting centers in the state. However, that plan did not include the designation of satellite voting centers in the adjacent states of Texas, Mississippi, and Georgia where the majority of the evacuees were deposited. The basis of the Justice Department's decision to preclear the New Orleans election was that the procedures did not amount to "retrogression" of the voting status of blacks (meaning they did not worsen the position of minority voters) and that the state had no affirmative obligation to enhance such opportunities to vote. Their argument was that although "the state may well have done more under the circumstances," because the state did not cause the circumstances that diminished the opportunities for voters to vote, their efforts were sufficient under the law. These arguments, however, received strong objection by individuals such as U.S. Senator John Kerry, who wrote Attorney General Alberto Gonzalez, saying: "I do not understand how the Department of Justice could approve a plan that threatens the civil rights of hundreds of thousands of citizens rather than postponing the upcoming elections until a time when a suitable plan could be developed."[12] The lack of an adequate reply by the Justice Department leaves open the question of why neither the state nor the federal government adopted a more aggressive plan under the circumstances and thus why they abdicated their ultimate responsibility to enhance the quality of democracy under which its citizens ostensibly live.

VOTER MOBILIZATION: THE "TWO-TRACK" STRATEGY

The election was approached by blacks under the two assumptions cited above, which was that organizations would work to assist those who desire to vote by attempting, by legal means, to remove the barriers to voting and by mounting assistance projects to provide information, transportation, and other resources to potential voters, while at the same time mounting "election protection" efforts.

This effort required 50–150 volunteers, who were needed to greet voters coming from both out of state and outside the city and to direct them to appropriate voting locations, that, in most cases, had been established. Many of these volunteers, including those who came from the surrounding colleges such as Xavier, Southern

University, and Dillard, would also man transport systems, child care facilities and feeding stations. Polling station workers would be sent to high-performing voting districts in the Uptown section, Gently and Pontchartrain Park, to distribute information and handle questions about election procedures, especially where there was some doubt about people's voting eligibility status. Volunteers were also going door to door in Algiers and the West Bank of Orleans Parish and other neighboring areas to "pull" voters out. The central base of operations was the Greater Urban League headquarters on Canal Street, which served as the main field office for poll monitors. There was also a legal command center staffed by attorneys to answer questions coming in from the field.[13]

Other organizations including the churches, the National Coalition on Black Civic Participation, the NAACP, and others sponsored transportation for displaced residents wishing to vote—buses of voters came from Atlanta, Baton Rouge, and other cities. These buses were organized through Ebenezer Baptist Church and Greater St. Stephens Full Gospel Ministry, among others.

In order to stimulate voter turnout and to guard against voting rights violations, the Reverend Jesse Jackson organized a march in New Orleans on April 1. It was a spirited affair that drew more than 5,000 people, who marched over a route that included the Gretna Bridge. In the ensuing rally after the march, speakers repeatedly referred to the juxtaposition of the New Orleans election and the pending Voting Rights Act reauthorization sitting before Congress. Reverend Jackson said: "All citizens have the right to protect their right to return. We are looking to establish jobs, training, and contracts. Now it is up to the state and local officials to carry the ball in the state legislature. No matter what we sacrifice today, [this march] is not as long as the march from Selma to Montgomery."[14] The march to Gretna was organized, in part, to highlight the fact that on September 1, 2005, when several hundred black people who were flooded out of their homes fled over that bridge, attempting to access higher ground in the town of Gretna, they were met by armed law enforcement officials who forced them to go back. This march, along with a previous one led by former Georgia Representative Cynthia McKinney, may have created the circumstances that led to the U.S. Department of Justice's decision to review the blockade. At the time of this writing, the state's investigation of the incident had been turned over to the Orleans Parish District Attorney for possible legal action.

At the same time, it was an opportunity for the legal and civil rights organizations that provided the infrastructure of support to meet, and in that setting, they were able to confirm their sphere of operation, tighten cooperation with various groups by clarifying their roles, and discuss overhanging legal problems and logistical considerations. The meetings were not only important but reflected an impressive display of black organizations working together to achieve the common goal of facilitating the right of people to vote.

THE VOTE

When Ray Nagin ran in the 2002, he did so as a former Republican cable company executive—a political outsider who was relatively unknown but stood out in a crowded field with an anti-corruption, clean-government campaign, and with strong ties to the city's white business establishment. He managed to win the primary election with 29 percent of the vote against Richard Pennington, then went on to defeat him in the runoff 59 to 41 percent.

The 2006 race for Mayor of New Orleans featured twenty-two candidates from various walks of life in the city, including Kimberly Butler, then-Clerk of the Criminal Court and chief elections official. However, the press focused on the two leading candidates, Mayor Ray Nagin and principal challenger Mitch Landrieu, with an outside chance given to Ron Forman, head of the city zoo and an aggressive fund-raiser with strong business connections. Landrieu was the brother of the Mary Landrieu, who is the current U.S. Senator, and son of the last white mayor of New Orleans, Moon Landrieu, whose term ended in 1978. In the April primary election, Nagin received 38 percent of the vote to 28 percent for Landrieu, with Forman running a distant third at 17 percent. In the May 20 runoff, Nagin defeated Landrieu by 52 to 48 percent.[15]

The April 22 election attracted 298,000 voters, 63 percent of whom were black, so the issue at hand was essentially whether enough of the black voters together with a coalition of whites wanted change. That black voters desired change was indicated by the fact that Nagin appeared to garner the lion's share of the black vote even though many of the poorest citizens could not get out of the city and ended up at the Superdome. Then, in an effort to recoup his standing with black voters in the runoff, Nagin suggested that in the reconstruction, New Orleans would become itself, a "chocolate city." The extent of support that Nagin received following Katrina stands in stark contrast to the 2002 election, in which he received a mere 40 percent of black voters support.[16]

It is also possible to suggest that the impact of the black electorate on the selection of leadership in the city was formidable to the extent that Mayor Nagin took into consideration their sensitivity to issues such as the size or "footprint" of the city, by the inclusion of the Ninth Ward in the priority areas for rebuilding. Before the election runoff, Mayor Nagin accepted the recommendations of a reconstruction task-force that he had created, known as the Bring New Orleans Back Commission, that a moratorium be placed on the rebuilding of flood-prone areas, including the lower Ninth Ward. However, opposition to this recommendation forced Nagin to alter his position and ultimately agree that all areas of the City were a priority for rebuilding.[17]

In fact, race played an even stronger role in the runoff election, given the expectation that the election between Landrieu and Nagin would be influenced, in large part, by the proposals for rebuilding that they put forth. Neither candidate, however, made any bold, effective, or clear proposals that could be attacked by the other. Instead, both candidates were purposefully vague, and Landrieu was loath to attack Nagin too sharply for risk of alienating black voters, since his own fortunes depended upon construction of a biracial coalition.[18]

Black voter turnout in the April election was 36 percent or 108,153, a modest performance in light of the 45 percent turnout in the 2002 election, and the estimated 30 percent turnout in the black sections lent strong credibility to the view that blacks had to overcome substantial barriers to vote.[19] Overall turnout in the May runoff election was estimated to be about 43 percent overall and about 50 percent among black voters, a phenomenal number given the obstacles that voters had to overcome. Nagin's voting bloc was comprised of 80 percent of blacks in majority-black voting precincts and 20 percent of the white vote, largely from the Uptown area, while Landrieu won the rest of the city's white precincts.

However, the fact that black voters came back to New Orleans from wherever they were in the country to provide Nagin a winning margin proved that their vote was not so much about Nagin as it was against the attempt to wrest control of the city from black leadership and by so doing, not only to disenfranchise them but to disempower them politically, and thus to affect their permanent displacement. This would give their opponents the opportunity to refashion New Orleans as they wished, free of the domination of a black majority. As such, the black vote was less, then, about Nagin, and more about their own investment in the city, its history and culture, and preserving a future in which they continued to see themselves as vital players through an act of political accountability.

THE ELECTION'S IMPACT ON THE RIGHT TO RETURN

In effect, the post-Katrina election was black New Orleans's modern-day civil rights movement, a moment in time when they chose to upset expectations that they would not turn out strongly and, as such, also upset calculations that the city would be reconstructed without having to take their views under consideration. As a member of the Katrina Survivors Association in Dallas shouted in response to the group's chairwoman Katie Neason's cry of "Let's go do it" as she and others filed out of a bus that brought them to New Orleans to vote, "We're back, we're back. . .and we're going to be at the table."[20] On the other hand, some white conservative votes for Nagin were born of the same calculation as when he was first elected—that he would be more amenable to their control than Landrieu, who had his own base of support through the State legislature and his family's political status Therefore, to support Nagin would both localize city power under their control and do so through the a black leader, thus avoiding rancorous racial politics.

This was something of a miscalculation of the power of black voters, many of whom had characterized Nagin as "Ray Reagan" in his first administration and who were acutely aware of their interests in the rebuilding process. Proof of this is that they have influenced and redefined Nagin's leadership in the post-election period when the contests for control of the plans for rebuilding have been tended both by legislators in Baton Rouge and in the city.

The Louisiana State Legislature created a Louisiana Recovery Authority (LRA), which constituted the vehicle through which funds would be channeled to the city and other areas for the rebuilding process. And, although the LRA was to work with the city at the primary level of planning, it has in fact developed its own policy postures. Thus, the connection between the Bush administration and the LRA is a linkage that has involved the White House in a process governing the control and expenditure of arguably the largest sums of urban redevelopment funds in American history.

Many of the wishes of the grassroots citizens in the city, then, would be dependent upon the behavior of their representatives on the city council, a lineup that was also changed somewhat by the elections. Nagin had governed in his first term with strong support from a coalition that included Oliver Thomas, the highly regarded council chair; Jay Batt, a white businessman; and Jackie Clarkson, an at-large member.[21] However, both Batt and Clarkson were defeated. Among the four members who often opposed Nagin, only one was defeated. The net effect of this was to shore up a level of accountability by the council to citizen demands.

Some evidence of this newfound accountability is made evident by the city council's recent efforts to establish its own planning process, one involving community hearings and deliberations, to create a unified plan for land use, infrastructure enhancement, and wetlands rebuilding. With Ron Forman having lost the election, the wealthier whites in New Orleans felt they had no direct involvement in the planning process and thus sought and attracted the financial support of the Rockefeller Foundation to develop a more comprehensive planning mechanism. However, that mechanism was perceived by grassroots community activists to be a substitute colonial system in which blacks were reduced to subordinate role players—they effectively intervened in the city's attempt to establish a memorandum of understanding between the foundation and the city council.

In another example, the mayor proposed wage increases for police but only first-year firefighters on the logic that firefighters, unlike other city civil service employees, received a 2 percent annual pay increase mandated by state law. However, a majority of the city council voted an increase for all members of the police and fire Departments that was eventually upheld by the mayor.[22]

Finally, in April of 2006, an August 29 deadline was established by the city government for homeowners to clean, gut, and board up damaged homes. However, after the election, the council voted to extend that deadline for several more months, in effect acknowledging the complexities surrounding verification of land ownership, contacting home owners, and completion of other administrative procedures. The council also voted to exempt the entire Lower Ninth Ward from that process, largely in response to the views of activists who argued that the city had allowed derelict homeowners to abdicate their responsibility to clean up their properties before the hurricanes, thus the unusual circumstances should result in a greater degree of leniency.[23]

IMPACT OF THE "TWO-TRACK STRATEGY"

The "two-track strategy" largely worked in this case, first through the mobilization of various interest groups and civic organizations that came together with city officials to express their electoral choices in a surprisingly effective manner. As indicated above, blacks were able to vote in significant numbers not only by physically returning to the city but also by overcoming the difficulties associated with the first-time voting, absentee voting, and provisional balloting procedures. The fact that the state legislature and the state administration did little to greatly ease the voting process in light of the context within which it occurred says much about the political obstacles encountered by displaced voters in a state largely dominated by a Republican establishment (even though a U.S. Senator and the Governor are Democrats). The election also illustrates the continued and effective resistance mounted by a largely black Democratic establishment.

The objections to the election raised by the Reverend Jesse Jackson, the Reverend Al Sharpton, the NAACP, the National Urban League, and others were also important as a backdrop to the federal legislative process that was occurring during Congress's reauthorization of the Voting Rights Act. As indicated above, the Justice Department precleared the election and in doing so said that it was not unusual for the jurisdiction in question to prematurely implement changes in the election procedures while the Justice Department was considering the case.

Unusual actions like this from the Justice Department raised questions about whether or not Republican legislators would attempt to legalize them in the process of the Voting Rights Act reauthorization. Hearings pertinent to reauthorization were conducted by the House Judiciary Committee, under the leadership of Representative F. James Sensenbrenner, from October 25, 2005, to November 9, 2005.[24] Two things occurred to mitigate the decision to bring court action after the election. The first factor was that the election turnout of blacks appeared not to have been damaged to the extent feared, although blacks were disproportionately hurt by the electoral structure. Nevertheless, their vote was large and influential in the final analysis, promoting the feeling that justice had been served by their own actions.

Then, the Voting Rights Act was eventually reauthorized on July 13, 2006, by a vote in the House of Representatives of 390 to 33, with most of the votes in opposition coming from Southern Republicans. Moreover, the amendments to the Act corrected the flaws in the process of preclearance that were the result of the Supreme Court's rulings in *Bossier Parish* and the *Georgia v. Ashcroft* cases that set a confusing standard for evaluating of whether changes in districts harmed or helped minority voters. The Voting Rights Act was subsequently passed in the Senate by a nearly unanimous vote of 98 to 0 and signed into law by President Bush.

CONCLUSION

The danger, however, in the positive impact that voters have had to date on the decisions of the mayor and city council is that the center may not hold. Time is an issue because the further the election and the hurricane's impact recede into the past, if a critical mass of residents does not return, black power will slip ineluctably into the reservoirs of those who are able to wait and profit from the barriers to the right of return. Black New Orleanians may find themselves stripped of the spirit of resistance and turning from their history to something new. The potential exists in that a mayor with a background rooted in the white power structure, still somewhat unsure of the strength of his fidelity to the black community, is still vulnerable to the previous forces that initially produced him.

Moreover, the possibility of the negative pull of time occurring is great also because of the massive barriers to the right to return, rebuild, and restore life posed by the challenges that displaced residents continue to face. This analysis suggests at least three formidable problems. The first is the inequalities of power in the various levels of control of spending and the regulations devised for spending by the White House through FEMA, the state through the Louisiana Redevelopment Authority, and the city through the local planning boards. The FEMA/LRA nexus would appear to be the most important, given the $110 billion that been allotted for

Louisiana reconstruction and the $12 billion allotted for the city of New Orleans itself, while the city government will have to adjust to rules that they design. Those rules doubtless could have the imprimatur of wealthy, locally connected private interests as well. In fact, the second contradiction is the superior ability of private interests to compete for land, political legitimacy, and other resources in order to promote their interests in developing a business-friendly area. That has been the driving force behind the reluctance of city leaders to vigorously support programs that would provide for the speedy return of poor and low-income residents who require affordable housing for purchase and rental. Third and finally, the Bush administration's Justice Department will have responsibility for the implementation of the Voting Rights Act until 2008, and it has exhibited a troubling pattern of laggard enforcement of the law. It has precleared changes in voting procedures such as the Georgia photo identification requirement and changes in other states that have had the capacity to disenfranchise black voters. So, despite the passage of the VRA, its enforcement will have to be monitored. If it is not, the political paradigm in New Orleans could change quickly if voting procedures, approved through the preclearance process, continue to make voter participation difficult. That could promote city officials who could enact legislation that opposes not only the specific issues involved in the right of return but that also has the affect of undoing the right itself.

Since Dahl's time, tools such as public opinion polling have also illustrated with a degree of specificity the wishes of the electorate. In this case, the Iraq-Katrina relationship exists as reflected by a September 2005 CNN/USA Today/Gallup poll, which finds that while 93 percent of respondents say that Katrina is the worst natural disaster in their lifetime, 63 percent say that New Orleans should be rebuilt.[25] However, a November 2005 poll of ethnic groups by *New California Media* found that blacks disapproved of the Iraq war the most of all groups. In particular, 77 percent of blacks, 69 percent of Hispanics, 60 percent of Asians, and a plurality of whites opposed the war. Moreover, blacks also privileged the elimination of poverty by the highest score (blacks, 58 percent; Hispanics, 43 percent; Asians, 38 percent; whites, 36 percent).[26] Nevertheless, a *USA Today* poll also found a twenty-point decline among those who said they "would definitely return" to New Orleans between October of 2005 and August 2006.[27]

So I recommend a minor correction of Robert Dahl's thesis that elections rarely give specific mandates to the winners by suggesting that when the issue faced by voters in an election is large enough, it constitutes a rather specific target, such that the result of the voting may indeed be informative. In this case, the perception of blacks was the specific issue of the extent to which they would be allowed to return and to participate in the life of a reconstructed New Orleans. And the political answer was resoundingly positive.

NOTES

1. Robert Dahl, *A Preface To Democratic Theory* (Chicago: University of Chicago Press, 1956), 131.
2. Ibid, 132–34.
3. Ronald Walters, *Freedom Is Not Enough: Black Candidates, Black Voters and American Presidential Elections* (Lanham, MD: Rowman and Littlefield, 2005).

4. See also Acts 2005, No. 220, para 4, Jan. 1, 2006; Acts 2005, No. 431, para 1, Jan. 1, 2006; Acts 2006, 1st Ex. Sess. No. 2, para 1, Feb. 23, 2006, at the following URLs: http://www.legis.state.la.us/Iss_doc/Iss_house/RS/18/Copy%20of%20RS%2018 and http://www.legis.state.la.us/Iss_doc/Iss_house/RS/18/Doc%2081629.html.

5. Telephone discussion, Ms. Kimberly Butler with the author, February 23, 2006.

6. James Parks, "Homes Gone, Lives Destroyed: Now Katrina Evacuees Try to Keep Right to Vote," PoliticalAffairs.net, April 12, 2006, Blog.aflcio.org.

7. Letter, State Senator Cleo Fields to John Tanner, Voting Section Chief, Civil Rights Division, U.S. Department of Justice, March 7, 2006.

8. "Voting Registration Update: Procedures," *Independent Electoral Commission of Iraq*, http://www.Iraqvote.org/ (accessed June 2005); Idem, "Iraq Election Polling Sites in United States Announced," idem, (accessed December 11, 2005).

9. Clerk of the Criminal Court, Election Administration, "Polling Place Changes," April 22, 2006. Presents changes of April 22 with the status of voting precincts as of March 20. This data was compared to ward racial data from the Department of State, Parish Report of Registered Voters as of March 7, 2006, Orleans Parish. My analysis indicates that of the wards with 99–100 percent changes in voting precincts (Wards 3, 5, 7, 8, 9, 10, 16), five of the seven had majority black populations.

10. "FEMA Voter Data Dispute," *The Institute for Southern Studies Newsletter* (February 23, 2006).

11. Peter Whoriskey, "Election Plan for New Orleans Approved," *The Washington Post*, March 17, 2006.

12. Senator John Kerry, letter to John K. Tanner, Chief, Voting Rights Section, Civil Rights Division, U.S. Department of Justice, April 3, 2006.

13. Author's notes.

14. Author's notes.

15. Mayor City of New Orleans, Louisiana Secretary of State, Parish Election data, 2006.

16. Cathy Booth and Russell McCulley, "Will Ray Nagin Win Redemption in New Orleans?" *Time*, April 10, 2006, online edition.

17. Coleman Warner, "Candidates Playing It Safe on Land Use," New Orleans *Times-Picayune*, May 13, 2006, online edition.

18. Gordon Russell, "Mastermind Puts the Mayors in City Hall; Political Strategist's win Record Will go Down in N.O. History," New Orleans *Times-Picayune*, June 11, 2006, online edition.

19. Mayors Election, Primary Election Results, Department of State, New Orleans Parish, 2002.

20. Gwen Filosa, "Voters Determined to Take a Stand; Many Drive In to Make Sure Voices Are Heard," New Orleans *Times-Picayune*, May 21, 2006, online edition.

21. Bruce Eggler, "Re-elected to Council," New Orleans *Times-Picayune*, April 23, 2006.

22. Bruce Eggler, "Panel Defies Nagin on Raise," New Orleans *Times-Picayune*, August 25, 2006, online edition.

23. Bruce Eggler, "City Council Eases the Gutting Law," New Orleans *Times-Picayune*, August 26, 2006, online edition.

24. House Committee of the Judiciary, Subcommittee on the Constitution, "Voting Rights Act: The Judicial Evolution of the Retrogression Standard," Hearing, November 9, 2005.

25. CNN/USAToday/Gallup, "Poll: Most Americans believe New Orleans Will Never Recover," September 7, 2005.

26. Haider Rizvi, "Post-Katrina Poll Finds Americans Prioritizing Poverty over Terrorism," *Common Dreams News Center*, http://OneWorld.net (accessed November 26, 2005).

27. "USA Today/Gallup Hurricane Katrina Survivors Follow-up Poll," *USA Today*, USAToday.com/news/polls/tables/live/2006-08-20-Katrina-poll.htm.

CHAPTER 2

THE NEW ORLEANS THAT RACE BUILT: RACISM, DISASTER, AND URBAN SPATIAL RELATIONSHIPS

DARWIN BOND GRAHAM

Far down Desire Street, off St. Claude Avenue, the main thoroughfare through the Ninth Ward of New Orleans, Jackie stands outside her home on an unusually cold December day eating a hot lunch provided by Common Ground, a mutual aid organization formed immediately after Katrina struck. The entire contents of her house are piling up on the sidewalk beside her. All of the electronics—television, radio, fan, lamps, VCR, and a pair of heavy electric metal-framed medical beds—are sitting in one stack by the bumper of a wrecked Oldsmobile with four flat tires. Carpeting dragged from her house is sitting in another humungous mound. The largest stack of waterlogged belongings consists of miscellaneous furnishings—clothing, books, CDs, drapes, toys, and kitchen utensils. In a few hours a new pile will rise beside all of this; the mold-infested drywall and plaster that must be torn from her home's interior, floor to ceiling.

Inside Jackie's house there are multiple brownish-yellow lines about two, three, five, and seven feet up on all the walls. Each line marks a level at which the toxic floodwaters settled for several days. The flooding took weeks to subside completely. On its way down it settled at many different levels for extended periods of time leaving graduated marks. The whole building smells of mold and mud. A crew of students from California (of which the author is a member) has been helping Jackie gut her house for several days. They don protective suites, respirators, goggles, and gloves. The entire house interior needs to be bleached and dried out before reconstruction can start, before Jackie can think about moving back in. In the front room of her home she's collected a dozen porcelain sculptures, keepsakes, and as many photographs and personal papers as she's been able to salvage. They're all coated in the same powdery brown film.

A haunting reminder of the floods that overtook New Orleans sits beached in her side yard. Constructed from a refrigerator, empty barrel, several ice chests, and fastened together with metal straps and nails, the improvised boat probably could have seated four to five adults and several children. A wide piece of scrap lumber about four feet long, a paddle, lies inside of it.

Jackie's neighborhood is dead quiet. The twin holly bushes that used to shine bright green with red berries on each side of the stairs up to the front porch are dead and withered. What's left of the lawn is a brown, shaggy mass of dead blades. Jackie's lived here a long time. She says the neighborhood has had more than its fair share of troubles in recent times, but that it's a good place in spite of the bleak pictures painted by demographic statistics, press reports, and sociological works that condemn communities like this as oppressive slums caused purely by spatial concentration and segregation of a poor, predominantly black population. Community cohesion and people's attachment to place have been profoundly powerful in the Ninth Ward (and New Orleans in general). William Falk's research on "family and belonging in a southern black community," articulates the transcendent power of place for those who dwell in geographies like the Ninth Ward, a power that appears capable of nullifying the negativity of oppression and that provides hope, happiness, and a collective means to struggle for those who live there:

> The real way in which one is related to the place is what matters. Thus, for a supposedly "poor" place. . .the real testament of faith for local people is: this is where I live. This is where my history is grounded. My biography is here. I know everyone here, at least everyone who matters the most to me. Here, me and mine are left to ourselves. I may not know much that matters to some other people, but I know well this place and my place in it. It is a place consisting of memories good and bad, but all of them are mine, things that cannot be bought or taken away from me. I am quite literally grounded in this place. It is in me; I am in it.[1]

Although it is poor, segregated, scorned by outsiders, criminalized, redlined, over-policed, and now finally flooded to Noahic proportions, New Orleans remains an assemblage of terrains imbued with so much meaning, nesting so much struggle and power, history and community. That residents of the Ninth Ward like Jackie are returning to rebuild, even without pledges from city leaders that their neighborhood will be provided with sewerage, gas, and electricity, is a testament to the power of place.

But will New Orleans be bought now? Will portions of it be taken away from those who have called it home for so long? "Land grab" is on the tip of every poor New Orleanian's tongue. Will community be taken away from those parts of New Orleans least able to protect themselves, or those neighborhoods most heavily targeted for sacrifice and redevelopment?

According to Jackie, the Ninth Ward "used to be a very white neighborhood when my family first moved in. There were a lot of white families and some elderly folks. My family was probably the first black one to move in on this particular block. Soon after it became all black." Her description matches the basic dynamics that sustain racial segregation; nonwhite urban migration and concentration, spillover into accessible and marginalized neighborhoods, reactionary white flight from many of these integrating areas, and the consequent social-urban ecological decline that results from environmental racism and neglect.[2] The Ninth Ward, where Jackie has lived most of her life, started out as a neighborhood of white-ethnic immigrants and a large number of black families, all working class. The modest homes with ample backyard space, wide streets fit for automobiles, numerous churches, and small corner shops, was the American dream for many. The properties were cheap and prone to minor flooding during a hard rain, but nothing catastrophic, as long as the levees held. The Federal government, state of Louisiana, and the city and levee boards mostly upheld this promise through the twentieth century. The city certainly experienced floods, but nothing on the scale of Katrina. The worst were hurricanes Betsy and Camille, but these were not catastrophic by any means.

Many older residents of the Ninth Ward have memories of Betsy, when their homes were flooded and they had to evacuate. New Orleans City Council member Oliver Thomas, whom Douglas Brinkley interviewed for his book about Katrina, recalled that "as a child, the scariest thing in my life was Hurricane Betsy."[3] Even though Betsy flooded parts of New Orleans, it wasn't anything approaching catastrophic. In the 1960s, it was in everyone's interest to make sure the levees held because back then much more than just the poor could be washed away. Industry, shipping, the growing tourism trade, and the workforce that these economic activities required meant that the levees and other infrastructure were supported in kind.[4]

By the 1960s, the Ninth Ward was fast becoming a segregated black ghetto. By the time Hurricane Betsy flooded parts of the city in several feet of water, the Ninth Ward and many of the most vulnerable neighborhoods had become virtually all black.[5] In 1965, when President Johnson wandered into a church in the Ninth Ward days after Hurricane Betsy, ankle deep in water, he announced "This is your President, I'm here to help you!" However, he saw only black faces staring back at him from the damp flashlight-illuminated interior. Days later he would write in his diary that the shelter he visited was a "mass of human suffering," and "most of those inside were Negro."[6] The differential impact that Betsy had on New Orleans's blacks made a lasting impression, but it ultimately served as little incentive for future investments in the area's infrastructure—to say nothing about redressing the environmental racism that was creating vast regions of vulnerability in and around New Orleans (and beyond). Indeed, global economic restructuring and the concomitant decline of the welfare state would soon lead to the opposite outcome; less social investment would be made in New Orleans's infrastructure. This, combined with racial residential segregation, is part of the cause of the Katrina catastrophe.

One precipitating event that set segregation into full swing in New Orleans was the integration of local schools. New Orleans schools were some of the first in the Deep South to be opened up by civil rights activists. After McDonogh #19 on St. Claude Avenue in the Lower Ninth Ward was integrated, many of the remaining white residents fled the city, believing that if living around blacks wasn't bad enough, having one's child attend school alongside them was the final straw. McDonogh #19 is now known as Louis Armstrong Elementary School. Like other schools across the nation that were desegregated, many in New Orleans took on the names of prominent black political leaders and artists. One can only imagine what Armstrong would think of this.

The school's original namesake came from John McDonogh, a wealthy local slave owner and businessman who made a point of educating his slaves and financing a recolonization society that transported black New Orleanians to the newly established nation of Liberia. On his deathbed he willed enough money to open dozens of schools in the greater New Orleans area. Some still bear his name. Most were the sites of civil rights struggles aimed at integration. William Frantz Elementary is in the same neighborhood that Jackie lives in. It was here in 1960 that six year-old Ruby Bridges and her mother drove to school one morning. They were greeted by an angry mob of parents. Up until that date the school had been reserved for whites only.[7] It wasn't long till the school, and the neighborhood, became nearly all black.[8] Many of the McDonogh high schools went through the same process (although some of them were open to blacks much earlier). That McDonogh's name is so prominent in the greater New Orleans educational system tells us something about education and racial inequality in Louisiana. The man worked his slaves in order to build up real estate around the city and Algiers, and some of these properties would be turned into schools for slaves, the emancipated, and the poor.[9]

McDonogh was a peculiar but also typical benevolent white patriarch in the Louisiana slavocracy. Leighton Ciravolo, a sympathetic biographer of McDonogh, describes his fatherly ante bellum style in the following terms: "McDonogh, being a deliberate thinker, realized that people work best when there is hope ahead. This continuous labor builds up a work ethic. He in fact treated his slaves more as students than like the property they were considered legally. The philosophy involved is a modern, fundamental idea of proper education: if students see prospects in their lives, they will develop self-esteem and will become energetic, useful members of society."[10]

McDonogh's efforts to "better" Afro-Americans was in no small part a means toward his re-colonization plans—sending slaves back to Africa. Recolonization was an early attempt by some white Americans to "balance" the races and distribute them to their proper spatial locations. In the first half of the nineteenth century, this meant sending emancipated blacks to found Liberia, a nation where black men and women could live better lives, according to men like McDonogh. It would right the wrong—the wrong not necessarily being slavery but rather the mixing of races in the New World. The proper place for blacks was ultimately viewed as Africa, far from the North American continent, which according to the McDonoghs of the nation was destined to become a white civilization.

This notion of relocating blacks to places where they will supposedly be better off—regardless of what black communities might think for themselves—is a thread that runs throughout the history of race relations in the United States. Although spatial notions of the proper places for certain races took particular forms in New Orleans, somewhat complicated by the city's truly unique history and traditions, the city was by no means an exception.

Exactly where this proper place for the black population was supposed to be would change dramatically after the Civil War and through the Jim Crow era, and again with the coming of World Wars I and II. Recolonization schemes would eventually be abandoned, maps of the Atlantic were shelved for maps of the city, and the spatial fix began take a more purely localized geographical form. Blacks were here to stay, so other means of geographically managing race relations would need to be sought. In white-dominated New Orleans, a city unique for always having had a large slave and free black population, this would mean living with slaves and mulattos, free and bound, manumission and recolonization plans, violent pogroms, marginalization into the "bottom of the bowl," and eventually a localized spatial division of the races in a hyper-segregated pattern whereby race and class correlates to topographical elevation. Residential and other localized types of segregation of one sort or another was always a reality, but by the twentieth century it became the only spatial division of races that mattered.

Patterns of racial segregation and spatial location in the socio-political ecology of New Orleans were by no means crafted through a logical process of intentional outcomes. There was no committee or organization seeing to it that race was manifested in space. The process was much more convoluted. Many actions were intentional, hostile, or outright violent. But many more were subtle, subconscious, and assumed.

Peirce Lewis describes New Orleans from the Civil War almost until the early 1960s as having a regime of racial segregation defined by its micro-spatial ordering. Prior to the civil rights movement and the end of legally sanctioned segregation, blacks and whites lived much closer to one another. Residential segregation was ordered in a complex patchwork in which clusters of predominantly black blocks would be encapsulated within super-blocks. On the main avenues and outer fringes of these super-blocks lived whites and more affluent citizens. According to Lewis, "such patches of black population by no stretch of the imagination can be called 'ghettos'. . .these Negro neighborhoods of New Orleans were quite small and multinucleated, with fuzzy boundaries."[11] This peculiar residential pattern was due in part to the custom of building slaves' and servants' quarters in back of the master's house. Jim Crow segregation in New Orleans adopted this spatial design, and there was very little pressure to change it until the black freedom movement of the 1960s.

As black communities began to question the norms and laws of Jim Crow in the middle of the twentieth century, this type of segregation had begun to give way to large contiguous areas within the city populated by blacks at rates approaching 80 to 90 percent. These areas consolidated the black population most intensely in the lowest, most vulnerable, and marginalized portions of the city.

For New Orleans, like much of the Southern United States, the "proper" spatial-racial regime was a problem of enormous proportions for the dominant-class and working-class whites who sought to separate themselves from blacks, a problem that had to be continually reworked as the region developed from a plantation economy to sharecropping and shipping, and finally the industrial production, transport, and tourism economy. To earlier New Orleanians, miscegenation had been identified as the most pernicious effect of an improper proximity amongst the races, but so, too, were rebellion, maroonage, crime, and loss of morality associated with spatial concentrations of blacks. What had always made New Orleans's problems of racial geography even more intractable for the ruling planter class was the relative abundance of free blacks and mulattos from early on. Later, racial geographies would be rearticulated to deal with the changing economic and political circumstances of blacks and whites in the city.

One way of understanding the Katrina catastrophe is that it is the product of the latest racial geographic regime in the Gulf Coast, one not all that different from the immediate past, but certainly very extreme in the degree to which nonwhites and the poor have found themselves relegated to extremely vulnerable positions within the socio-political ecology (and with very few resources to deal with this vulnerability). Hurricane Katrina's effects were structured by the spatial reordering of New Orleans since the mid twentieth century that came about directly through racial segregation, and its consequences are being played out under the new politics of race—a politics in which racial hierarchies and inequalities are being rearticulated and fashioned in novel ways that mask the racist causes and consequences.[12]

From the 1950s on, white New Orleanians moved just outside the city limits, separating themselves from the Ninth Ward and other increasingly black neighborhoods. Many whites migrated east, just a mile or two into St. Bernard Parish, separating their new homes from the old with a set of railroad tracks, a military reserve, and an ample sized patch of green space. Many more moved into Jefferson Parish to the west. The push and pull forces that motivated white flight were many. They included racist yearnings for a segregated society that was being fundamentally challenged, but also economic forces and technological breakthroughs that made suburbanization possible. St. Bernard Parish became a destination that promised secure jobs, a large affordable home, and distance from the ever-darkening "chocolate city."

Beyond city limits in the Parish, St. Claude Avenue becomes the St. Bernard Highway, and Claiborne Avenue becomes Judge Perez Drive. The homes in St. Bernard are much newer than those in the Ninth Ward. Many of them are two stories tall and almost every last one is made of sturdy brick. The houses in the Ninth Ward are mostly small wood-framed structures. In St. Bernard, the yards are larger, and the streets twist and turn in the curvaceous fashion that is a hallmark of the modern suburban development. The trees are taller, and one can presume that, before the floods, the grass was greener on this side. St. Bernard Parish was flooded just as deep and long as the neighboring Ninth Ward. Only the area's sturdier infrastructure kept it from taking as bad a beating from the winds and rain—that and the fact that the worst effect of flooding was only felt in the Lower Ninth Ward where levee failure created a wall of rushing water that leveled several square blocks, picked up and moved homes and cars, and crushed anything in its path.

Arabi and Chalmette, the first two towns in St. Bernard adjacent to the Ninth Ward, were 96 and 92 percent white before Katrina hit. Both towns have a poverty rate below the national average of 12 percent. Homes in Arabi are valued at a median of $77,000, with comparable houses in Chalmette worth about $10,000 more (for the New Orleans metro region, this is very high). In New Orleans's Ninth Ward, 36 percent of Jackie's neighbors live below the poverty line. Some blocks have astronomical rates of impoverishment with more than half of all residents living in poverty. The Ninth Ward is virtually all black today except for the southern section around the edge of the French Quarter, which is not coincidentally on higher ground and separated from the rest of the ward by St. Claude Avenue's four lanes and a median. Home values are much lower in the Ninth Ward even though ownership rates are incredibly high.[13] This is spatial segregation on the local level, and it is characteristic of the larger pattern that can be discerned at the citywide level.

For the first half of the eighteenth century, St. Bernard Parish was a French colony (as was all of Louisiana). Planters established themselves on the high ground near the banks of the Mississippi and used African slaves to raise sugarcane and indigo in the rich soils deposited over several millennia of delta building. The natural levees up and down river from New Orleans were developed in the same manner. Thus, New Orleans began and thrived as a port city surrounded by vast plantations.[14] (Even today the region's most important economic activity takes place on "plantations" outside the city—many of the Port of Southern Louisiana's facilities are named after the properties that used exist there; bulk goods are shipped from Angelina Plantation, Hope Plantation, Goldmine Plantation, and so forth. Some of the refineries and chemical plants along the "cancer alley" corridor are also located on industrial lots still called plantations.)

The slave-based economy flourished through thirty years of Spanish rule, the second brief French dominion, and American control until the Civil War. All the while, New Orleans expanded rapidly, becoming the largest Southern city (the sixth largest in the United States) by 1860. After the war, many emancipated slaves left the rural parishes of Louisiana for New Orleans. Others stayed as sharecroppers and small farm owners alongside Spanish and German immigrants and the propertied upper class. Until the end of World War II, St. Bernard remained a rural backwater—its total population a mere 7,000, according to the 1940 Census. In 1950 its population exploded. Several large factories, including an aluminum works, a sugar refinery, and an oil refinery, provided an economic base for the working-class population that moved in. The majority of this growth took place right on the edge of New Orleans's city limits. Builders provided spacious homes for the new suburbanites. State policies and practices within the real estate and banking industries promoted racial segregation at the expense of nonwhites. Blacks were kept out via norms and laws.

By 1970, more than 50,000 people lived in St. Bernard Parish, most of them closer toward the New Orleans end. White families moved into the new residential developments, finding employment in the factories that were moving out of New Orleans. Jefferson Parish, on the opposite side of the city, experienced a similar pattern of growth. Both parishes experienced rapid economic and population growth. Nonwhites, meanwhile, were restricted inside the city limits of New Orleans, unable to gain employment in the booming exurban industries or purchase a home in the rapidly expanding (and financially appreciating) suburban neighborhoods. These new dynamic places were not theirs to enter. They were forced to rely on whatever they could manage within the city limits. This was the consolidation of the new spatial regime of race that jelled after World War II and solidified after the 1960s. As industries left and investments dwindled, times became tougher for those working-class and poor populations restricted geographically (and by nonspatial means, of course). Redlining was rampant. Property values in black neighborhoods stagnated, jobs became increasingly scarce, and major redevelopment projects, like the I-10 freeway, cut through the social fabrics of these communities, laying waste to business districts and cultural centers like the Claiborne neutral grounds.

Left in the Ninth Ward and other formerly mixed working-class districts, many poor black New Orleanians were able to purchase their homes and build surprisingly vibrant communities and some degree of wealth, all in spite of their situation. However, the overall costs outweighed the benefits. White racism was rearticulated in new forms during and after the civil rights movement. Across much of America, geographical segregation that created powerfully different life chances became a paramount means of maintaining the racial hierarchy that had previously been constructed through other less spatially deterministic forms. In New Orleans this created an especially vulnerable poor black population, but also much vulnerability amongst middle-class blacks and whites who fled the inner city only to settle in swamp lands no more fit for habitation than the Ninth Ward or the "bottom of the bowl."[15] The socio-political ecology resulting from the push for racial integration and reactionary suburbanization (propelled by the white population's possessive investment in whiteness that promised a maintenance of privilege in the suburbs) built a cityscape under sea level, surrounded by a maelstrom for a river on one side, lakes to the north, and rapidly disappearing marshes to the south and southeast. In 1965, Hurricane Betsy exposed the harm wrought by this regime of racial segregation and disaccumulation structured into the geography of greater New Orleans. Betsy was a preview. Katrina, a disaster of far larger magnitude, has irrefutably affirmed this point.[16]

As of July of 2006, Jackie's house sits quietly on Desire Street, gutted, its former contents hauled away, but little more progress has been made on it. Her neighbors are filtering back in slowly, many of them just to survey the damage and salvage a few belongings. The street is mostly deserted, however. Most inhabitants of the Ninth Ward have few savings to spend on reconstruction. Few of them have flood insurance. What wealth they had was literally embodied in their homes. Now it's in no condition to use as any kind of equity. Other streets have more activity. Some are bustling with trucks, hammers, generators, and crews of men and women gutting interiors and fixing rooftops. The Ninth Ward's pace of reconstruction has been slow, but it appears to be on track. More and more lights can be seen flickering in windows at night. Many families have returned and parked trailers in their yards. Children roam the streets on bicycles, making occasional forays out to Claiborne Avenue to the Family Dollar Store to buy candy and soda.

Once, New Orleans was named the "Crescent City" because it was literally shaped like a slender moon, following a big bend in the Mississippi. Most of the houses, businesses, churches, and schools were built on the high ground closest to the riverbank. Then it waxed in size, filling in the low-lying areas east and west, bounded only by Lake Pontchartrain above. The way things are going politically, it looks as though New Orleans might once again become the Crescent City in shape as well as name. Half of the city's population may never be able to return. Many of their homes, businesses, and places of worship might be leveled and turned into parks or more expensive housing. Residents now refer to the wealthy parts of town that compose the bright edge of this crescent as the "sliver by the river." To carry the lunar metaphor even further, one can imagine the New Orleans that existed pre-Katrina as a crescent moon, a crescent only because of its black core. The city's black majority has long been the core of its cultural and economic vitality. But this core is unseen, like the shaded side of the real crescent moon. Politically speaking, many black and poor residents, returned or otherwise, believe that the city's powerful want to tear out this black core but keep its cultural products—music, food, language, history, and art. It will be the aesthetic sort of crescent moon, a crescent without a shaded core; the convex portion will be punched out, and sitting on it, perched on the bottom lip of the hook-shaped moon, will be the representation of that black-Creole, selling the city, selling voodoo, sin, gumbo, and the blues to the world at large.

The politics of rebuilding New Orleans are incredibly complex, but not without precedent. Major disasters throughout American history have more often than not been followed by intense political jockeying among economic and political elites in order to retain control of local and regional government or wrest it from one another. Redevelopment schemes have always emphasized the various and competing spatial interests of different groups cut along lines of class, race, ethnicity, and gender. Capital has often responded to widespread destruction in a Schumpeterian fashion by taking advantage of the liquidation of finances, firms, and the physical landscape to refashion whole new designs for locally based accumulation. Social movements in the forms of place-based mobilizations, racial coalitions, and labor organizations have arisen to meet the innumerable challenges facing different populations after disaster events.

Early on, the most powerful lobby in post-Katrina New Orleans was Mayor Ray Nagin's Bring New Orleans Back Commission (BNOB). Composed mostly of local business elites (real-estate developers in particular) with a vested interest in growth-machine politics,[17] the Commission also seated prominent artists, several black businessmen, and a few putative community leaders. Its vision for the future of New Orleans promotes the city's "culture" and music, improving government effectiveness, and radically reshaping the urban landscape and overall footprint. Although the BNOB Commission has disintegrated due to lack of funding or political stability, its vision for New Orleans represents the thinking of city elites, real estate developers, many powerful civic leaders, and large portions of the middle and upper classes (predominantly whites).

Early in 2006, several commission members publicly affirmed the right of all neighborhoods to be rebuilt on several occasions, but the Commission made no commitments to provide resources for residents to do the rebuilding. BNOB leader Joseph Canizaro even happens to be the chairman of the Urban Land Institute, a think tank that advised the city in Katrina's wake that many neighborhoods like the Ninth Ward should definitely not be rebuilt. Apparently, the Commission has not held any recent public meetings or moved forward with any of its proposals. But this does not necessarily mean that its vision is not being pushed forward by the new official planning process or through the haphazard reconstruction efforts underway.

According to the *Times-Picayune*, as of July 2006, the city has only managed to craft a "plan to create a plan to rebuild New Orleans."[18] This latest official framework for reconstruction, called the Unified New Orleans Neighborhood Plan, is described as, "a single comprehensive land-use planning process, which will coordinate the recovery of more than seventy distinct neighborhoods."[19] So far so good, but the plan appears short on substance, despite being funded to the sum of $4.5 million from the Rockefeller Foundation and Greater New Orleans Foundation. The Louisiana Recovery Authority has decided to throw its weight behind this scheme, making it the current front-runner among planning schemes. Mayor Nagin and other officials have said that any planning already conducted by third parties will be incorporated into this master plan for the city, giving those neighborhoods that have used private resources to rebuild a further lead in the chaotic process.

Initially the BNOB Commission proposed giving neighborhoods a window of one year to rebuild, and that if at the end of that timeframe a community had failed to show its "viability," it would be demolished, entirely re-planned, and redeveloped. As of July 2006 it is not clear whether this mandate is being upheld. Nagin was forced to distance himself from this proposal after multiple communities voiced their opposition to it. If this idea is adopted in some form or another, exactly who will define progress, and what will be considered a neighborhood "comeback," are subjective determinations that remain to be seen. As of today, it appears that the massive public outcry (coming from diverse quarters of the city) defeated this point on the wish list of large-scale real estate capital.

As one walks through the streets of New Orleans, it's hard to tell what the storm wrought and what neglected conditions were already present. Much of the city had already been slummed by racist real estate practices and the realities of poverty. Current planning policies are heavily weighted toward those neighborhoods inhabited by more affluent residents and those not hit as hard by the storm, where organizing a community recovery effort will occur smoothly or will not be needed at all.

The BNOB Commission's final report, released on January 11, 2006, spells out the Commission's ambitious plans in detail.[20] Its proposals are worth studying because they represent a wish list for the city's power elites and will carry weight regardless of the fact that BNOB has been all but disbanded.

Their recommendations for rebuilding embrace many of the design philosophies that can be found in new urbanism, a school of thought within the professions of urban planning and architecture that emphasizes the aesthetics of city life (without its harsh realities). To rebuild the city's neighborhoods, they state, reconstruction must be "built on neighborhood history and culture; respectful of historic block patterns, architecture, and landscape; mixed income communities with a diversity of housing types; parks and open space connected to a city-wide system; city-wide accessibility through transit; neighborhood centers that provide a high quality of daily life." It all sounds very appealing, but it's premature to assume that the actual reconstruction process will abide by these principles in any equitable shape or manner. Statements such as this also tend to avoid issues of race or economic justice by using language such as "culture," "diversity" and "mixed-income" solutions.

The Commission's report is laden with empty appeals to the disadvantaged alongside benign-sounding code words for demolition and gentrification. Some are probably there only to appease the gods of ecological sustainability and justice—while other recommendations actually mean what they say and will be carried out in the fullest. The BNOB Commission's plans, as well as the other efforts that will influence the Unified New Orleans Neighborhood Plan, must all be synthesized by city authorities before they can unlock the funds currently held by the Louisiana Recovery Authority.

Residents across the city responded with outrage and suspicion to the officials' early plans, especially Nagin's comments about issuing a moratorium on building permits in sections of the city like the Ninth Ward. (Nagin later withdrew this idea and allowed building to commence). One aspect of the BNOB Commission's thinking that may be incorporated into the final planning (explicitly or tacitly) is its identification of four "opportunities for neighborhood rebuilding" based on location and the extent of resulting damage from the hurricane and floods. "Immediate opportunity areas" are those that sustained little or no damage and can be improved without further debate. Their future is assured. Not surprisingly, these areas mostly cover the Uptown, Garden District, French Quarter, Bywater along the Mississippi, and northernmost portions of the Lakeview and Gentilly communities—affluent and predominantly white neighborhoods. True to their label, residents in these neighborhoods immediately began repairing their homes. Contractor vehicles can be spotted on every street. Power tools are buzzing and hammers are pounding away at homes that suffered mostly minor wind damage, fallen trees, or leaky roofs.[21] Planning in these zones of the city is somewhat redundant because of the minor damage they sustained and the overabundance of resources at the disposal of residents and other property owners.

The second major classification for neighborhoods is termed "infill development areas." The Commission recommends that these areas be consolidated under public and private ownership, that development plans be drafted, and proposals solicited from demolition and construction firms. The areas in question would probably be consolidated under the auspices of some sort of redevelopment agency. (The BNOB Commission proposed the creation of a Crescent City Redevelopment Corporation [CCRC], which would sell land back to private developers). Again, not surprisingly, neighborhoods deemed "infill development areas" are predominantly black and poor sections of the city. They include large swaths of the Ninth Ward, Bywater, Gentilly, Mid-City, Tremé and Central City, and the McDonogh, Whitney, and Fischer neighborhoods across the river in the Algiers district. Under the Unified New Orleans Neighborhood Plan, the mechanism by which large tracts of land are seized and redeveloped by real estate capitalists is likely to take a different form, but some means of consolidating properties is in the works.

The BNOB Commission was also asked to look ahead and create an overall vision for the kind of city they'd like to see. Among the more novel proposals is one to build a network of high-speed light rail commuter lines linking New Orleans with its suburbs—Jefferson, St. Bernard, St. Tammany Parish, and the Slidell region. Most critical studies of rapid transit around the nation have concluded that the benefits of such a system will primarily go to those on the ends of the lines—the suburban bedroom communities and the central business district. Those in the middle, the neighborhoods the rail lines will cut through, will see very little utility in such a system.[22] The benefits to the real estate lobby and large employers concentrated in the downtown area could be immense, as parcel values for office and retail developments would likely multiply in value, and a light rail system would guarantee a white-collar professional suburban workforce. Whether neighborhoods like the Ninth Ward or the Tremé live to see this future is an open question. Whether these neighborhoods will retain the communities—the residents, families, and local businesses that give them their culture and traditions—is even more uncertain.

The plans for the regional rail system gives some insight into the larger goals and vision of the interests originally behind the BNOB Commission and now at work in other organizations and agencies or behind the scenes as private actors to influence the Unified New Orleans Neighborhood Plan. The New Orleans they envision is a global city, surrounded by an affluent suburban professional workforce. The central business district will be the center of the region. Its office towers will fill every morning with professionals and white-collar workers from outlying areas. It will be an information economy instead of what New Orleans has been, a working class city with a large population of supernumeraries (the structurally unemployed). Much of the city will become parkland. The metropolitan region's population will shrink by half. Families will also move back into the central city districts. On top of what remains, redevelopment will create neighborhoods in the architectural style of the past. It will attempt to recapture the sense of community lost in the past half century by deindustrialization and suburbanization. Condos will fill in the old warehouses, the racially cleansed housing projects like Lafitte, and the new towers built upon old barren blocks. Townhouses will parallel city streets just beyond the downtown. The streets will be sanitized for the newcomers. In some respects, census data confirms that New Orleans had already begun to move in this direction.

This vision for New Orleans is motivated in part by a desire among exurbanites to return to the city and rediscover its magic. New Orleans, like many cities, is being seen as an appealing place to live. But sadly, it's a desire that will be fulfilled only in the displacement of those who currently reside in central cities, those who were abandoned in Tremé and the Ninth Ward decades ago. In New Orleans it's a return that will be made possible by exploiting a disaster. In other cities it's simply called gentrification. But in post-Katrina New Orleans, it will be called "logical," the "return of nature," "opportunity," and "market forces." Or, in the now-infamous words of U.S. Representative Richard Baker, God did it.[23]

The disaster that hit New Orleans in the form of a hurricane and flood wasn't a freak event. It wasn't a break with the normal as much as it was a punctuated moment of continuity in the longue durée of this city. Katrina is business as usual, packed into an exclamation point. The eradication of poorer neighborhoods like the Ninth Ward has long been the dream of many of New Orleans's more affluent and privileged.

One of the primary ways racism and socio-economic exploitation works is along spatial lines. Racialized and poor populations are related to by the dominant society most concretely in terms of geographical proximity. The different regimes that subjugate poor and nonwhite communities must calibrate a proper spatial distribution of populations. Post-Katrina New Orleans appears to be unfolding toward a new type of spatial ordering unlike anything New Orleans has known in the past. There are like-minded people in every American city who just want the ghettos or the "wrong side of the tracks" to disappear, and they don't care how or where they go, as long as it's somewhere else. It's what helped create the conditions for this latest disaster in New Orleans. But other complex factors (economic, cultural, and happenstance) push and pull on the demographic order.

This peculiar American creed of racism and cruel indifference toward the poor characterizes much of our urban history. It was there when plantations spread along the banks of the Mississippi and slashed into the delta swamps. It didn't end with the Civil War or the civil rights movement. It raged in 1866 when ex-Confederates and police officers massacred black and white radical Republicans at their New Orleans convention (which was also fundamentally a struggle over who would rebuild New Orleans and how it would be rebuilt following a civic disaster of another sort). It was there when public housing went awry and when the I-10 highway destroyed the Claiborne neutral grounds. It was there when white parents pulled their kids out of integrating schools and fled to St. Bernard and Jefferson Parishes. What floats to the top now in this time of uncertainty and indecision is a desire that many have for a new New Orleans. Leaving the city a half-century ago was a privilege made possible largely along lines of race. But leaving the city took its toll on the spirits of many. The desire for an urban way of life always remained in the hearts of some suburban-ites. That sort of magical urban place was calling. Coming back to the city after Katrina is the new privilege. The Bring New Orleans Back Commission articulated the first statement of this vision. As the process progresses and takes a new form, we still find that there is little or no room in this vision for the Ninth Ward and neigh-borhoods like it. There is certainly no room for public housing like the St. Bernard projects. The Department of Housing and Urban Development has now called for its demolition along with the Lafitte, C. J. Peete, and B. W. Cooper Homes.[24]

After a jazz benefit concert for the reconstruction of New Orleans, poet Amiri Baraka explained, "What Bush wants is to make New Orleans like his mother—shriveled and colorless."[25] Whether or not Bush wants this, the city has certainly shrunk in numbers. Without a serious federal commitment, far beyond what has been allocated to date, and without strident activism at all levels to ensure that fed-eral funds and policies are administered fairly, any effort to rebuild the city will end in an outcome that can only be characterized as racial cleansing and the eradication of the poor. (We are already seeing this in the year since, but we are also seeing diverse social movement activity.) This is the other side to the politics of rebuilding New Orleans. It's the grassroots response. It's agitating, organizing, rallying, docu-menting, building, cleaning, and speaking out. For every speech that Nagin gives, for every new official planning scheme announced, there are a thousand volunteer work crews helping to revive shell-shocked neighborhoods. For every visit Bush makes to the Gulf Coast, there are thousands of college students, activists, former residents, and citizens making the trip to the Big Easy to provide assistance to indige-nously led groups like Common Ground, the People's Hurricane Relief Fund, C3, United Front for Affordable Housing, Survivor's Village, and others.

Back on Desire Street in the Ninth Ward, Jackie's house stands empty. It's July, the midst of a new hurricane season. The house's insides are completely gutted. If she can muster up the money to install new drywall and carpeting, fix the holes in the roof, and to buy the bare necessities of life like a bed and kitchenware, she could move back in before long. If she does, she might be alone. However, judging from the activ-ity on her street in the past few months she'll probably be moving back in alongside some neighbors. Every day, more and more residents are coming home. Many are still just stopping by to survey the scene, but quite a few are breaking out their tools and putting good, old-fashioned sweat equity back into their humble houses. Whether or not they can make enough of a difference and help to rekindle the shattered social bonds and a sense of community is uncertain, but the future relies upon them, not on some city official who measures and deems a "viable comeback" or on some city- or state-sanctioned planner who imposes wonderful new plans upon them.[26] The point here, if you ask many people, is to rebuild the community so that it can resist the inevitable attempts that the powerful will make to destroy it, attempts that always seem to be made when a "natural" disaster provides an alibi.

NOTES

1. William S. Falk, *Rooted in Place: Family and Belonging in a Southern Black Community* (New Brunswick, NJ: Rutgers University Press, 2004).

2. Social scientists have been reconsidering the dynamics of racial succession in neighborhoods, noting that ecological approaches that focus entirely on spatial variables like the number of black residents on a block, the particular patterns of black residence in relation to whites, and so forth are inadequate to explain how it is that certain neighborhoods become all white or non-white (for instance, see Kevin Fox Gotham, "Beyond Invasion and Succession: School Segregation, Real Estate Blockbusting, and the Political Economy of Neighborhood Racial Transition," *City & Community, 2002* 1, no. 1 [March 2002]: 83–111). Nevertheless, the decline of black neighborhoods like the Ninth Ward are due to structural racism in American society, no matter how complex it is.

3. Douglass Brinkley, *The Great Deluge: Hurricane Katrina, New Orleans, and the Mississippi Gulf Coast* (New York: Harper Collins, 2006) 45.

4. Additionally, the perceived legitimacy—however shaky—of the welfare state during this era (and up until at least the 1980s) meant that social investments and social consumption protecting places like the Ninth Ward from catastrophic flooding were uncontroversial and ensured. The levees were funded, and so were welfare programs for the growing population of dislocated workers (supernumeraries) pushed out of work by rapidly transforming monopoly sector industries, most of whom were black Americans who lived in places like the Ninth. See James O'Connor's *Fiscal Crisis of the State* (New York: St. Martin's Press, 1973), for a detailed theoretical treatment of how this political-economy that sustained social investment/consumption in levees, city services, and welfare (broadly defined) is inevitably outpaced by a fiscal crisis that leads to cutbacks, the sort of which weakened New Orleans infrastructure and put much of its population in a position of critical vulnerability.

5. For a more historical treatment of environmental racism in New Orleans (particularly the segregation of poor blacks into the "bottom of the bowl" and the back swamps), see Craig E. Colten, *An Unnatural Metropolis: Wresting New Orleans from Nature* (Baton Rouge: Louisiana State University Press, 2005); also, Peirce F. Lewis, *New Orleans: The Making of An Urban Landscape* (Cambridge: Ballinger, 1976).

6. Quoted in David Remnick, "High Water: How Presidents and Citizens React to Disaster," *The New Yorker*, October 3, 2005.

7. Ruby Bridges, *Through My Eyes: Articles and Interviews Compiled and Edited by Margo Lundell* (New York: Scholastic Press, 1999).

8. Morton Inger, *Politics and Reality in an American City; The New Orleans School Crisis of 1960* (New York: Center for Urban Education, 1969).

9. As of July 2006, many of the public schools in New Orleans remain closed. In their place have sprouted up dozens of charter schools. Prior to Katrina, the New Orleans public school system resembled many other inner-city districts; it was chronically under-funded and highly segregated. Bill Quigly, lawyer and professor at Loyola University of New Orleans, says, "Public education in New Orleans is mostly demolished and what remains is being privatized. The city is now the nation's laboratory for

charter schools—publicly funded schools run by private bodies. Before Katrina the local elected school board had control over 115 schools—they now control four. The majority of the remaining schools are now charters. The metro area public schools will get $213 million less next school year in state money because tens of thousands of public school students were displaced last year." Bill Quigly, "Ten Months After Katrina: Gutting New Orleans," *Dissident Voice*, July 1, 2006, http://www.dissidentvoice.org.

10. G. Leighton Ciravolo, *The Legacy of John McDonogh* (Lafayette: Center for Louisiana Studies, 2002).

11. Peirce F. Lewis, *New Orleans: The Making of An Urban Landscape* (Cambridge: Ballinger, 1976).

12. Howard Winant, *The New Politics of Race: Globalism, Difference, Justice* (Minneapolis: University of Minnesota Press, 2004.

13. U.S. Census, http://www.census.gov.

14. Thomas Ingersoll describes New Orleans as an urban core surrounded by rural plantations, a situation of such proximity and almost conflicting roles that the city became a microcosm of America as a whole, torn between rural and urban life, slavery and wage labor, and the industrialism of the factory and the plantation. Thomas N. Ingersoll, *Mammon and Manon in Early New Orleans: The First Slave Society in the Deep South 1718–1819* (Knoxville: University of Tennessee Press, 1999).

15. The "bottom of the bowl" refers to some of the lowest lying grounds in the city—literally in the center of New Orleans, far below sea level.

16. Note that I am not arguing that Katrina was a greater disaster because of the storm's characteristics. Rather, Katrina was a greater disaster in the sense that the disaster was the sum of several parts, both natural (the storm) and social (the political system, economic structure, race relations, poverty rate, condition of infrastructure, et cetera).

17. John Logan and Harvey Molotch, *Urban Fortunes: The Political Economy of Place* (Berkeley: University of California Press, 1987).

18. "Nagin Sets Guidelines to Plan Rebuilding," New Orleans *Times-Picayune*, July 5, 2006, http://www.nola.com/newslogs/breaking/index.ssf?/mtlogs/nola_localbreaking news/archives/2006_07_05.html.

19. Susan Saulny, "New Orleans Sets a Way to Plan Its Rebuilding." *The New York Times*, July 6, 2006.

20. Wallace Roberts & Todd, LLC, "Action Plan for New Orleans: The New American City," *Bring New Orleans Back Commission, Urban Planning Committee*, January 11, 2006.

21. As I write this piece I sit on the porch of a double-shotgun home in the Faubourg-Marigny neighborhood. I moved in a week ago. According to my neighbor, who is a longtime resident, the flood waters only covered the street by a few inches. The worst damage in this section of the city was to walls and roofs from the wind.

22. See Joseph A. Rodriguez, "Rapid Transit and Community Power: West Oakland Residents Confront BART," *Antipode* 31, no. 2 (1999). Rodriguez summarizes some of the literature critical of BART's claims that it would benefit low-income communities along its lines and also describes the black community's political opposition to various aspects of the BART project, including its overall purpose, impact on black business districts, and union racism that locked nonwhites out of from high-paying construction and operation jobs.

23. Baker's words, first reported by the *Wall Street Journal* were, "We finally cleaned up public housing in New Orleans. We couldn't do it, but God did" ("Washington Wire," *Wall Street Journal*, September 9, 2005).

24. Filosa, Gwen. "HUD Demolition Plan Protested: Residents Say They're Being Shut Out of City." New Orleans *Times-Picayune*, June 16, 2006.

25. Larry Blumenfeld, "America's New Jazz Museum! (No Poor Black People Allowed)," *Salon*, July 7, 2006, http://dir.salon.com/story/ent/feature/2005/10/12/jazz/index.html.

26. However, one planning/architectural firm working on the Ninth Ward through the New Orleans Neighborhood Rebuilding Plan is Stull and Lee, an African American firm with experience in designing racially conscious architecture that is sensitive to poverty and economic development in places like West Palm Beach, Florida, for instance. The official planning process may not be all that bad for the Ninth Ward after all. See Stull and Lee Inc., "A Community Vision for the Future of the Lower Ninth Ward: Presentation of Initial Sketch Plan Alternatives," June 17, 2006, http://www.nolanrp.com (accessed July 6, 2006).

Chapter 3

Race-ing the Post-Katrina Political Landscape: An Analysis of the 2006 New Orleans Election

Kristen Clarke

Introduction

The elections that were conducted in New Orleans in the spring of 2006 were historic in a number of regards. These elections were the first conducted following a major natural disaster and with significant numbers of residents displaced from the city. In many respects, these elections tested our nation's commitment to ensuring broad levels of equal and open participation in the political process while raising profound questions regarding the impact of delayed elections on the democratic conscience. Ultimately, these elections will figure significantly into the rebuilding and reconstruction process that is now unfolding. This essay will describe the positive and negative aspects of the elections while analyzing their larger relevance to black political power in the Gulf.

Preelection Struggle

There was no shortage of controversy surrounding the spring 2006 elections in New Orleans. In the weeks and days leading up to the elections, many black leaders called for a delay to give people enough time to get back to the city to cast their ballots.[1] Others called for satellite voting—the establishment of strategically placed centers around the country that would have allowed displaced voters to cast their ballots from their temporary places of residence. Given the displacement patterns that emerged following Katrina, the placement of satellite voting centers in cities such as Houston, Atlanta, Memphis, Jackson, and Dallas would certainly have helped make participation easier for significant numbers of displaced voters. Although a federal judge was unwilling to order it,[2] the body that could have crafted and implemented legislation authorizing the satellite voting centers was the Louisiana State Legislature. Ultimately, the legislature voted against this ambitious undertaking, arguing in part that they lacked the ability to establish secure computer databases out of state and that they were uncertain that officials in the respective jurisdictions would comply with their mandate.[3] Instead, legislation was passed that allowed for limited satellite voting in ten parishes around the state: Caddo, Ouachita, Rapides, Calcasieu, Lafayette, East Baton Rouge, St. Tammany, Tangipahoa, Jefferson, and Orleans.[4] These centers, based at the registrar's offices in each of the ten parishes, helped make voting easier for the large numbers of African American voters who remained displaced in-state.

Despite the problems and challenges with these elections, there are some notable and appreciable aspects to the process. For one thing, these elections were conducted despite significant damage to the city's infrastructure. Numerous polling locations were destroyed or damaged beyond repair.[5] The identification of alternative locations that would provide voters the opportunity to cast ballots was likely not an easy feat. In addition, many poll workers needed to be recruited and trained in order to ensure that there were sufficient numbers of personnel working the polls during the elections. Notice of all changes regarding the elections needed to publicized to voters who were now dispersed throughout the country—a task that must have been particularly difficult given the fact that many displaced voters were in a constant state of flux. This reality was further complicated by the unwillingness of the Federal Emergency Management Agency (FEMA) to share updated address lists on the grounds that it would breach privacy restrictions. Although these tasks were accomplished under difficult circumstances, providing more time would have benefited voters.

THE LACK OF POLITICAL WILL TO ENSURE BROAD PARTICIPATION IN THE NEW ORLEANS ELECTION

In many respects, the recent elections in New Orleans will serve as a guidepost for future elections conducted under emergency circumstances. With the specter of intensifying international conflict lying on the country's shoulders and a world that is increasingly vulnerable to natural disaster, Hurricane Katrina will certainly not be the last tragedy to befall the country. The elections illustrates how much of a delay in the election schedule the public is willing to tolerate and reveals how aggressive officials are willing to be to ensure that displaced voters are able to cast their ballots. In general, it appears that the public will not tolerate a long delay in elections, in part because disasters embroil human emotions and spark political debate. Indeed, in the days following Katrina, blame was levied on local, state, and federal officials of all stripes, and there were immediate calls for the removal of FEMA Director Michael Brown, Governor Kathleen Blanco, and Mayor Ray Nagin at various points.[6] It is difficult to suspend elections under these circumstances because emotions are running incredibly high and the public is eager to have some say about the type of leadership they want to help them get through the difficult period that inevitably lies ahead. Indeed, at least two lawsuits were filed by those seeking to expedite the elections. However, these elections also show that elected officials are unwilling to take aggressive steps to facilitate voting under these circumstances, perhaps because the scope of the tragedy did not reach far enough and officials have a hard time grasping the severity of the situation. The failure of the state to do more explains why many displaced African American voters may have wanted to participate in these elections but were unable to.

Certainly, there are a number of other steps that could have been taken to make it easier for displaced voters to participate in what were arguably the most important elections in the city's storied history.[7] The out-of-state satellite voting centers are certainly one such measure that would have allowed those who had been unable to find a way back into the city to cast their ballot from their temporary domicile. Taking this concept one step further, an extensive period of early voting could have been made available at registrars' offices around the country, giving displaced voters a wide window of time during which to cast their ballots. Preaddressed, prestamped ballots could have been made available at post offices around the country with a verification process established at the Orleans Parish Registrar's Office for all ballots returned by mail. Problems with mailing lists aside, ballots could have been automatically mailed to all displaced voters, thus eliminating the requirement that voters first request their ballots and wait for their arrival. Finally, election officials could have conducted extensive voter outreach by negotiating with FEMA officials to make ballots available at FEMA Centers and during well-attended public meeting held around the country to educate displaced voters about the receipt of social services and benefits.

SHIFTING ALLEGIANCES AND OVERCOMING OBSTACLES

In 2002, Ray Nagin ran a campaign that clearly appealed to white voters and business interests in New Orleans. He received marginal support from the black community (approximately 45 percent), with the majority of African American voters supporting his opponent, former New Orleans Police Chief Richard Pennington. Nagin's lowest levels of support were in black enclaves of the city such as the Ninth Ward. However, white voters turned out to the polls in exceptionally high numbers, helping ultimately tip the scales in Nagin's favor. During his time in office preceding Hurricanes Katrina and Rita, Nagin did not enjoy a great reputation among the city's black community. He continued to be regarded a mayor on behalf of white and monied interests in the city. However, the tide shifted in Nagin's favor in the months following Katrina when it appeared that he was the only viable African American candidate to throw his hat in the ring for the mayoral race. Early polls suggested that among the large pool of candidates, Nagin, Louisiana Lieutenant Governor Mitch Landrieu, and Ron Foreman were the frontrunners.

Analysts suggested that Landrieu was the more moderate and progressive of the three candidates; he was running on a platform that would have likely brought a more immediate impact on poor New Orleanians and increased the likelihood that poor, black voters might be able to make a speedy return to the city. Nagin's victory, however, meant that an African American continued to hold the most powerful position in the city and reduced fears among many displaced black residents that the city would be rebuilt to their exclusion. Although Nagin came under sharp criticism for his now-infamous "Chocolate City" comment, some black voters interpreted this as a firm, if perhaps belated, promise on his part to maintain the racial status quo.[8]

Nagin's victory in the most recent election reflects the fact that race continues to play a significant role in the political conscience of the Deep South—a reality that has not been altered by the hurricanes. African American voters in New Orleans are deeply concerned about holding on to their political gains and fear that the hurricanes may upset a delicately forged power balance in the city. Many black voters, particularly in the Ninth Ward, cast aside their profound disdain for Nagin out of a desire to keep the city's twenty-eight–year-old black mayoral legacy intact.[9] Likewise, white voters have seen the mass destruction of the city as an opportunity to rebuild a "new" New Orleans. Indeed, some white voters, the vast majority of whom are middle or upper class, have expressed a desire to see major reform to the social fabric of the city—a sentiment that some interpret as a desire for exclusion of the black poor.[10] White voters saw these elections as a unique opportunity to put a new face on the city, while black voters feared what that face might look like.[11] With these heightened stakes as a backdrop and despite the obstacles that displaced voters face, many endured long bus rides back to the city, found their way to one of the in-state satellite voting centers or cast absentee ballots in order to ensure that their voices would be heard. Although a sizeable number were not able to overcome the obstacles, the political will exercised by many African Americans evidences a real commitment to not only rebuild their homes but to maintain the long sought political gains that existed pre-Katrina. In this sense, Katrina presents interesting prospects for heightened black civic participation as communities throughout the Gulf are forced to confront the glaring specter of continued racism and assess the power balance that exists within their respective communities.

NOTES

1. "Protesters Call for New Orleans Election Delay: Jackson, Sharpton among Critics Claiming Vote Not Fair to Evacuees," Associated Press, April 1, 2006.
2. See *Wallace v. Blanco*—litigation brought by the NAACP Legal Defense and Educational Fund in the Eastern District of Louisiana seeking remedial relief to help ease the burdens on displaced voters.
3. "Bill fails to Change Rules for New Orleans Election," Associated Press, April 2006.
4. Charles Lussier, "New Orleans Election Off to a Slow Start," *Baton Rouge Advocate*, April 11, 2006, http://naacpldf.org/. . ./epp_katrina/Early_Voting_and_Satellite_Locations.pdf.
5. "Committee OKs New Orleans Election Plan," Associated Press, January 23, 2006.
6. Elizabeth Bumiller, "Casualty of Firestorm: Outrage, Bush and FEMA Chief," *The New York Times*, September 10, 2005; Brown Puts Blame on Louisiana Officials," *CNN.com*, September 28, 2005, http://www.cnn.com/2005/POLITICS/09/27/katrina .brown/index.html.
7. Kristen Clarke-Avery and M. David Gelfand, "Voting Rights Challenges in a Post-Katrina World: With Constituents Dispersed, and Voting Districts Underpopulated,

How Should New Orleans Hold Elections?" *findlaw*, October 11, 2005, http://writ
.news.findlaw.com/commentary/20051011_gelfand.html.

8. John Pope, "Evoking King, Nagin Calls New Orleans Chocolate City: Speech Addresses
   Fear of Losing Black Culture," New Orleans *Times Picayune*, January 17, 2006.

9. John Pope, "Black Precincts Buttress Nagin Victory," New Orleans *Times-Picayune*,
   April 24, 2006.

10. Michael Eric Dyson, *Come Hell or High Water: Hurricane Katrina and the Color of
    Disaster* (New York: Perseus Books, 2006), 204–205.

11. Manning Marable, "New Orleans Reconsidered: Race or Class," *Along the Color Line*,
    January 2006; Ceci Connolly, "Ninth Ward: History Yes, But a Future? Race and Class
    Frame Debate on Rebuilding New Orleans District," *Washington Post*, October 3, 2005.

CHAPTER 4

PROPERTY AND SECURITY, POLITICAL CHAMELEONS, AND DYSFUNCTIONAL REGIME: A NEW ORLEANS STORY

D. OSEI ROBERTSON

INTRODUCTION

We from a town where
everybody drowned
. . .
Everybody crying,
. . .
There's no doubt in my mind it was *George Bush*.

—New Orleans Hip Hop Artist Lil' Wayne[1]

On August 29, 2005, a series of events were put into motion that would change my life forever. The week before, I had been watching the news, observing a massive hurricane named Katrina move toward the Gulf of Mexico. This was an annual ritual, however. New Orleans and many other communities along the Gulf Coast deal with hurricane season every year. My main concern was for my family; they needed to leave. So I called my mother and father and urged them both to leave, which they did by August 27. Up to this point things were pretty much protocol; this was not the first time they had fled their homes as a precautionary measure. I thought I knew the drill—they would leave for a few days, there would be some minor flooding and roof damage, and the power would probably be out several days. On August 29 this annual routine turned into a nightmare. When the initial reports came in, it appeared that New Orleans had dodged a bullet. Small flooding and power outages constituted the preliminary reports. At approximately 1:00 a.m. on August 30, I logged onto a radio feed from a local CBS affiliate. Mayor Nagin was reporting that 80 percent of the city was experiencing major flooding, with up to ten feet in some areas. This couldn't be happening. I called my father right away, and told him the news. I was afraid to call my mother, I was afraid to tell her what had probably happened to her house. By daybreak all news agencies were beginning to report the depths of the disaster. This was not the usual drill, and people would not be returning in a few days. Three major levees had been breached—the 17th Street canal, the London Avenue canal, and the Industrial canal. The place where I was born and raised was under water.

As a native of the city, I found it very interesting to observe news analysts, scholars, and politicians making various assessments of the outcomes associated with Katrina. Some writers wish to highlight the impact of racism, while others focus on class-based discrimination.[2] Although these dynamics are important parts of the story, many accounts fail to capture some of the deeper challenges that confront the city. New Orleans has been a dying city for the past fifteen to twenty years. To grasp the uphill battle that we face, it is essential to understand the political and economic context in the city before Katrina and how these factors will affect future rebuilding plans.

Before beginning this analysis, however, I must first admit to bending a number of (traditional) social science rules. The usual stance of objectivity is one that I cannot claim. My vantage point is far from clear, clouded by reflections of my home. No matter where you lay your head, home is home—that place where your formulative experiences reside. Thus, because this event has affected my family and me, I must acknowledge that I am far from a distant observer. The role I occupy in this analysis is more like that of a participant observer selectively referencing "hidden transcripts."[3] In light of this position, I occasionally employ narrative and cultural devices in the analysis in an effort to tie the personal to the general electoral and political-economic dynamics this chapter explores. In addition to experiences and observations drawn from family and friends, I draw on traditional political-economic sources that include public opinion data, socio-economic statistics, and local periodicals.

This is a study of New Orleans before and after Katrina. It begins with an examination of the electoral and attitudinal patterns in the city over the past couple of decades. This section concentrates on the mayoral office and African American occupancy of City Hall since the administration of Ernest "Dutch" Morial, with particular attention on Mayor Nagin's first electoral campaign. The next section continues this theme by examining the transformation and shift Nagin made in his bid for reelection following the hurricane. Following an analysis of mayoral politics, the second section explores the city's political-economic policies and performance since the 1980s. New Orleans was a struggling city prior to Katrina, and this section details some of the root causes of the city's political and economic woes. In the final section, I offer some conclusions regarding the city's future.

NEW ORLEANS: AN OVERVIEW

To accurately analyze the challenges the city of New Orleans faces in the future, it is important to be clear about its political, economic, and cultural history. Culturally speaking, New Orleans is unique; it is one of the most fascinating places in the country. The historical blend of Spanish, French, African, and Native American cultures has created a city rich in musical, artistic, and spiritual traditions. It was New Orleans that both gave birth to jazz and vanguarded its resurgence in the early 1980s. The New Orleans Jazz and Heritage Festival, the Bayou Classic, the Essence Festival, and Mardi Gras are all national events. Layered underneath the public view of such events as Mardi Gras are traces of very old racial patterns. For example, few outsiders recognize that there are two different Mardi Gras events that take place on Fat Tuesday. Whites historically have gathered on St. Charles Avenue and Canal Street, while blacks historically have gathered at the intersection of Orleans and Claiborne Avenues, where the Zulu parade ends. Additionally, the Mardi Gras Indians function as a living tribute to the associations between slaves and Native Americans.

Politically, New Orleans has had African American mayors since the late 1970s. Following the enactment of the 1965 Voting Rights Act, several American cities witnessed a large surge in the number of African American voters, which increased the electoral strength of blacks. By the mid-1970s, it was not a question of whether an African American would be mayor, but when. Ernest "Dutch" Morial would follow an electoral path to City Hall comparable to that of Atlanta's Maynard Jackson. Neither candidate was selected to run for office by the establishment. In New Orleans, Sidney Barthelemy was gradually being groomed by insiders for the seat, while Morial ran for election without strong ties to former mayor Moon Landrieu and his associates. Morial was from the Seventh Ward, the historical center of the Creole community in New Orleans, a community that exemplified what historian Arnold Hirsch has labeled "creole radicalism."[4] Though race has always been a part of American politics, in certain areas of the country there has historically been a deeper intra-group tension focused on complexion. Beyond aesthetic considerations, the main concern here is the general belief that Creoles (the earliest free blacks of mixed genealogy drawing from French, Spanish, and Native American peoples) were granted access to society's institutions prior to darker blacks. Although the influence of the traditional Creole community has waned in recent decades, traces of these intra-group patterns still linger today.

FLIPPING CHANNELS

In this day and age of cable and satellite TV, we often flip though channels, searching for our favorite movies, shows, news, or sporting events. Out of two hundred channels, for example, we may only regularly look at thirty or fewer channels, skipping several channels in search of our favorites. On the local level, many cable providers have public access channels that feature local activists, entertainers, and news. In the nineties, the local cable provider in New Orleans (Cox Communications) had a regular information program where managers and executives would update the viewing audience on various service issues as they related to Cox. It goes without saying that this was a flip-through channel: you passed it when looking for something else to watch. I recall asking a friend to tell me who was running for mayor in 2002, and he mentioned a few names. The only ones I recognized were career politicians Paulette Irons (a local state senator), Troy Carter (a city councilman), Jim Singleton (City Council Chair), and Richard Pennington (then-Chief of Police). But when my friend mentioned another name, Ray Nagin, my response was, "Who is that?"

He said, "You know, the guy from Cox, the one always on their public access channel."

Like the city itself, politics in New Orleans has long followed a traditional path in that there are relatively few unknowns operating in the political landscape. Following "Dutch" Morial, mayors in the city were either longtime politicians or had links to political families. Some people felt that Sidney Barthelemy, because of his associations with the Landrieu administration, would be the city's first African American mayor, so his election following "Dutch" was of little surprise. Marc Morial was the son of Dutch, and it was merely a matter of time before he ran for office. So when a businessman who was a political outsider announced his candidacy, I thought he had little or no chance. Little did I know that he would become a central figure in one of the most traumatic and challenging times that the city has ever faced.

Many scholars of black politics distinguish between the first wave of BEOs (Black Elected Officials) that came to office in the late 1960s and early 1970s and a new crop of BEOs who began winning office in the late 1980s and early 1990s. The early BEOs, such as Carl Stokes, Kennith Gibson, and Maynard Jackson, gained political office through explicit appeals to their black voting base. "Most early black mayors," Persons contends, "explicitly articulated a social reform agenda in their campaigns emphasizing issues of police brutality, the hiring of blacks in municipal jobs, increasing low-income housing choices," and other areas associated with working-class and lower-class blacks.[5] These city leaders were, in theory, carrying on the struggles of the civil rights movement within the institutional confines of local governance. None of these mayors were radicals or called for massive redistribution efforts while in office, but they at least gave the impression that they were concerned with the collective good of African Americans. Regarding this tradition, Walters asserts, "The essence of the problem is that the aggressive legacy of the civil rights movement contains the inherent notion of confrontation and pressure as a methodology for forcing major American institutions to legitimize black demands for justice."[6] This characterization of the first wave of BEOs has been contrasted with groups of candidates emerging in the late 1980s who neither in substance nor style seem to reflect the same concerns of those earlier elected officials. In many cases these women or men are seeking to "normalize black politics or to subordinate black demands to the normal workings of the political process and, thereby, delude it of its urgency and oppositional character."[7] In response to this growing trend, some scholars have attempted to conceptualize this process under the analytical heading of deracialization. Deracialization is a concept that analysts have advanced to account for the ways in which race has been downplayed by African Americans running for office in recent years. Specifically, it refers to the process whereby black candidates campaign without overtly emphasizing race or making direct appeals to issues that are associated with African Americans in an effort to draw white support. McCormick and Jones are among a group of scholars who have sought to fully articulate this concept. They define deracialization as a electoral strategy as "conducting a campaign in a stylistic fashion that defuses the polarizing effects of race by avoiding explicit reference to race-specific issues, while at the same time emphasizing those issues that are perceived as racially transcendent, thus mobilizing a broad segment of the electorate for purposes of capturing or maintaining public office."[8]

In addition to this general concept, the authors suggest that deracialized campaigns may be divided into three parts: political style, mobilization, and issues. *Political style* refers to the manner in which candidates project themselves in a campaign, particularly in terms of presenting a nonthreatening or safe image to potential voters. The confrontational or protest image represented by first generation of BEOs is replaced with a candidate focused on racial reconciliation. *Mobilization* tactics refer to a candidate's direct or explicit efforts to target African American voters, as candidates operating from a deracialized context make generic multicultural appeals to voters seeking to avoid alienating potential white voters. *Issues* directly associated with African Americans, such as police brutality or allocating minority contracts, are rejected in favor of racially neutral policy advocacy.

Deracialized campaigns are largely dependent on two factors: the demographic makeup of the jurisdiction and the racial background of candidates. If a city, for example, has a majority minority population, one can generally assume that a minority candidate will easily win an election if they are running against a white opponent. But when there are several minority candidates, the minority vote is split, thereby increasing the chances that whites, or other minorities, will provide the critical threshold to win the election.[9] For example, following Dutch Morial's two terms in office, Sidney Barthelemy won his first term in office with 85 percent of the white vote. It is in these scenarios that the chances of a deracialized campaign increase. Looking at the 2002 election, one could argue that Nagin's campaign reflected these factors in terms of political style, mobilization, and issues.

In terms of political style, Nagin took the honest businessman approach, pointing to the fact that he had worked with various different groups in the business community. For example, he often pointed to his role as president of the local hockey team—the New Orleans Brass. As one of his supporters noted, "The Brass is a difficult group to work with. . . .There are about twelve partners, and they're from five parishes, and they're black and white, young and old. The only thing they have in common is that they're all successful. Ray holds the group together with his humility, his pleasant demeanor and his unquestioned integrity."[10] In just about every discussion of Nagin leading up to election he was described as an effective businessman with an "off-the-cuff" speaking style, and these were characterizations offered by whites. Because Nagin was not a career politician, he did not have connections to many of traditional political organizations in the city such as SOUL, BOLD, COUP, ACORN, or LIFE.[11] "I don't owe anything to a BOLD or a SOUL or a COUP or a LIFE, all that alphabet-soup stuff," he asserted in one interview.[12] All of these organizations are affiliated with political insiders, and if he made any direct appeals to ACORN or SOUL, for example, this could have been viewed as a direct appeal to African American voters. Moreover, as the *Times-Picayune* endorsement of Nagin noted, "He's not tight with the city's alphabet soup of vote-getting organizations."[13]

In terms of policy issues, Nagin advocated a classical pro-business approach. Urban regime theorists often speak of different types of policies advocated by city administrators. Sites points to three types of regimes that scholars commonly employ in their analytical assessments: caretaker, pro-growth, and progressive. Those regimes that support classical liberal doctrines often advocate pro-growth strategies, in which the emphasis is on "market-oriented development, using incentives or public subsidies to promote the kind of economic growth favored by downtown interest."[14] Those regimes that seek to stabilize or reform government advocate caretaker (maintenance) positions in which developmental issues are largely avoided, "concentrating instead on fiscal stability and improvements in the provision of routine services."[15] And finally, those regimes that "seek to limit downtown expansion in favor of more community-oriented development" are viewed as progressive or social-reform regimes. Nagin advocated the path of pro-growth, with little or no concern for progressive or redistributive alternatives. One of the highlights of his platform was the suggestion that Louis Armstrong Airport should be sold or leased, and then the city could use the proceeds for rebuilding infrastructure. Another indicator of his racially neutral and pro-growth stance involved a proposed City Council referendum to raise the minimum wage $1 above the federal rate. He was the only candidate who opposed the referendum outright, while his opponents expressed support with reservations.[16] In terms of style, mobilization, and policies Nagin was able to present himself as a nonthreatening candidate who focused on economic issues from a market or pro-growth perspective. There would be no threats of redistributive policies on his watch, and because of his lack of political ties, concerns about allocational abuses (awarding political appointments and contracts based on connections rather than merit) were minimal.

A total of fifteen candidates were running for mayor, and prognosticators pointed to Pennington and Irons as the early frontrunners. One poll taken in early December 2001 indicated that voters favored Pennington (23.3 percent), followed by Irons (21.1 percent), Singleton (13.6 percent), and Carter (10.7 percent).[17] At this point in the campaign Nagin was not on the political radar screen, and by early January he was receiving barely 5 percent in the polls. A series of allegations directed at frontrunner Irons caused an opening in the crowded landscape, with Nagin being the primary beneficiary,[18] and by late January Nagin had increased his support to 11.9 percent. In a matter of two months he went from dark horse to frontrunner.

In the February 2 primary election, Nagin received the most votes, 29 percent versus Pennington's 23 percent. When Nagin made it into the runoff, endorsements and campaign contributions began to roll in. Several members of the political establishment threw their support behind Nagin, including Council Chair Jim Singleton, Councilman Oliver Thomas, and Councilman Troy Carter, while Pennington received the support of ACORN and the support of Mayor Morial.[19] Allegations of corrupt business dealings were directed at Nagin, while Pennington had to answer questions about rumors of domestic abuse. Nevertheless, in a campaign that became very dirty, even by Louisiana standards, Nagin was able to hold on to his lead and be elected to office. He won his first term because he secured a larger percentage of the white vote, more than 80 percent. He was clearly a candidate embraced more by whites than blacks. In particular he was supported by the very influential business community. Looking at election results from 2002, Nagin won those areas heavily populated by whites (Appendix A). In the Lakeview and Audubon Park/St. Charles parts of the city, Nagin received almost five times as much support as Pennington. For example, in the Fourteenth Ward, which represents the center of old wealth in the city, Nagin received 5,782 votes compared to 1,635 for Pennington. In the Lakeview area (Fourth Ward), which may be the most segregated part of the city,[20] Nagin had a similar winning margin of 5,949 versus Pennington's 1,471. It may be noted, however, that Nagin made a strong showing throughout the city and won the election by more than 20,000 votes. In three of the largest wards that are predominantly black—the Seventh, Eighth, and Ninth Wards—Nagin won the majority of votes.

I Once Was Lost, but Now I'm Found: Nagin's Return Home

"I can't stand him," I recall my mother saying in the immediate months following the hurricane. She was talking about Mayor Nagin. Since he had been elected in 2002, she would from time to time mention her general dislike for the new mayor. He was a new face, not part of the traditional old guard, and he secured his election by appealing to whites through his pro-business rhetoric. In the first election he won 85 percent of the white vote, compared with only 35 percent of the black vote. So when my mother mentioned her dislike for the mayor, it was not an isolated viewpoint. To a certain degree, the city was at a crossroads. The traditional Creole elite had no new candidates because both Barthelemy and Morial had secured successive terms. Although Pennington had been endorsed by outgoing mayor Marc Morial and Congressmen William Jefferson, the combination of last-minute scandals and Nagin's alliance with the business community allowed him to be elected. Nagin entered office with a classic reform/pro-business agenda. In his inaugural address he suggested that the city needed to be a place where "the business community is heavily involved and invested in the growth and success of the entire New Orleans metropolitan area," and he spoke of the need to put an end to "past parochial politics."[21] His reform agenda mainly centered on ending the high degree of corruption associated with previous administrations. Like the scandals that can emerge in Congress when a new party takes power, the target of many of Nagin's probes were former associates of Mayor Morial; he also enacted a dramatic firing of hundreds of city workers (partially due to budgetary issues). Nagin did little to endear himself to the old guard, those associated with either the Barthelemy or the Morial regimes. He was not the typical Seventh Ward Creole. Nevertheless, incumbency in the New Orleans, as it is in most cities, was viewed as virtually a shoo-in. That was until Katrina hit. Instead of an easy reelection campaign, Nagin found himself in one of the most interesting reelection campaigns in recent state history.

As mentioned earlier, Nagin won his first term by running a classic deracialized campaign, gaining that critical threshold of white support needed to win an election between black candidates. Following the hurricane there was general perception among blacks that Nagin shared a large responsibility for the incompetence of government in the first ten days after the storm. Local hip-hop artist Juvenile summed up the feeling of many in one of his songs: "Your mayor ain't your friend he's the enemy / Just to get your vote a saint is what he pretend to be / F__ him."[22] This is not surprising, considering the horrific experience the people of New Orleans went through during and after the storm; for one not to be angry would be a surprise. But given the large amount of public outcry following the storm, one would assume that a Nagin reelection would be virtually impossible, but this was not the case.

KATRINA AND THE POLITICAL CONTEXT

Hanes Walton argues that any analysis of the field of black politics must first seek to identify the particular political context, spatial and temporal, in which the subjects are operating. He contends that black politics is not linear, nor does it replicate itself cross-nationally; instead, black politics is highly sensitive to the local context in which it is operating. Furthermore, "the political context variable is. . .a thesis that postulates that political behavior at either the individual or the group level is not independent of the political environments (a particular time and a particular period) in which it occurs."[23] Walton's positions are reinforced by Kaufmann in her analysis of mayoral elections in Los Angeles and New York. Drawing from socio-psycholog-ical group interest theory, she contends that in times of racial conflict voting will be "polarized" along racial lines, whereas in the absence of racial conflict voting patterns will follow traditional cues such as party identification and ideology. "When elec-tions take place in a conflictual environment," Kaufmann contends, "voting behav-ior will likely reflect the temporal salience of these interests. However, when racial conflict recedes, voting behavior will likely revert to more normal political consider-ations."[24] Despite different theoretical origins, both Walton and Kaufmann high-light the salience of the political context, particularly at the local level. Before we consider the campaign efforts of the mayor, it is critical to recognize the immediate demographic byproduct (and therefore political opportunity) of the storm that immediately affected the political context.

Katrina affected the political context in New Orleans in two key ways: attitudi-nally and demographically. The post-Katrina response by government agencies elicited a critical response on the part of blacks and whites. Although several national and local commentators attributed government mismanagement to Mayor Nagin, public opinion data suggest that both blacks and whites attributed more culpability to Governor Blanco and President Bush than to the mayor. Mayor Nagin had a 54 percent approval rating, compared with 33 percent for Governor Blanco, 23 percent for President Bush, and 22 percent for the Federal Emergency Management Agency.[25] Looking more closely at the approval ratings of key actors, some racial dif-ferences are revealed. Nearly 60 percent of blacks approved, compared with 47 per-cent of whites. Only 24 percent of those who approved of Governor Blanco were white, versus 41 percent of blacks. Overall, blacks were more prone to direct respon-sibility at the federal government than at local or state politicians, except in the his-torically sensitive area of law enforcement. In the days immediately following the storm, a number of politicians seemed to be more concerned with "law and order" or "property and security" than with "search and rescue." These cognitive cues were also reinforced by the physical byproducts of the storm.

Demographically, initial surveys indicated that the city was now majority (roughly 60 percent) white. Over twenty candidates threw their hats in the electoral ring, with only two other African American candidates (the Reverend Tom Watson and Civil Clerk Kimberly Butler) in the race. There was a general feeling that white elites wished to exploit the forced mass exodus of blacks into a political advantage. One state representative even commented that the storm had accomplished what they were trying to do for years—clean up public housing (translation: get rid of poor blacks). Scholars of black politics have long discussed issues of gentrification and the increase in the number of low-income blacks being forced from central cities because of massive redevelopment projects. New Orleans has the potential to be the largest gentrification project in history, and there was a growing sentiment among African Americans that this was being planned. This demographic factor had a large impact on the public perception of blacks. Thus the political context that Walton speaks of had been significantly altered. Keeping in mind the contextual factors that affected Nagin's attempt for reelection, we need to ask the next critical question: what type of campaign measures did Nagin undertake to secure reelection?

Chocolate City: The Anatomy of a Political Crossover

"What's up with Nagin?" I recall asking one of my friends after Nagin's comments on Martin Luther King Day. On MLK Day most networks were broadcasting sound bites of the Mayor stating that New Orleans would be a "Chocolate City" again. I wanted to know if I was missing something in the national media; maybe it was different in New Orleans. Mayor Nagin talking about New Orleans being a Chocolate City was as surprising to me as it must have been for some whites. "I don't know," my friend responded, "I think he might have been drunk." He was joking, but this explanation seemed to be more plausible than the thought of a contemporary politician making an explicitly racial comment about the demographics of his jurisdiction. This was far from the political style that characterized him during his first term; he rarely made any direct appeals to African Americans. As the previous section noted, his electoral victory was largely due to white support; Pennington won a larger percentage of the black vote than Nagin. If Nagin represented a classic case of a deracialized campaign the first go-round, he was risking a key component of his electoral base by making these comments. Of whites, 52 percent indicated that they were offended by his statement, compared with, predictably, only 17 percent of blacks stating they were offended.[26] The majority of blacks (66 percent) and a significant number of whites (42 percent) responded that they were not offended but thought he could have said it better. Contemporary politicians, both black and white, often speak in general terms absent of specific racial markers. When a politician says they are concerned about "soccer moms" or middle-class issues, they are indirectly referring to issues that largely concern whites. Conversely, when a politician starts mentioning "cleaning up housing" or more "law and order," these are cognitive cues that convey to some whites their desire to eliminate any perceived threat posed by blacks.

The issue is not whether or not Nagin should have mentioned his concern that African Americans return to New Orleans, but how he said it.

What initially was seen as political misstep, may have been, in fact, a well-calculated move. The demographic shift toward a white majority was clear immediately after the hurricane, and any questions about how this would affect the political landscape were answered when the candidates were announced. While there had occasionally been a white challenger since Dutch Morial's mayorship, there had never been more than two other white candidates in a campaign. With several white candidates running for mayor, it was clear that whites saw this as an opportunity to retake City Hall. Rather than deny the existence of this new political context, Nagin sought to abandon any hope of matching the large percentage of whites who supported him during his first election and instead focused on attracting a greater proportion of African American voters. This process would entail both national and local linkages.

Although there is a significant degree of fragmentation within the black community, most blacks accept that they have a shared history and culture and view racism as still a salient component of American society (although in more indirect ways). Whether one chooses to label it as a cultural-historical dynamic or a variant of group identity (socio-psychological), there is a general sense of being black. As is often the case following serious incidents with racial undertones, most African Americans viewed the post-Katrina problems as indicators of racism. A small indicator of this are the lyrics offered by Mos Def on an underground remix released days after the storm:

> Its dollar day in New Orleans
>
> . . .
>
> and Mr President he bout that cash
> he got a policy for handling the niggas as trash. . .[27]

Despite the claims of some "post"-oriented critics, cultural forms have always revealed a certain essence about African Americans, and the lyrics from this song seem to reflect what a large number of blacks nationally were thinking following the storm. These positions are empirically reinforced by some of the surveys taken following the storm. According to a *New York Times/CBS News* Poll, two-thirds of African Americans said that race was a factor in the government's response, compared to only a third among whites.[28] Data from a Pew Center poll also supports this view. Of blacks, 66 percent felt the government's response would have been faster had the victims been white, compared with only 17 percent of whites. Predictably the event was a painful reminder to African Americans that racism still exists in this country, with 77 percent of blacks responding that "racial inequality still a major problem." As in other polls conducted during this period, the majority of African Americans held President Bush responsible for the government's incompetence following the storm, with 85 percent of blacks responding that he could have done more. The looting following the storm, additionally, was also viewed differently by blacks and whites. Of blacks, 57 percent felt that the looters were just "ordinary people who were desperate," versus 38 percent of whites. Beyond the difference between white and black responses to the hurricane, these data also indicate that there was a national concern about the collective fate of blacks in the months to come. In terms of electoral considerations, several national organizations indicated their concern about the upcoming elections. They suggested that Louisiana officials were attempting to disenfranchise blacks by holding early elections. Initially, the Louisiana Secretary of State sought to postpone the election indefinitely, but a group of whites challenged this ruling in court. The court ruled in favor of the plaintiffs, and the election was scheduled for April 22, 2006. The NAACP and Urban League attempted to appeal the decision, but to no avail. These events were important because they indicate there was a certain level of national support for the candidacy of Nagin; in particular, there was the feeling nationwide that whites were trying to strip blacks of their voting rights.

Locally, Mayor Nagin focused his attention on winning the support of African Americans, those still in New Orleans and those scattered across the country. He made a 180-degree turn in terms of style, mobilization, and issues, abandoning the deracialized tactics that had defined his first campaign. Both locally and nationally there was a huge uproar about his chocolate city remarks. At a Martin Luther King Day march the Mayor said, "We ask black people: it's time. It's time for us to come together. It's time for us to rebuild New Orleans—the one that should be a chocolate New Orleans. And I don't care what people are saying Uptown or wherever they are. This city will be chocolate at the end of the day."[29]

Despite the fact the comments were largely unfiltered in terms of the language he used, the Mayor did touch a sensitive issue among African Americans from the city—that whites were going to use the hurricane to significantly reduce the number blacks in the city. Not only did Nagin make frequent mentions of race in his Martin Luther King Day speech, but he also did so in speeches throughout the city and in places like Houston and Atlanta, where a large number of New Orleans residents had relocated. In one talk in Houston, the Mayor noted that none of the other candidates looked like him (i.e., black), while several campaign billboards in the city referred to Nagin as "our mayor."

In terms of mobilization, Nagin gained the support of ACORN, a community-based organization seeking to address the needs of working- and lower-class African Americans. This group served a key role in both voter registration efforts and busing out-of-state residents to the polls. On a national level, the National Coalition of Black Civic Participation (an umbrella group consisting of the National Urban League, the NAACP Legal Defense and Educational Fund, the Lawyers Committee on Civil Rights, and the Louisiana '06 Coalition) assisted in holding mayoral debates in Houston and Atlanta and provided poll monitors in several sites.[30] In terms of policy articulations, Mayor Nagin resisted recommendations by several groups to restrict building in areas such as the Ninth Ward. Many groups called for a selective repopulation of the city, starting with (disproportionately white) areas that had lower levels of damage. The mayor rejected this suggestion, announcing all areas could be repopulated but at the citizens' own risk. De facto gentrification via a process of large-scale seizures of property is one of the key concerns of African Americans in the post-Katrina context, as the Ninth Ward east of Industrial Canal (New Orleans East) has long been viewed among the worst land (in terms of flood

potential) in the city. However, to condemn the Ninth Ward would be political suicide for a candidate counting on African American votes to be reelected. The Ninth
Ward is the largest ward in the city (in the 2002 elections over 40,000 votes came
from this area), and New Orleans East has a large number of middle- and upper-class
blacks. Overall, Nagin courted these areas, suggesting to African Americans on a
consistent basis that he was their only hope for the future.

Mayor Nagin's reelection strategy worked, because of him and in spite of him. A
total of twenty-two candidates were on the April 22 primary ballot, and Mayor Nagin
and Lieutenant Governor Mitch Landrieu finished first and second in voting. Nagin
received 38 percent, Landrieu received 29 percent, Audubon Institute Director Ron
Forman received 17 percent, Republican Lawyer Robert Courhig received 10 percent, and the rest of candidates each received less than 2 percent (the other two
notable African American candidates, the Reverend Tom Watson and Civil Clerk
Kimberly Butler received 1 percent each). Although Nagin finished first in the primary (securing nearly all precincts that were black), nearly 60 percent of voters voted
for Landrieu, Forman, and Courhig. Scholars of voting behavior have long noted that
runoffs can potentially serve as a form of vote dilution, and in races where the bulk
of candidates are white, the general assumption is that their votes will be aggregated
behind one candidate, thereby making it more difficult for the minority candidate to
win. In theory, if Landrieu was able to secure the support of Forman and Courhig,
winning the May 20 runoff should have been easy. Landrieu was able to secure the
endorsement of Forman, but Courhig threw his support behind Nagin. Within a
racialized context many observers thought the contest would come down to who
would win the largest percentage of the crossover vote, but in the runoff both candidate, received roughly 20 percent from crossover voters. The key to Nagin's victory
was the overwhelming support among African Americans. The mayor received near
unanimous support from local and national black leaders and organizations. More
importantly, he was able to increase voter turnout among blacks, whose turnout
increased from 53 percent in the primary to 55 percent in the runoff. Looking at the
election returns (Appendix B), Nagin won virtually all precincts where blacks constituted well over 80 percent of the population; conversely, Landrieu won nearly all
precincts where whites represented over 80 percent of the population. In wards Four
and Five, which represent the Lakeview/City Park area, Landrieu led Nagin by a ratio
of almost 3 to 1 (3,876 versus 1,365 in the Fourth Ward, for example). In the influential Uptown/Audubon Park area (the Fourteenth Ward) Landrieu had his highest
turnout of any wards with 5,246 versus Nagin's 1,733. When one looks at the areas
with a large number of blacks, a similar trend emerges. In the now-infamous Ninth
Ward, Nagin received 13,972 votes compared with Landrieu's 6,163, while he also
won a solid majority in the Seventh Ward (5,230 versus 3,442). The white vote that
gave Nagin his victory in 2002 was almost completely against him in 2006. For
example, in two areas that have a high concentration of wealthy whites, the Fourth
and Fourteenth Wards, in 2002 Nagin won by a ratio of almost four to one (11,371
compared to Pennington's 3107). In the 2006 election his opponent won those areas
by almost a three-to-one ratio (9,122 versus 3,098 for Nagin). Overall, African
Americans were able to reelect Nagin in one of closest elections in the city's history
with Nagin, receiving 52 percent of the vote versus Landrieu's 48 percent. I know of

no other contemporary politician who has been able to switch bases and still win reelection. For their strategic insight the Nagin camp must be applauded, but for one to view this election as simply about Ray Nagin would be thoroughly shortsighted.

When election night came, my mother asked me who I thought would win (I'm the political scientist, right?), and I said Landrieu. If the demographics the media were reporting of a new white majority were right, I didn't see any way Nagin could win. She seemed down when I said that, in an "oh, well" kind of way. When she found out Nagin won the following morning, she was visibly happy. "Yeah, I voted for him," she hastily admitted. This was the same person that "couldn't stand him" six months ago. She was by no means a Nagin convert, but she did recognize (as most other blacks from New Orleans) the significance of this election. One could argue that African Americans were making a statement: they were not about to let whites take over the city. Scholars of local politics acknowledge the limited power that most mayors have in implementing policy without the support of the federal government and local business communities. Moreover, contemporary mayors rarely favor any type of progressive or redistributive policies that favor minorities or low-income groups. Most are forced to adopt a pro-growth or caretaker posture if they wish to achieve anything. After nearly three decades of black mayors, there has been very little improvement in the city, and most blacks are well aware of this. However, on a symbolic level, blacks were not ready to concede City Hall to a white mayor. In an environment devoid of post-Katrina racial polarization, this may have been possible. But the political context of Katrina created a nonphysical conflictual environment between blacks and whites over the future of New Orleans, and in such an environment racial symbolism can be very powerful. Nagin deserves credit for skillfully manipulating the context, but the true victors (at least symbolically) were the black voters of New Orleans.

(DYSFUNCTIONAL) REGIME POLITICS: A NEW ORLEANS STORY

From the dirty dirty, ya heard me,
u either shoot or get shot
From where the unemployment line
be as long as your block. . .

—New Orleans hip-hop artist Juvenile, "NOLA Clap"[31]

To only focus on the electoral aspects of New Orleans would be terribly shortsighted. A critical problem in black politics is that African Americans frequently evaluate minority candidates employing only descriptive and symbolic considerations, whereas substantive evaluations often elude the evaluation process.[32] Instead of this process, Manning Marable suggests, "accountability must be measured objectively according to a list of priorities, and not determined by political rhetoric at election time."[33] Despite any momentary euphoria caused by Nagin's successful reelection, the most substantial problems that confront New Orleans were present prior to Katrina's devastation. New Orleans native and political theorist Adolph Reed Jr.

asserted shortly after the storm, "What happened in New Orleans is the culmination of twenty-five years of disparagement of any idea of public responsibility."[34] The lyrics quoted above were released in 2004, and they reflect (although somewhat exaggerated) the problems of high poverty and crime that gripped New Orleans. The city has been plagued by a number of problems that reflect mismanagement or failed policy in key economic and educational areas. Public schools in the city have been deteriorating since the early 1980s, while there are only two "Fortune 500" companies in the metropolitan area. As a result of the city's underperformance, it has witnessed one of the highest population declines of a mid-sized city in the region (-3.2 percent) between 2000 and 2003.[35] With the exception of the tourist industry, the city had been steadily declining economically for the past twenty-five years. This decline can be attributed (locally) to two primary reasons: a dysfunctional regime and a one-dimensional focus on tourism. Elite greed and corruption has lead to a dysfunctional regime that lacks the ability to attract or develop corporate interest locally, while the main source of revenue—tourism—has not been connected with any developmental projects for the general population. Before examining these underlying factors that plague the city, however, it is first important to have baseline understanding of the city's economic performance.

POLITICAL ECONOMY IN BRIEF

When one looks at various economic indices in New Orleans, the picture of a struggling economy emerges. The decline is more apparent when the city is compared with other cities in the South. The median income in New Orleans is $27,133, more than seven thousand less than Houston ($36,616) and Atlanta ($34,770), and only slightly more than Birmingham ($26,735). Poverty and underemployment are also major issues in the city, as nearly 30 percent of individuals make less than $15,000 a year, while another 15 percent of the population makes less than $25,000 a year. When compared to the regional jewels of Houston and Atlanta, New Orleans lags significantly behind. In terms of low-income households, New Orleans has a larger percentage than Atlanta and Houston at 36.3 percent, that is more than 10 percent higher than Houston (which does have a significantly larger population) and higher than Birmingham, which has a smaller population. A similar gap emerges when comparing high-income groups, with only 12 percent of households in the high income group in New Orleans, compared with 20.9 percent in Atlanta and 17.1 percent in Houston.

Income inequality is only one part of the equation in examining New Orleans's economic woes; the composition of industry is another major problem. Cities like Atlanta and Houston have a major corporate presence; this is the true engine of economic growth, not tourism. New Orleans only has two Fortune 500 companies, while Atlanta has seventeen, and Houston has twenty-two. I have lived in both New Orleans and Atlanta, and the numbers fail to capture the huge discrepancy between the two cities in terms of economic development. When I moved to Atlanta in the late 1990s, I was struck by the level of development and economic growth. Driving around the perimeter (I-285), one witnesses regional or home offices for nearly every major corporation in the country; large cranes and construction are part of daily life in both the central city and the suburbs. Atlanta's economic boom has led to a dramatic increase in its population. One indicator of this growth is building permits. Although Atlanta is generally viewed as much larger than New Orleans, much of its population resides in suburban areas. In 2003 the population of the city of Atlanta was 419,729, while the total population of the metropolitan area was 4,529,371. Conversely, the population of the city of New Orleans is 467,934, while the total metropolitan area is 1,336,348. Keeping these demographics in mind, the gaps between the two cities in terms of housing growth is substantial. In 1980, New Orleans issued 2,047 housing building permits compared with Atlanta at 1,285. By the mid 1980s this had reversed, with the number of new building permits significantly dropping in New Orleans while there was a steady increase in Atlanta.

Housing growth peaked in New Orleans in 1983 when 3,199 new permits were issued (1,103 in Atlanta); after that the city began a steady decline. In 1986 there were 2,346 building permits issued in Atlanta with only 637 issued in New Orleans. The recession that affected the national economy in the late 1980s and early 1990s had a predicable impact on housing growth in both cities, but by 1993 Atlanta was well on its way to a recovery while New Orleans would struggle until 1996 before it returned to pre-1987 levels. The mean number of housing permits over this twenty-four year period is 2,534 for Atlanta and 893 for New Orleans. In the last year for which data was available (2003), there were 917 permits issued in New Orleans, compared with 6,893 in Atlanta. Within this time period New Orleans has yet to issue more than 1,000 permits in one year and falls well short of the 1983 total of 3,199. In nearly every major category of social and economic development the city of New Orleans falls well behind its regional neighbors.

Some of New Orleans's economic problems, though, can be partially attributed to the general decline in federal support to cities that occurred after the 1970s. Under the heading of "New Federalism," presidents since the 1970s have significantly reduced federal aid to cities.[36] Republican occupancy of the oval office has led to two policy patterns that have significantly reduced the amount of revenues distributed to local governments. First, the Republican belief that the private sector is a better source for economic development and growth than are federally funded programs has had a significant impact on urban development. Beginning with Reagan and continuing to the second President Bush, lawmakers have championed the virtues of market-based growth. "The Reagan Administration," Judd and Swanstrom note, "set about sharply reducing federal urban aid proclaiming that 'the private market is more efficient than federal program administrators in allocating dollars.'"[37] This was also related to the second policy shift of devolution, "a strategy in which the national government would grant more authority over a range of policies currently under national government authority," although in many instances the states bore a substantial portion of the cost of these programs.[38] The move from a centralized to decentralized view of the government on the part of Republican presidents has lead to an environment in which state and local governments have more authority but a decline in intergovernmental grants. As a result of this policy shift, between 1980 and 1990 there was a 46 percent net reduction in federal grants to city governments.[39] When the federal government does provide grants, Rich asserts, "the vast majority of federal urban assistance today is not being administered by city governments or local agencies aligned with City Hall."[40] To solely locate New Orleans's economic struggles within an intergovernmental context, however, would be to overlook the gains that other cities have made since the early 1980s. As the previous data has indicated, places like Atlanta and Houston continued to exhibit strong economic performance in the same environment that New Orleans struggled with. Therefore, to understand the source of the city's troubles it is necessary to examine the performance of the local regime.

REGIME POLITICS

Scholars of local politics often use the term *urban regime* to refer to the interaction of political and business interests that facilitate the growth and development of cities. At the core of this relationship are elected officials who are responsible for steering institutional mechanisms and business actors who command large amounts of financial capital. According to Stone, an urban regime refers to the "informal arrangements by which public bodies and private interest function together in order to be able to make and carry out governing decisions."[41] While civic associations (such as community groups) may sometimes enter into the equation, the central area of concern is the interaction between business and government officials. The end of big government, exemplified in the Great Society programs of the 1960s and 1970s, led to an increasing strain on municipal governments, thereby making the support of private business critical in the vitality of modern cities. Conversely, despite the large amounts of capital that financial elites can throw around, they are still dependent on elected officials for navigating institutional regulations (property taxes, land grants, public works, et cetera).

Some scholars have noted that urban regime theory, as articulated by Stone and others, fails to account for macroeconomic shifts that oftentimes constrain the activities of local government.[42] Another important criticism of urban regime theory is the failure to account for racial and ethnic dynamics in some cases.[43] In terms of an urban regime with a black majority, Reed notes that the combination of macro political-economic pressures and the intra-group hegemony of the black middle and upper class has produced a generation of black mayors who proclaim a progressive agenda while maintaining an unwavering commitment to pro-growth strategies. He asserts that "the new regime of race relations. . .has exerted a demobilization effect on black politics precisely by virtue of its capacities for delivering benefits and, perhaps more important, defining what benefits political action can be legitimately be used to pursue."[44] The main factor that Reed and others point to is not anti-communal policy, but it is the ability of BEOs to frame their public discourse to shape the public's perception that "downtown development" strategies are good for lower-class blacks. Employment promises (as a byproduct of these development projects), combined with selective administrative appointments, have allowed many BEOs to give the impression that they were involved in policy promulgation and implementation that would aid the majority of the population, when in fact only elites receive any substantial returns from these directives.[45] When one looks at the actions of the urban regime in New Orleans, this pattern is clearly evident. Unlike the elites in municipalities like Atlanta, New Orleans elites cannot claim aggregate economic growth. The following section highlights some of the characteristics of this dysfunctional regime.

Outside of tourism, New Orleans generates the bulk of its revenues from small businesses and the health industry. This is not to say there have not been opportunities for corporate development; rather, the other key issues—corruption and greed—have served to alienate potential investors. For example, high levels of corruption have existed since the days of Governor Huey Long, and they were exemplified in the conviction of former Governor Edwin Edwards. Mardi Gras is another example: for all the fanfare it receives nationally, few outsiders recognize that on certain levels it is an elite ritual; many of the largest parades, like Bacchus and Endymion, are sponsored by private social organizations whose members (nearly all white) are composed of some of the wealthiest people in the city. In an earlier study, Schexnider noted that "wealth and power in New Orleans are concentrated in a small but powerful oligarchy that unduly influences the direction of economic development and public policy."[46] Citing from a study by James Bobo, he continues, "Significantly, there is a small but powerful, highly structured upper class and the middle income class is much smaller in metropolitan New Orleans, compared to all SMSA's [standard metropolitan statistical areas], Atlanta, Dallas, and Houston. In few, if any, major metropolitan areas is income so inequitably distributed. And an inequitable income distribution is in part self-perpetuating."[47]

Contemporary talks of reform often overlook the influence of this group, but as one writer noted prior to Mayor Nagin's first term, these individuals share an equal responsibility for the city's decline over the past fifty years: "No one denies that corrupt politics has held Louisiana back, yet for all their protestations about reform, many members of the business community have formed a co-dependent relationship with the very people they publicly revile."[48]

Some observers have characterized the economic plans of the city as "ad hoc" or "surprise" planning. One columnist in the *Times-Picayune* has noted, "New Orleans lurches forward in an unplanned, ad hoc manner with private-sector development gathering support and public works receiving approval largely based on what interest group, developer or politician influences the planning process at a particular time. While more progressive cities in this country have master plans, New Orleans has what I like to call 'planning by surprise.' You wake up in the morning, you read *The Times-Picayune*, you see what the plan is."[49]

Some may contend that these backroom deals are a normal part of politics. Peterson, for example, contends that "issues of patronage and corruption dominate local politics not because local officials are particularly venal or devious, but because employment issues are one of the limited set of matters readily resolved at the local level."[50] Despite the commonality of this practice, nevertheless, in New Orleans allocational policies are so exaggerated that competence, economic growth, and the general impact on the city are completely overlooked in exchange for political connections. One example of this involved the management of Louis Armstrong Airport during Marc Morial's second term. Although it is located in Jefferson Parish, the airport is governed by New Orleans officials, and every five years a contract worth approximately $20 million is awarded to a firm to manage the facilities. In a story published by the *Louisiana Weekly*, it was reported that Mayor Morial was attempting to bypass the committee setup to evaluate candidates and select a personal ally. It noted:

> Bob Tucker, friend of New Orleans Mayor Marc Morial, has endeavored to win the $20,000,000 management contract at the Airport for some time. Of the original six applicants for the private operations contract, only Tucker's AMC and Parson's Aviation remain. A private poll of the Aviation Board allegedly reveals that the majority of the members want Parson's, an international firm that operates airports all over the world, to get the contract. The members were reportedly impressed by Parson's experience and expertise. However, sources reveal to The Louisiana Weekly that "friends of Tucker and Morial" have blocked Parson's bid.[51]

These types of occurrences were so commonplace that they rarely attracted public attention; allocational corruption was viewed almost as a part of the city's political culture.

Reed notes that one of the main characteristics of urban regimes with black majorities is the ability of black elites to mold the public's expectations of success and failure. Like other mayors in the past, Marc Morial was able to shape public perception of his performance in a positive light largely due to reductions in crime and new tourism projects. Howell and McLean report that Morial enjoyed a 90 and 91 percent approval rating among blacks in 1996 and 1998, respectively, with a 59 percent approval rating among whites.[52] In addition to improvements in law enforcement, Morial was able to point to a number of large projects under his watch, including an expansion of the Ernest "Dutch" Morial Convention Center, the construction of an amusement park—JazzLand—in New Orleans East, and the construction of a new sports arena built to attract an National Basketball Association (NBA) franchise. Among blacks and whites these efforts led to positive evaluations, yet, as the data in the previous section reveals, New Orleans lagged significantly behind places like Houston and Atlanta in median income and income distribution among the population during his administration. The regime's emphasis on insider politics, without any efforts to expand the city's middle class, led to a situation of incremental economic growth exaggerated under the pretext of tourist events and downtown building projects.

TOURISM IN THE BIG EASY

New Orleans was the place everyone liked to go see, visit, and party. A place that people from other places wanted to go to act silly, it was viewed largely as a party town. It had good food, good music, and relaxed laws when it came to social pleasures. This is a key part of the problem. Because New Orleans combines a rich artistic history with an open tolerance for various types of human behavior viewed much more negatively in other areas of the country, its economy has largely been built around tourism. Tourism contributes $5–8 billion to the city and state economies annually, as well as 30 percent of its tax revenue.[53] In addition to its natural endowment of fine food and music, the city and state have built an infrastructure aimed at harnessing the tourist industry. The Superdome and Convention Center were built to draw large sporting events (the Sugar Bowl, the Super Bowl, the Bayou Classic, and the Essence Festival) and various conventions and conferences. The problem is not that tourism serves no benefit to the local economies, but rather the nature of that benefit. New Orleans has long been a top (adult) entertainment destination, not only for native events such as Mardi Gras or the Jazz & Heritage Festival, but also for sports. Prior to conflicts between the state and Saints owner Tom Benson, the city had hosted the most Super Bowls, along with Miami and Los Anteles (it is also home for the college Sugar Bowl). Over the past twenty years nearly all major development projects have centered on tourism. The state has financed the building of the Morial Convention Center (with subsequent expansions) and the New Orleans Arena (in order to attract an NBA team). Like other cities, the "downtown" area has also been subject of investments such as the Riverwalk, Jax Brewery, and the Aquarium of the Americas. Many of the earlier downtown projects were an outgrowth the World's Fair, which the city hosted in 1984.

The state and city benefit heavily from these venues and events. As noted earlier, the city and state received more than $5 billion from tourism in 2004. The issue is linkage: does an economy centered on tourism benefit the general population? Building a new sports arena or aquarium cost hundreds of millions of dollars, finances that could greatly benefit a public school system that has been struggling for decades. Politicians often speak about the benefits of these projects to the city and state, but after years of investment there has been virtually no improvement in the city's other areas. The reality, as Eisinger notes, is that "the amount of fiscal and political resources and the level of energy that local elites must devote to the revitalization of large entertainment projects are so great that more mundane urban problems and needs must often be subordinated or ignored."[54] Several studies have noted that cities that engage in these gathering industries rarely see the returns they predicted from these investments, or the revenues received are not funneled towards more pressing needs. Moreover, as Judd and Swanstrom indicate, "the downtown corporate economy has few links with small business and job creation in neighborhoods. Central city residents often lack the skills and education to qualify for the knowledge-intensive jobs that locate in central business districts. . . .Many of the jobs in the tourist industry that do go to central city dwellers are unpleasant, low-skilled jobs, with little or no prospects for advancement."[55]

Although the tourist industry can generate significant revenues for the city and state, it does not develop the type of corporate infrastructure that defines the country's big cites. It has failed to attract industry like Atlanta, lacks the financial and industrial legacies of New York and Chicago, and, more importantly, it has failed to capitalize on its comparative advantage in the petroleum industry (due to its close location to the Gulf of Mexico), losing out to Houston. This is not to suggest that these cities have linked these industries to progressive or redistributive policy agendas, as these areas have also neglected their lower-income populations.[56] These industries, however, do lead to aggregate city growth, whether it be in finance, technology, or manufacturing. The only people I ever hear mentioning that they want to move to New Orleans are artists or musicians; everyone else would mention it as a place they would like to go and party. "The Big Easy" may be a nice name for a movie or a song, but in terms of politics and economics, it symbolizes that the city is a joke. Houston, Chicago, or Atlanta—you go to these places to do business; people go to New Orleans to get drunk. Sometimes I wonder who is more intoxicated—the visitors or the local elites who celebrate these excesses as an accomplishment.

CONCLUSION: NEW ORLEANS—THE CITY THAT CARE FORGOT

There is a lot unknown in terms of the rebuilding process and the future of New Orleans. The city is incrementally struggling to get back on its feet. When I ask friends about their future plans, I get mixed responses. A longtime friend who lived in one of the few areas that had minimal damage (the West Bank) is planning on moving his family to Atlanta. Both he and his wife worked for the public school system prior to the storm, but since the hurricane an already deplorable school system is currently in shambles as the state seeks to reorganize it. My friend lost his job, while his wife was offered another position with the school board at the "very competitive" salary of $28,000. This is a typical middle-class family like those most cities seek to attract, yet New Orleans is driving these residents away. Unlike my friends who are younger and more mobile, my immediate family wishes to return home. "After the hurricane season" is the familiar response I hear from them, as no one wants to deals with the anxiety of being in the city if a hurricane enters the Gulf. This is the unknown part of the equation, but as the example of my friend illustrates, much of the city's future is dependent on the plans and policies that will be implemented in the near future. More than a year after Katrina, politicians at the local and state level continue to move at a snail's pace. There are at least four different commissions appointed by the governor, the mayor, and City Council that are charged with articulating policy options. Each plan has a different vision of the city's future. One major issue is whether the city should focus its resources on the rebuilding process. Outside planners have favored the areas that suffered the least amount of damage, those at higher elevations. However, these areas are disproportionately inhabited by whites. After winning an election through strong African American support, the mayor would be committing political suicide to condemn large parts of the city. At the state level, the governor has been slow to distribute funds to individuals, particularly the Road Home Grants. Needless to say, the frictions that existed between Blanco and Nagin before the storm continue to be an issue. Yet within this political stalemate, the people of New Orleans continue to suffer. Visiting there in August 2006, I was struck by this weird combination of despair and hope. The bulk of the city still looks like a ghost town, but there are pockets where the spirit of New Orleans is alive and well. But the question remains: Will New Orleans ever be the same?

NOTES

1. Lil' Wayne, "George Bush," *DJ Drama and Lil' Wayne Mixtape* (Sum. 2006).
2. Consider the following: Michael Parenti, "How the Free Market Killed New Orleans," *The Humanist* (November/December 2005): 16–18; Saladin Muhammad, "Hurricane

Katrina: The Black Nation's 9/11! A Strategic Perspective for Self-Determination," *Socialism and Democracy* 20 (July 2006): 3–17; Henry Giroux, "Reading Hurricane Katrina: Race, Class, and the Biopolitics of Disposability," *College Literature* 33 (Summer 2006): 171–96; and David Troutt, *After the Storm: Black Intellectuals Explore the Meaning of Hurricane Katrina* (New York: New Press, 2006).

3. Several scholars have used terms like "hidden transcripts" and "standpoint epistemologies" to highlight the difference between public (superordinate) and private (subaltern) discourses. These authors note the importance of examining personal narratives and cultural production as a means of countering the dominant discourse, and understanding the resistance practices of the oppressed. My usage of the concept focuses mainly on tapping into perspectives that may have been overlooked in the dominant discourse on Hurricane Katrina. See Patricia Hill Collins, *Fighting Words: Black Women and the Search for Justice* (Minneapolis: University of Minnesota Press, 1998); James Scott, *Domination and the Arts of Resistance: Hidden Transcripts* (New Haven: Yale University Press, 1990); Manning Marable, *Living Black History: How Reimaging the African-American Past Can Remake America's Racial Future* (New York: Basic Books, 2006); Robin Kelley, *Freedom Dreams: The Black Radical Imagination* (Boston: Beacon Press, 2002); Gayatri Spivak, "Can the Subaltern Speak?" in *Colonial Discourse and Post-Colonial Theory: A Reader*, ed. P. Williams and L. Chrisman (New York: Columbia University Press, 1994); Bat-Ami Bar On, "Marginality and Epistemic Privilege," in *Feminist Epistemologies*, ed. L. Alcoff and E. Potter (New York: Routledge, 1993); and John Michael, "Making a Stand: Standpoint Epistemologies, Political Positions, Proposition 187," *Telos*, no. 108 (Summer 1996): 93–103.

4. See Arnold Hirsch, "Harold and Dutch Revisited: A Comparative Look at the First Black Mayors of Chicago and New Orleans," in *African-American Mayors: Race, Politics, and the American City*, ed. D. Colburn and J. Adler (Urbana: University of Illinois Press, 2001), 109–10; also consider Gwendolyn Hall, *Africans in Colonial Louisiana: The Development of Afro-Creole Culture in the Eighteenth Century* (Baton Rouge: Louisiana State University Press, 1992).

5. Georgia Persons, "Black Mayoralties and the New Black Politics: From Insurgency to Racial Reconciliation," in *Dilemmas of Black Politics: Issues of Leadership and Strategy*, ed. G. Persons (New York: Harper Collins, 1993), 46.

6. Ronald Walters, "Two Political Traditions: Black Politics in the 1990s," *National Political Science Review* 3 (1990): 199.

7. Ibid.

8. Joseph McCormick II and Charles Jones, "The Conceptualization of Deracialization: Thinking Through the Dilemma," in *Dilemmas of Black Politics: Issues of Leadership and Strategy*, 76. Additional discussions of this concept can also be found in Huey Perry, introduction to *Race, Politics, and Governance in the United States*, ed. Huey Perry (Gainesville: University of Florida Press, 1996); and Roger Oden, "The Election of Carol Mosley-Braun in Illinois," in *Race, Politics, and Governance*.

9. Another type of deracialized campaign occurs when African Americans are running against whites in cities or districts where whites constitute the majority. In these situations the African-American candidates make strong appeals to white voters and "usually do not make strong racial appeals in their campaigns. This is the essence of a deracialized campaign." Examples of this type of deracialized campaign include the election of Douglas Wilder as governor of Virginia and David Dinkins as Mayor of New York. See Perry, introduction to *Race, Politics, and Governance*, 5.

10. Roy Rodney, cited in "Nagin Counts on Compromise, Integrity," New Orleans *Times-Picayune*, January 12, 2002, p. 1.

11. There are a number of PACs (political action committees) that have been operating in New Orleans since the mid 1960s. The Community Organization for Urban Politics (COUP) was associated with Seventh Ward conservative Creoles, and key leaders have included the city's second African-American mayor, Sidney Barthelemy. The Southern Organization for Unified Leadership (SOUL) was an outgrowth of the activities of the Congress Of Racial Equality in New Orleans and was initially associated with the emerging black consciousness of the 1960s. The Black Organization for Leadership Development (BOLD) emerged out of an association of uptown blacks affiliated with former mayor Moon Landrieu. The Louisiana Independent Federation of Electors (LIFE) is an organization of former Mayor Marc Morial and his supporters, while the Association of Community Organizations for Reform Now (ACORN) is a group formed in 1990s by a number of community activists. For discussions of SOUL, COUP, and BOLD, see Arnold Hirsch, "Simply a Matter of Black and White: The Transformation of Race and Politics in Twentieth-Century New Orleans," in *Creole New Orleans: Race and Americanization*, ed. A. Hirsch and J. Logsdon (Baton Rouge: Louisiana State University Press, 1992).

12. "Nagin Counts on Compromise, Integrity," New Orleans *Times-Picayune*, January 12, 2002, p. 1.

13. Editorial, "Ray Nagin for Mayor," New Orleans *Times-Picayune*, January 17, 2002, p. 6.

14. William Sites, "The Limits of Urban Regime Theory: New York City Under Koch, Dinkins, and Giuliani," in *The Politics of Urban America: A Reader*, ed. D. Judd and P. Kantor (New York: Longman, 2002), 215.

15. Ibid.

16. See Frank Donze, "7 Mayoral Rivals Stick to Their Scripts in Debate," New Orleans *Times-Picayune*, January 31, 2002.

17. Stephanie Grace, "Nagin Shoots Up, Irons Loses Steam in Newest Poll," New Orleans *Times-Picayune*, January 23, 2002, p. 1.

18. The slide of Irons in relationship to Nagin is important for one reason: she was the candidate receiving the most support from whites in the beginning of the mayoral campaign. She was subject to a number of allegations; among the more notable accusations were that she was employed in two different state government positions (Dual Office Holding Law) and that she had claimed that her brother had been innocently shot years ago when in fact he had been shot by the police after robbing a store. See Christopher Tidmore, "Inside Political Track," *Louisiana Weekly* January 14, 2002; idem., January 21, 2002; "Integrity Questions Are Mounting for Mayoral Candidate: Ad, Job Inquiries Hit Irons Campaign," New Orleans *Times-Picayune*, January 12, 2002, p. 1; "Mayoral Attack Hits the Airwaves," New Orleans *Times-Picayune*, January 19, 2002.

19. See Christopher Tidmore, "Mayor's Race Heats Up as Nagin Gets Backers," *Louisiana Weekly*, February 11, 2002, p. 1; Frank Donze and Stephanie Grace, "ACORN Plants Its Support in Pennington," New Orleans *Times-Picayune*, February 22, 2002, p. 1.

20. Although there is housing segregation in New Orleans, the city is not rigidly segregated in a spatial sense. The average distance that separates a white neighborhood from a black neighborhood is often only a few blocks. For example, the wealthiest part of the city, the St. Charles Audubon Park area, is thoroughly integrated, with many lower-income and high-crime neighborhoods.

21. C. Ray Nagin, "New Orleans Is on the Move Again: New Orleans' Mayor C. Ray Nagin Inaugural Address," New Orleans *Times-Picayune*, May 7, 2002, p. 6.

22. Juvenile, "Get Ya Hustle On," *Reality Check* (Atlantic Records, March 2006).

23. Hanes Walton, *African American Power and Politics: The Political Context Variable* (New York: Columbia University Press, 1997), 7. Also see idem,"The Nature of Black

Politics and Black Political Behavior," in *Black Politics and Black Political Behavior*, ed. Hanes Walton (Westport, CT: Praeger, 1994); and idem., *Black Politics: A Theoretical and Structural Analysis* (Philadelphia: Lippincott, 1972).

24. Karen Kaufmann, *Group Conflict & Mayoral Voting Behavior in American Cties* (Ann Arbor: University of Michigan Press, 2004), 5. Additional discussion of racial models employed by voting and attitudinal analysts can be found in Katherine Tate, *From Protest To Politics* (New York: Russell Sage Foundation/Harvard University Press, 1993); Michael Dawson, *Behind The Mule: Race and Class in African-American Politics* (Princeton, NJ: Princeton University Press, 1995); Dianne Pinderhughes, *Race and Ethnicity in Chicago Politics: A Reexamination of Pluralist Theory* (Urbana: University of Illinois Press, 1987); Cathy Cohen and Michael Dawson, "Neighborhood Poverty and African American Politics," *American Political Science Review* 87 (June 1993): 286–302; and Richard Allen, Michael Dawson, and Ron Brown, "A Schema-Based Approach to Modeling an African-American Racial Belief System," *American Political Science Review* 83 (June 1989): 421–41

25. The specific question was, "Do you approve of the job–person– in responding to the effects of Hurricane Katrina?" Gallup Poll conducted February 18–26, 2006.

26. Gallup Poll conducted February 18–26, 2006.

27. Mos Def, "Katrina Remix," Underground (net) Release (Approx. Sept. 2005).

28. Todd Purdam et al., "Support for Bush Continues to Drop as More Question His Leadership Skills, Poll Shows," *The New York Times*, September 15, 2005.

29. Mayor Nagin, cited in John Pope, "Evoking King, Nagin Calls N.O. 'Chocolate City,' Speech Addresses Fears of Losing Black Culture," New Orleans *Times-Picayune*, January 17, 2006, p. 1.

30. See George Curry, "Political Cross-dressing in New Orleans," *Louisiana Weekly*, May 29, 2006.

31. UTP, "Nolia Clap," *NOLA CLAP* (2004).

32. The terms *descriptive, symbolic*, and *substantive*, as modifiers of *representation*, are often used by congressional scholars as an analytical tool to evaluate the representativeness of elected officials to their constituents. "Descriptive representation," according to Walton and Smith, "is the extent to which the legislature looks like the people in a demographic sense. Symbolic representation concerns the extent to which people have confidence or trust in the legislature, and substantive representation asks whether the laws passed by the legislature correspond to the policy interests or preferences of the people." See Hanes Walton and Robert Smith, *American Politics and the African American Quest for Universal Freedom* (New York: Longman, 2000), 178; Katherine Tate, *Black Faces in the Mirror: African Americans and Their Representatives in the U.S. Congress* (Princeton, NJ: Princeton University Press, 2003), 13.

33. Manning Marable, "Violence, Resistance, and the Struggle for Black Empowerment," in *Speaking Truth to Power: Essays on Race, Resistance, and Radicalism* (Boulder, CO: Westview, 1996), 132.

34. Aldoph Reed, Jr.,"Classifying the Hurricane," *The Nation*, October 3, 2005, online edition.

35. U.S. Census Bureau.

36. For discussions of the contemporary federalism see: Paul Peterson, "From the Price of Federalism," in *The Enduring Debate: Classic and Contemporary Readings in American Politics*, ed. D. Canon, J. Coleman, and K. Mayer (New York: Norton, 2006); Craig Volden, "The Politics of Competitive Federalism: A Race to the Bottom in Welfare Benefits," *American Journal of Political Science* 46 (April 2002): 352–63.

37. Dennis Judd and Todd Swanstrom, *City Politics: Private Power and Public Policy* (New York: Longman, 1998), 236.

38. Theodore Lowi, Benjamin Ginsberg, and Kenneth Shepsle, *American Government: Freedom and Power* (New York: Norton, 2006), 65.

39. Judd and Swanstron, *City Politics*, 239.

40. Michael Rich, "The Intergovernmental Environment," *Cities, Politics, and Policy: A Comparative Analysis*, ed. J. Pelissero (Washington, DC: CQ Press, 2003), 48.

41. Clarence Stone, *Regime Politics: Governing Atlanta: 1946–1988* (Lawrence: University Press of Kansas, 1989), 6. Also consider Russell Murphy, "Politics, Political Science, and Urban Governance: A Literature and a Legacy," *Annual Review of Political Science* (2002): 63–85.

42. Regarding this omission, Davies contends that "the heart of the problem is the limited theorizing of the way economic forces affect local political institutions and the balance of power within them. It is not enough to acknowledge the influence of the market economy on local political processes, it is also necessary to explain how fluctuations in the economy enable and constrain political options." Most regime theorists reject the type of economic determinism associated with Marxism; however, critics maintain that these proponents have failed to account for market forces at macro level and their potentially constraining effects on municipal governance. The recessions of the 1980s, combined with a federal government determined to reduce the amount of funding at the local level, clearly had an impact on the ability of mayors to govern. The key issue, these authors raise, is to what degree these macro processes influence or limit decision making on the local level. These pro-growth market strategies also contradict any efforts toward egalitarian social-communal development. Imbroscio characterizing Paul Peterson's analysis of local politics, says, "He argued that the mobility of economic resources (resulting from the nature of capitalism), combined with the intergovernmental (subnational) competition for investment (resulting from the nature of federalism) left American cities constrained by economic pressures. These constraints limited the range of political choice. Cities were compelled, by necessity, to pursue developmental policies (that enhance the economic position of the city) while eschewing policies to redistribute wealth (that harm this economic position)." Therefore, one witnesses both an analytical limitation that leads to an ideological contradiction on the part of regime theorists. See Jonathan Davies, "Urban Regime Theory: A Normative-Empirical Critique," *Journal of Urban Affairs* 24, no.1 (2002): 13; David Imbroscio, "Overcoming the Neglect of Economics in Urban Regime Theory," *Journal of Urban Affairs* 25, no. 3 (2003): 272.

43. The issue is not the assignment of racial categories to key actors, but the failure to examine how race sometimes serves as a structuring principle, that can lead institutional biases. These institutional biases play themselves out in a class-based analysis that overlooks intragroup activities outside the systemic context. See Cynthia Horan, "Racializing Regime Politics," *Journal of Urban Affairs* 24, no 1 (2002): 25.

44. Adolph Reed Jr., "Sources of Demobilization in the New Black Political Regime: Incorporation, Ideological Capitulation, and Radical Failure in the Post-Segregation Era," in *Stirrings in the Jug: Black Politics in the Post-Segregation Era* (Minneapolis: University of Minnesota Press, 1999), 121.

45. A key consideration from Reed's perspective is the tactical ability of BEOs in relationship to their constituents. He notes that "Jackson's success at reinvention suggest one mechanism through which the black administration can mediate the tension between electoral and governing constituencies. By virtue of his representation of his position as the 'black' position, he added an epicycle of racial self-defense to blacks' consideration of policy options; to that extent, Jackson shifted the basis for black policy debate away from substantive concern with the potential outcomes and toward protection of

the racial image and status, as embodied in the idiosyncratic agenda of the black official. In this way it becomes possible for black officials to maintain support from their
black constituents and the development elites that systematically disadvantage them."
Ibid., 177.

46. Alvin Schexnider, "Political Mobilization in the South: The Election of a Black Mayor
in New Orleans," in *The New Black Politics: The Search for Political Power*, ed. M.
Preston et al. (New York: Longman, 1982), 223. Also consider Huey Perry, "Black
Political and Mayoral Leadership in Birmingham and New Orleans," *National Political
Science Review* 2 (1990): 154–61.

47. Schexnider, "Political Mobilization," 223.

48. Christopher Tidmore, "It's Not Just the Politicians' Fault," *Louisiana Weekly*, February
25, 2002.

49. William Borah, "City Must Have a Real Master Plan," New Orleans *Times-Picayune*,
May 21, 1998.

50. Paul Peterson, *City Limits* (Chicago: University of Chicago Press, 1981), 154.

51. Christopher Tidmore, "Inside Political Track," *Louisiana Weekly*, January 21, 2002.

52. Susan Howell and William McClean, "Performance and Race in Evaluating Minority
Mayors," *Public Opinion Quarterly* 65, no. 3 (2001): 327–33.

53. See Bringing New Orleans Back Commission (BNOBC), "Post-Katrina Economic
Redevelopment Plan," Version 4.1, January 2006; BNOBC, "Rebuilding New Orleans,"
January 2006.

54. Peter Eisinger, "The Politics of Bread and Circuses: Building the City for the Visitor
Class," in *The Politics of Urban America: A Reader*, 257. Also see John Kasarda, "Urban
Change and Minority Opportunities," in *The New Urban Reality*, ed. P. Peterson
(Washington, DC: Brookings Institution, 1985).

55. Dennis Judd and Todd Swanstrom, *City Politics: Private Power and Public Policy* (New
York: Longman, 1998), 368.

56. While Atlanta, for example, has been a model city in terms of rapid political-eco-
nomic growth, this is not to suggest that low income and minority groups' relative
socio-economic position has improved. For discussions of this contradictory perform-
ance see Manley Banks, "A Changing Electorate in a Majority Black City: The
Emergence of a Neo-Conservative Black Urban Regime in Contemporary Atlanta,"
Journal of Urban Affairs 22, no. 3 (265–78); Rob Gurwitt, "How to Win Friends and
Repair a City," in *State and Local Government 2005–2006*, ed. K. Smith (Washington,
DC: CQ Press, 2006), 173–77; Mack Jones, "Black Mayoral Leadership in Atlanta:
A Comment," *National Political Science Review* 2 (1990): 138–44; Reed, "A Critique
of Neoprogressivism in Theorizing about Local Development Policy: A Case from
Atlanta," in *Stirrings in the Jug*; and Stone, *Regime Politics*.

Hurricane Katrina as an Elaboration on an Ongoing Theme: Racialized Spaces in Louisiana[*]

K. Animashaun Ducre

Introduction

A search of the WORLDCAT database yielded 786 sources related to Hurricane Katrina in summer 2006.[1] Over half of the sources cited in WORLDCAT are books (419) related to the impacts of the disaster, as well as rebuilding the city of New Orleans and the Gulf. The wealth of opinion and intellectual work on Katrina reveals a particular assumption about the storm. The common sentiment is that Hurricane Katrina was an extraordinary event. While it is not my intent to belittle the significant pain and suffering of Katrina victims and their families, I contend that Katrina does not mark a significant departure, as many have suggested. Rather, Hurricane Katrina should be examined as an eloboration on an ongoing theme, another significant note in the sordid history of American racism. In fact, the state of Louisiana emerges as an important site for chronicling America's race relations, beginning in 1865 with the Louisiana Black Code, the *Plessy v. Ferguson* decision in 1896, the rise of the oil and gas industry in Louisiana, and the treatment of African Americans during the Great Mississippi Flood of 1927. Given this chronology of events and against

*Direct correspondence to: K. Animashaun Ducre, Department of African American Studies, Syracuse University, 200 Sims Hall, Syracuse, New York 13244.

this historical backdrop, the chaos that ensued in Katrina's aftermath is unremarkable. Furthermore, placing the history of Louisiana within this historical context can explain why many in the Gulf Coast continue to suffer until this day.

RACIALIZED SPACES IN LOUISIANA: RACE, SPACE, AND POWER

> There was little question that Katrina had sparked renewed debate about race, class, and institutional approaches towards vulnerable population groups in the US. . . .Whether or not one believed racist charges were well-founded (and clearly a majority of our members did not) the Select Committee agreed it should recognize and discuss the socioeconomic and racial backdrop against which Katrina unfolded.[2]

This statement, issued by the Congressional Select Committee, reflects the persistent amnesia regarding racism in the United States. In their 500-plus–page report, race is only mentioned briefly. Unlike Congress, many see race as an important dimension in understanding Katrina. The concept of racialized spaces is a useful framework to support the idea of Hurricane Katrina as a postscript. This concept evolved prior to Katrina as a possible explanation for the emergence of environmental racism and injustice. While not a native Louisianan, I had spent most of my adult life in and around the state, first as an undergraduate at Tulane University and later as a community organizer in the environmental justice movement until the late 1990s. Louisiana's culture seems distinct to me. I figured that there must be something unique about its history that facilitated contemporary examples of racism and injustice. For example, I was a student when David Duke ran his campaign for governor and later an unsuccessful bid for the Presidency. He only lost by a narrow margin to former Governor Edwin Edwards. I can also recall the controversial acquittal of Rodney Peairs of Baton Rouge, who shot a Japanese exchange student, Yoshihiro Hattori, on Halloween in 1992.[3] I also recall the marching of the Ku Klux Klan on the town square in Homer, in northern Louisiana. Homer was also the proposed host site of a uranium enrichment facility—a proposal that was later challenged on the basis of environmental racism. In fact, one of the most decisive battles for environmental justice occurred in southeastern Louisiana in the city of Convent. There, residents launched a successful campaign against a proposed plastics complex.[4]

The state of Louisiana emerges as an important site for a discussion of race relations centering on race, space, and power. This intersection of race, space, and power is what I refer to as *racialized spaces*. The racialized spaces hypothesis is defined as "historic practice and spatial designation of a particular area for racial and ethnic minorities as a means of containment and social control. This practice serves to reinforce preconceived notions of Otherness or, result in the creation of a culturally inferior Other."[5]

The key themes of this concept concern history, containment, and social control. The first element is history. History acknowledges the genealogy of patterns that push and pull the poor and people of color under racial segregation, class segregation, or both. Examples include racial violence, deed restrictions, restrictive covenants, blockbusting, redlining, slum clearance, creation of public housing, and discriminatory mortgage lending. The second element, containment, emphasizes the geographic isolation of Others within racialized spaces, from and by those with power. The final element is social control, which refers to the methods that maintain racialized spaces, like surveillance and institutionalization. Quite literally, social control is the element designed to keep *folks in their places*. All of these tenets can be applied to the historical and socio-spatial context of African American life in the state of Louisiana. Moreover, the concept of racialized spaces helps place the events surrounding Hurricane Katrina and the resulting levee failure into context.

In the early 1970s, Blauner was the first academic to suggest that the severe concentration and socioeconomic disparities in black neighborhoods resulted in internal colonialism.[6] The racialized spaces framework relates to the internal colonization thesis. France was the first colonial power in the area that encompasses present-day Louisiana. The first settlement started in 1699, and the importation of African slaves began as early as 1717. The largest consignment of African slaves came in June 1719, with Africans originating from Senegal-Gambia.[7] By 1724, African slaves outnumbered the French colonial population, necessitating the need for stricter controls. The first sets of controls, the "Code Noir," or Black Codes, were instituted. The codes would serve as the basis for the American institution of the Black Codes in 1865.

The Louisiana Purchase was perhaps the most lucrative land deal in U.S. history. The American government paid almost four cents an acre for 900,000 square miles of territory, expanding the size of its empire by 23 percent. In 1808, federal law prohibited the importation of African slaves into the United States, but interstate slave-trafficking was alive and well. There are also numerous accounts of illegal importation of African slaves post-1808. Louisiana is admitted into the Union as the eighteenth state in 1812. By 1849, there were 1,536 plantations, producing 96 percent of the nation's raw sugar, using the labor over 325,000 enslaved Africans.[8] And like many places in the South, its landscape and economy were devastated by Civil War and Union occupation.

While the plantation regime and slavery proved the cornerstone of the Gulf economy, its strength was always tested by nature. Major hurricanes hit in 1856 and 1915. Hurricane Betsy devastated the region in 1965. Major flooding occurred in 1884, 1897, 1902, 1903, 1912, 1913, 1922, 1973, 1983, and 1995. Despite these challenges, New Orleans rose to become a major port. By 1880, New Orleans was the second-largest port in the United States, falling short to New York.[9] By 1995, New Orleans was ranked the world's largest port by cargo volume. To remain competitive, New Orleans economy revolves around control and mastery of space by the exploitation of (black) labor, land, and capital.

The mastery over space by land relies on the daunting task of controlling the mighty Mississippi River and its tributaries. The United States Army Corps of Engineers was created in 1802 by the federal government. Soon after its establishment, one of the key tasks of this federally funded civil engineering agency was to ensure navigability of the Mississippi River for trade. From then on, the Army Corps of Engineers began a long and complex relationship with Louisiana's natural environment. Its central role in controlling land reflects the value of engineering and enterprise during the Progressive Era. Barry notes, "In the century of the engineers, the study of this writhing river began as a scientific enterprise. The resulting policy became a corruption of science."[10]

The mastery over space by labor centers on the role African Americans play in empire building. In applying the element of history from the racialized spaces framework, there is evidence that Louisiana initiated the formal era of racially discriminatory policies with landmark legislative initiatives like the Black Codes of 1865 and the events leading up to the Plessy decision in 1896.

Louisiana is credited as the pioneer of the South's new post–Civil War labor system, which led to the establishment of labor contracts through the Freedmen's Bureau.[11] After New Orleans fell to Union forces in 1862, a commander in the Department of the Gulf (the antecedent to the Freedmen's Bureau) was faced with the mass exodus of former slaves from area plantations, resulting in a severe labor shortage. The solution was a contractual agreement between the government and planters, supposedly on behalf of the freedmen. These labor contracts were reminiscent of ante-bellum policies. This type of sentiment would become the basis for the Black Codes of 1865.

In reaction to imminent emancipation, Louisiana, along with other Southern states, established laws, often referred to as Black Codes, to limit black mobility. The French had established *Code Noir* to govern the rights of slaves as early as 1724 in the Louisiana territory. Southern plantations were regarded as the priority in restoring the devastated, post-Civil War economy. While these codes guaranteed civil rights to freedmen such as the right to marry and own property, their general purpose was to ensure a supply of former slaves as workers on labor-intensive plantations. In addition to these codes, some scholars have suggested that the establishment of vagrancy and convict laws during the same era were also a means to restrict black mobility and guarantee labor.[12] Vagrancy laws succeeded in compelling black workers to choose between the field or imprisonment. In Louisiana, violation of a labor contract was tantamount to a criminal act. With regards to criminality, convict laws were established to create a policy for prison labor. Under the racialized spaces hypothesis, the earlier forms of labor contracts for freedmen and the enactment of the Black Codes can be viewed as the attempt to restore the racial order of white supremacy that existed before the Civil War. Blacks were *free* in a very limited sense. However, blacks were not without agency. There were attempts to organize under the Knights of Labor, but their efforts were met with racial violence.[13]

The most significant opportunity for black advancement was in the proposal for land redistribution. In August of 1865, the assistant commissioner for the Freedmen's Bureau of Louisiana announced plans for the redistribution of thousands of acres of abandoned and confiscated lands; however, President Johnson quickly nixed those plans.[14] The lands were restored to their prior owners, and so the chance to establish *place* for black Louisianans was abandoned.

While efforts to create new black spaces were thwarted, there was also political momentum to further diminish the remaining spaces where blacks could live, work, and play. This political momentum was codified into what is referred to as the "separate but equal" doctrine. While many recall and herald the 1954 *Brown v. Board of Education* decision to desegregate public schools and other facilities, the shameful decision in American history that catalyzed the institutionalization of the "separate but equal" doctrine was the 1896 Supreme Court decision of *Plessy v. Ferguson*.

Homer A. Plessy was born in 1863 in New Orleans to parents known in French creole as *gens de colour*, "free people of color." On June 7, 1892, Plessy, a shoemaker, sat in the whites-only railcar on the East Louisiana Railroad; he was arrested for violating the state law mandating separate cars for whites and blacks. His trial judge, John Ferguson, ordered him to pay fines for this violation, and the case was appealed to the U.S. Supreme Court. Lawyers for Plessy argued that his arrest violated his civil rights under the Thirteenth and Fourteenth Amendments. Delivering the opinion, Justice Brown wrote in 1896,

> The object of the [fourteenth] amendment was undoubtedly to enforce the absolute equality of the two races before the law, but in the nature of things it could not have been intended to abolish distinctions based on color, or to enforce social, as distinguished from political equality, or commingling of the two races upon terms unsatisfactory to either. Laws permitting, and even requiring, their separation in places where they are liable to be brought into contact do not necessarily imply the inferiority of either race to the other, and have been generally, if not universally, recognized within the competency of the state legislatures in the exercise of their public power. . .we cannot say that a law which authorizes or even requires the separation of the two races in public conveyances is unreasonable.[15]

The intersection between race and space here in this case is quite profound. This decision cements the establishment of black spaces and white spaces as a rule in race relations. Over a century later, most American neighborhoods, particularly New Orleans, are still divided along racial lines despite civil rights challenges to the contrary.[16]

The Great Mississippi River Flood of 1927 provides an interesting parallel to Hurricane Katrina, especially in its treatment of African Americans and the vulnerability of oppressed communities during natural disaster. It is reported that over 700,000 people were displaced during the floods and 330,000 were African Americans. The relief agencies respected Jim Crow policy and provided racially segregated shelters for those displaced. Barry chronicles the most vivid example of racial oppression, the fate of African American families in the Mississippi Delta during the flood.[17] Black men were forced to work at gunpoint to increase levee protection. Once the water broke in Greenville, Mississippi, 13,000 blacks were forced to live in squatter conditions on the actual levee. The city of New Orleans was spared in the 1927 flood, in part because a portion of the levee in St. Bernard Parish was blown up with dynamite to keep the city from being flooded. Thus, allegations that the predominantly African American neighborhood in the Ninth Ward of New Orleans was similarly sacrificed during Hurricane Katrina are not entirely outrageous.[18] At one time, the political elite of New Orleans did pull together to sacrifice the livelihood of St. Bernard Parish in the interest of New Orleans.[19]

Parallel to the development of racially segregated spaces was the transition of Louisiana's agricultural economy to a petrochemical-based economy. The first successful oil well was established in 1901 in Jennings, Louisiana, marking the beginning of a century-long relationship between land, oil, and capital in Louisiana. The other major event that signaled a shift in the economic base was the decision by Rockefeller to build a Standard Oil refinery in Baton Rouge, Louisiana in 1909.[20] Southeastern Louisiana was attractive to the petrochemical industry due to the abundance of natural resources like salt, freshwater, oil, and natural gas, as well as transportation routes by rail and waterway along the Mississippi River.[21]

The conversion of sugarcane fields to petrochemical facilities resulted in what many refer to as *cancer alley*, the eighty-five-mile stretch of the Mississippi River corridor with over 100 chemical complexes. Many reports indicate that there is a disproportionate impact on the African American population in this area.[22] Why? As former field workers of the sugarcane plantations, many of the African American neighborhoods found themselves sandwiched between these immense chemical complexes. In fact, Wright relates the corporate buyouts of historic African American communities. These communities were established as freetowns by former slaves after emancipation, and the buyouts represent the most extreme example of the impact of racialized spaces in Louisiana.[23] Reveilletown was bought out by Georgia Gulf, Good Hope was purchased and moved by St. Charles Refinery, and the community of Sunrise was taken over by Placid Refining Company. Additionally, Morrisonville was relocated by Dow Chemical, and the former homes of this historic black community have become *green space* by the Dow complex grounds. One of the most heart-wrenching sacrifices of this relocation is the historic cemetery of Morrisonville, which is still located on Dow's grounds. Former church members are allowed access to the gravesites, but the area is closed to the public. The pattern of environmental racism found in Louisiana provides even more evidence of racialized spaces in Louisiana. The environmental justice perspective is useful in understanding the racial disparities in the impacts of Katrina.[24]

KATRINA AS AN ELABORATION ON AN ONGOING THEME

> Racism stands apart by a practice of which it is a part and which it rationalizes: a practice that combines strategies of architecture and gardening with that of medicine—in the service of the construction of an artificial social order. . .racism manifests the conviction that a certain category of human beings cannot be incorporated into the rational order, whatever the effort. . . .The consequence is that *racism is inevitably associated with the strategy of estrangement.*[25]

The estimate of those displaced from Hurricane Katrina is akin to the number of African Americans who were displaced by massive restructuring of cities during urban renewal in the 1960s and 1970s. The mass exodus or removal of African Americans is not a new phenomenon, and this is another reason why Katrina should be viewed as an elaboration on the ongoing theme of American racism.

As with the rhetoric behind urban renewal policies, local politicians are not interested in promoting a new New Orleans with its former low-income residents. Also reminiscent of urban renewal, resources for neighborhood revitalization in New Orleans and all along the Gulf are prioritized according to race and class privilege. There has been a strong response from the academic community in the aftermath of the devastating hurricane season of 2005 and the horrors of Katrina. For example, many have spoken about the consequences of Katrina from various medical, social, behavioral, engineering, and religious perspectives. Others critique the political and militaristic systems that fostered the cataclysmic effects of Katrina on the Gulf Coast.

My contention in this chapter is that Katrina's aftermath should not be highlighted as something significantly different than what has happened in America's racial history prior to August 2005.

The racialized spaces hypothesis explicates the ways in which policy can be used to erect rigid racial boundaries. In the case of Louisiana, there has been evidence of these segregationist mechanisms. The most historic and significant event was the Plessy decision, which hailed from an incident on the East Louisiana Railroad. Other key early segregation policies centered on housing. Prior to 1930, the Louisiana state legislature enacted a series of laws to restrict housing and enforce segregation. In 1917, municipalities were granted the power to deny building permits for homes for both blacks and whites in areas principally inhabited by residents of a different race. This law was strengthened by a new statute in 1921 that provided penalties for housing blacks and whites in the same dwelling, with the exception of domestic help.[26]

In the case of Louisiana, containment is exemplified by the geographic and economic isolation of African Americans within the state. With the exception of New Orleans, the African American population of the state was historically situated in rural areas. This demography is no coincidence. Many strategies were employed to limit black mobility after emancipation. The first of such strategies included the Black Codes of 1865, vagrancy laws, and forms of debt peonage.[27] The final element of the racialized spaces hypothesis is social control. If the previous element of history and containment served to create and subordinate black space, the institution of Jim Crow reified those spaces.[28] Placed within this context of racialized spaces, there is no doubt that African Americans and the poor in New Orleans were burdened with disproportionate impact from the storm. The social position of African Americans throughout the history of Louisiana makes them vulnerable to the adverse impact of disasters, whether natural or manmade. Efforts to improve the fate of African Americans in the event of an emergency should begin with efforts to improve their social condition: better education, better homes, better jobs, and provide greater levels of access to opportunity. We cannot stop storms like Katrina, but we can ameliorate their impact. It is my contention that eradicating institutional racism and injustice would go far in lessening disproportionate burdens of disasters like Katrina.

BIBLIOGRAPHY

Allen, Barbara. "The Popular Geography of Illness in the Industrial Corridor." In *Transforming New Orleans and Its Environs*, edited by Craig Colten, 178–201. Pittsburgh: University of Pittsburgh Press, 2000.

Barry, John. *Rising Tide: The Great Mississippi Flood of 1927 and How It Changed America.* New York: Simon and Schuster, 1997.

Bauman, Zygmunt. *Modernity and the Holocaust.* Ithaca, NY: Cornell University Press, 1989.

Blauner, Robert. "Internal Colonialism and Ghetto Revolt." *Social Problems* 16, no. 4 (1969): 393–408.

Bullard, R. D. "Confronting Environmental Racism: The Case of Shintech and Convent, Louisiana." Testimony prepared for the Louisiana Department of Environmental Quality Hearing. Convent, LA: January 24, 1998.

Bullard, Robert D. *Unequal Protection: Environmental Justice and Communities of Color*. San Francisco: Sierra Club Books, 1994.

Burby, Raymond. "Baton Rouge: The Making (and Breaking) of a Petrochemical Paradise." In *Transforming New Orleans and Its Environs*, edited by Craig Colten, 160–77. Pittsburgh: University of Pittsburgh Press, 2000.

Cohen, William. *At Freedom's Edge: Black Mobility and the Southern White Quest for Racial Control*. Baton Rouge: Louisiana State University, 1991.

Colten, Craig. *An Unnatural Metropolis: Wresting New Orleans from Nature*. Baton Rouge: Louisiana State University Press, 2005.

de Jong, Greta. *A Different Day: African-American Struggles for Justice in Rural Louisiana, 1900–1970*. Chapel Hill: University of North Carolina Press.

Ducre, K. Animashaun "Racialized Spaces and the Emergence of Environmental Injustice." In *Echoes from the Poisoned Well: Global Memories of Environmental Injustice*, edited by Sylvia Hood Washington, Paul C. Rosier, and Heather Goodall, 109–24. Lanham, MD: Lexington, 2006.

Dugas, Caroll Joseph. "The Dismantling of De Jure Segregation in Louisiana, 1954–1974." PhD diss. Baton Rouge: Louisiana State University, 1989.

Dyson, Michael Eric. *Come Hell or High Water: Hurricane Katrina and the Color of Disaster*. New York: Basic Civitas Books, 2006.

Ellis, Catherine. "The Legacy of Jim Crow in Rural Louisiana." PhD diss. New York: Columbia University, 2000.

Hair, William Ivy. *Bourbonism and Agrarian Protest: Louisiana Politics 1877–1900*. Baton Rouge: Louisiana State University Press, 1969.

Halpern, Rick. "Solving the 'Labour Problem': Race, Work and the State in the Sugar Industries of Louisiana and Natal, 1870–1910." *Journal of Southern African Studies* 30, no. 1 (2004): 19–40.

Hines, Revathi I. "African Americans' Struggle for Environmental Justice and the Case of the Shintech Plant." *Journal of Black Studies* 31, no. 6 (2001): 777.

Jackson, Kenneth. *Crabgrass Frontier: The Suburbanization of the United States*. New York: Oxford University Press, 1985.

Louisiana Advisory Committee to the U.S. Commission on Civil Rights. "The Battle for Environmental Justice in Louisiana. . .Government, Industry, and the People." Washington, DC: Government Printing Office, 1993.

Maguire, Robert E. *Hustling to Survive: Social & Economic Change in a South Louisiana Creole Community*. Vol. 2, *Project Louisiane*. Quebec: Departement de Geographie de l'Université Laval, 1989.

Massey, Douglas S., and Nancy A. Denton. *American Apartheid: Segregation and the Making of the Underclass*. Cambridge, MA: Harvard University Press, 1993.

Miller, Randall. "The Freedmen's Bureau and Reconstruction: An Overview." In *The Freedmen's Bureau and Reconstruction:Reconsiderations*, edited by Paul Cimbala and Randall Miller, xiii–xxxii. New York: Fordham University Press, 1999.

Pastor, M., R. D. Bullard, James Boyce, Alice Fothergill, R. Morello-Frosch, and Beverley Wright. *In the Wake of the Storm: Environment, Disaster and Race after Katrina*. New York: Russell Sage Foundation, 2006.

Plessy v. Ferguson, 163 U.S. 537 (1896).

Powell, Bernice, and R. D. Bullard. *From Plantations to Plants: Report of the Emergency National Commission on Environmental and Economic Justice in St. James Parish, Louisiana*. Cleveland, OH: United Church of Christ Commission for Racial Justice, 1998.

Reid, T.R. "Japanese Media Disparage Acquittal in 'Freeze Case': Commentators See America as Sick Nation." *Washington Post*, May 25, 1993, p. A14.

Ripley, C. Peter. *Slaves and Freedmen in Civil War Louisiana.* Baton Rouge: Louisiana State University Press.

Roberts, J. Timmons, and Melissa Toffolon-Weiss. "Roots of Environmental Injustice in Louisiana." In *Chronicles from the Environmental Justice Frontline,* 29–41. New York: Cambridge University Press, 2001.

Rodrigue, John. "The Freedmen's Bureau and Wage Labor in Louisiana Sugar Region." In *The Freedmen's Bureau & Reconstruction: Reconsiderations,* edited by Paul Cimbala and Randall Miller, 193–218. New York: Fordham University Press, 1999.

U.S. Congress. Select Bipartisan Committee to Investigate the Preparation for and Response to Hurricane Katrina. *A Failure of Initiative: Final Report of the Select Bipartisan Committee to Investigate the Preparation for and Response to Hurricane Katrina.* 109th Congress, 2nd Session. Washington, DC: United States Government Printing Office, 2006.

Shallat, Todd. "In the Wake of Hurricane Betsy." In *Transforming New Orleans and Its Environs,* edited by Craig Colten, 121–37. Pittsburgh: University of Pittsburgh Press, 2000.

Sternberg, Mary Ann. *Along the River Road.* Baton Rouge: Louisiana State University Press, 2001.

NOTES

1. WORLDCat, *Search:Hurricane Katrina,* http://newfirstsearch.oclc.org.libezproxy2.syr .edu/WebZ/FSQUERY?format=BI:next=html/records.html:bad=html/records.html:n umrecs=10:sessionid=fsapp4–59032–eqwivfav-49sp8f:entitypagenum=2:0: searchtype=basic.

2. Select Bipartisan Committee to Investigate the Preparation for and Response to Hurricane Katrina, "A Failure of Initiative: Final Report of the Select Bipartisan Committee to Investigate the Preparation for and Response to Hurricane Katrina," 109th Cong. 2nd Session, 2006, pp. 19–20.

3. T. R. Reid, "Japanese Media Disparage Acquittal in 'Freeze Case': Commentators See America as Sick Nation," *Washington Post,* May 25, 1993.

4. R. D. Bullard, "Confronting Environmental Racism: The Case of Shintech and Convent, Louisiana." Testimony prepared for the Louisiana Department of Environmental Quality hearing, Convent, LA, January 24, 1998; Revathi I Hines, "African Americans' Struggle for Environmental Justice and the Case of the Shintech Plant," *Journal of Black Studies* 31, no. 6 (2001).

5. K. Animashaun Ducre, "Racialized Spaces and the Emergence of Environmental Injustice," in *Echoes from the Poisoned Well: Global Memories of Environmental Injustice,* ed. Sylvia Hood Washington, Paul C. Rosier, and Heather Goodall (Lanham: Lexington Book, 2006).

6. Robert Blauner, "Internal Colonialism and Ghetto Revolt," *Social Problems* 16, no. 4 (1969).

7. Mary Ann Sternberg, *Along the River Road* (Baton Rouge: Louisiana State University Press, 2001).

8. Robert E. Maguire, *Hustling to Survive: Social & Economic Change in a South Louisiana Creole Community,* vol. 2, *Project Louisiane* (Québec: Université Laval, 1989).

9. John Barry, *Rising Tide: The Great Mississippi Flood of 1927 and How It Changed America* (New York: Simon and Schuster, 1997).

10. Ibid.

11. William Cohen, *At Freedom's Edge: Black Mobility and the Southern White Quest for Racial Control* (Baton Rouge: Louisiana State University Press, 1991).

12. Ibid.; Greta de Jong, *A Different Day: African-American Struggles for Justice in Rural Louisiana, 1900–1970* (Chapel Hill: University of North Carolina Press, 2002), Randall Miller, "The Freedmen's Bureau and Reconstruction: An Overview," in *The Freedmen's Bureau and Reconstruction:Reconsiderations*, ed. Paul Cimbala and Randall Miller (New York: Fordham University Press, 1999), C. Peter Ripley, *Slaves and Freedmen in Civil War Louisiana* (Baton Rouge: Louisiana State University Press).

13. William Ivy Hair, *Bourbonism and Agrarian Protes: Louisiana Politics 1877-1900t* (Baton Rouge: Louisiana State University Press, 1969); Rick Halpern, "Solving the 'Labour Problem': Race, Work and the State in the Sugar Industries of Louisiana and Natal, 1870–1910," *Journal of Southern African Studies* 30, no. 1 (2004).

14. John Rodrigue, "The Freedmen's Bureau and Wage Labor in Louisiana Sugar Region," in *The Freedmen's Bureau & Reconstruction: Reconsiderations*.

15. *Homer A. Plessy v. John H. Ferguson*, 163 U.S. 537; 16 S. Ct. 1138; 41 L. Ed. 256; 1896 U.S. Lexis 3390 (1896).

16. Kenneth Jackson, *Crabgrass Frontier: The Suburbanization of the United States* (New York: Oxford University Press, 1985); Douglas S. Massey and Nancy A. Denton, *American Apartheid: Segregation and the Making of the Underclass* (Cambridge, MA: Harvard University Press, 1993).

17. Barry, *Rising Tide*.

18. Michael Eric Dyson, *Come Hell or High Water: Hurricane Katrina and the Color of Disaster* (New York: Basic Civitas Books, 2006).

19. Todd Shallat, "In the Wake of Hurricane Betsy," in *Transforming New Orleans and Its Environs*, ed. Craig Colten (Pittsburgh: University of Pittsburgh Press, 2000).

20. Raymond Burby, "Baton Rouge, LA: The Making (and Breaking) of a Petrochemical Paradise," in *Transforming New Orleans and Its Environs*, ed. Craig Colten (Pittsburgh: University of Pittsburgh Press, 2000); Craig Colten, *An Unnatural Metropolis: Wresting New Orleans from Nature* (Baton Rouge: Louisiana State University Press, 2005).

21. Barbara Allen, "The Popular Geography of Illness in the Industrial Corridor," in *Transforming New Orleans and Its Environs*.

22. Louisiana Advisory Committee to the U.S. Commission on Civil Rights, "The Battle for Environmental Justice in Louisiana. . .Government, Industry, and the People," Washington, DC: Government Printing Office, 1993); Bernice Powell and R. D. Bullard, "From Plantations to Plants: Report of the Emergency National Commission on Environmental and Economic Justice in St. James Parish, Louisiana" (Cleveland: United Church of Christ Commission for Racial Justice, 1998); J. Timmons Roberts and Melissa Toffolon-Weiss, "Roots of Environmental Injustice in Louisiana," in *Chronicles from the Environmental Justice Frontline* (New York: Cambridge University Press, 2001).

23. Robert D. Bullard, *Unequal Protection: Environmental Justice and Communities of Color* (San Francisco: Sierra Club, 1994).

24. M. Pastor et al., "In the Wake of the Storm: Environment, Disaster and Race after Katrina," (New York: Russell Sage Foundation, 2006).

25. Zygmunt Bauman, *Modernity and the Holocaust* (Ithaca, N.Y.: Cornell University Press, 1989), 65.

26. Caroll Joseph Dugas, "The Dismantling of De Jure Segregation in Louisiana, 1954–1974" (Louisiana State University Press, 1989).

27. Halpern, "Solving the 'Labour Problem.'"

28. Catherine Ellis, "The Legacy of Jim Crow in Rural Louisiana" (PhD diss., Columbia University, 2000).

CHAPTER 6

AN INTERVIEW WITH JUDGE IVAN L. R. LEMELLE

SUZETTE M. MALVEAUX

The Honorable Ivan L. R. Lemelle (E. D. La)* was born in 1950 and grew up in a small, rural town outside of Lafayette, Louisiana—Opelousas. He graduated from Xavier University and Loyola University, as well as the New Orleans School of Law, and then practiced law in the private and public sectors. He was appointed a United States Magistrate Judge in 1984 and then United States District Judge (Eastern District of Louisiana) in 1988 by President Clinton.

Judge Lemelle, one of many displaced citizens of New Orleans and the only African American district judge on the United States District Court in the Eastern District of Louisiana, presided over *Wallace v. Blanco*, a federal lawsuit challenging Louisiana's emergency election plan for failing to sufficiently protect the voting rights of the primarily African American New Orleans population displaced by Hurricanes Katrina and Rita. The NAACP Legal Defense and Education Fund (LDF) and local counsel from Louisiana argued that measures taken by the state legislature were inadequate and would disenfranchise temporary evacuees from participating in New Orleans's April 2006 primary and May 2006 runoff elections, in which the mayor and city council selections would be made. The LDF contended that the state's voting scheme was inadequate and threatened to disproportionately impact black voters, in violation of the Voting Rights Act. While the city's primary election did not take place on February 4 as originally planned, it was later slated for April 22, a date many civil rights organizations and leaders challenged as too soon. Judge Lemelle, while recognizing room for improvement in the electoral process, refused to delay the election.

Judge Lemelle shared his thoughts on why he permitted New Orleans's first municipal elections post-Katrina to go forward, how Katrina has impacted the city

*Judge Lemelle is the author's cousin. This interview took place on June 16, 2006.

and himself personally, and what lessons can be learned for the future. Judge Lemelle also reflected on what it is like to be one of the few African American federal judges in the United States Court of Appeals for the Fifth Circuit and on the pioneers that inspired him.

*   *   *

SM: Tell us about Opelousas, Louisiana and what that was like when you were young, growing up there.

JUDGE LEMELLE: I was born in Opelousas, Louisiana. Most of my early and high school years were spent in the public and Catholic school systems in Opelousas. Upon graduation, I went on to obtain degrees from Xavier University in New Orleans and the Loyola School of Law.

During those formative years in Opelousas, I still recall some of the used and marked books that were discarded by the white Catholic school and sent to us at the black Catholic school—Holy Ghost Catholic School. My parents emphasized education as the key to overcoming racial barriers.

SM: So what made you decide to become a judge?

JUDGE LEMELLE: A difficult decision made easy with the help of some heroes in the early civil rights era. Founding members of the law firm I practiced in, Judge Robert Collins, who later became the first African American federal judge in the Deep South. Judge Nils Douglas, a magistrate judge commissioner in state court. And Lolis Elie, attorney, activist, and frequent judge pro tem in the state courts, Ron Nabonne, Jeff Wilkerson, and Okla Jones II, all alumni like me from the same firm, provided encouragement and support. They thought the time was right for someone to be the first black magistrate judge in the federal courts in Louisiana. Through their persistence and with additional support from my family, I accepted the challenge and applied. The Eastern District of Louisiana has a grand history of brilliant jurists. I was presented with a chance to walk in the footsteps of J. Skelly Wright, Alvin Rubin, Fred Heebe, and many others.

SM: What kind of challenges have you had being the only African American judge in the Eastern District of Louisiana? And one of the few in the Fifth Circuit?

JUDGE LEMELLE: We have ten African American jurists in the Fifth Circuit. The Fifth Circuit, as you know, is one of the busiest circuits in the country.

The challenges that I face are no different than when Judge Collins was the first and only one, when Judge Okla Jones II was the second and only one. Court employees, especially minority members, would often present us concerns about working conditions, hiring and promotion practices. I was honored when our Chief Judge appointed me to be the chairman of our court's EEO [Equal Employment Opportunity] and affirmative action committee. Through the years, we were able to resolve a lot of complaints. People felt comfortable talking to us and not discouraged in bringing issues they might otherwise stay silent about. While not an issue for me, I could sense some misgivings in a few, not most, senior members of downtown majority firms who appeared in Title VII racial discrimination litigation. But I did not have the time or inclination to prove myself to anyone. Justice had to be served by an impartial application of law to facts, not by perceptions or one-sided thinking.

SM: Can you tell me about Hurricane Katrina? How has that impacted your life? You were in New Orleans. You've been in New Orleans for how long?

JUDGE LEMELLE: Since 1967. Along with our fellow citizens, we were displaced from our homes by the storms and [are] still coping with every imaginable and unexpected matter in the aftermath of our exodus. Our family home is in New Orleans East. Katrina's path came right over that area. It left us with an average flood depth of eight feet. My home had over six feet of water. It settled at four feet for about ten days before it finally drained out. My first floor was devastated. My second floor is intact but damaged by wind, hail, and looting that occurred in our neighborhood. But, you know, I was lucky. My family safely evacuated before that momentous day, August 29, 2005. We lost close friends as a direct result of the storm and a sister-in-law's husband as an indirect result of the same. Our material losses pale in light of the human casualties.

There was a total collapse of our infrastructure. New Orleans East is the largest land area of the city, composed of a significant number of middle-class African Americans, including my subdivision, Lake Willow. Solo practitioner lawyers, doctors, poor people, retirees, [and] teachers who evacuated ultimately found, in many instances, better opportunities elsewhere. But most still want to return home to New Orleans. Many just could not afford to come back to New Orleans.

As a Red Cross volunteer in Lafayette, Louisiana, where I eventually relocated to, I vividly recall meeting an eighty-plus-year-old couple. . .evacuees from Betsy, and again Katrina. The only reason they are alive was because they recalled from Betsy that a lot of people died from drowning or suffocation in the attics of their homes, trying to escape the flood. Accordingly, this couple kept an axe to cut a hole in the roof of the attic, in the event that there was another hurricane that forced them into the attic. And that's what they did in Katrina. They were rescued from their home's rooftop in New Orleans East by helicopter.

The wife told me, "Judge, I never saw him move so fast in chopping the hole in that attic!"

And he said to me, "You know, Judge, the only frightful moment I had was not when the flood came and when I was chopping that hole—because I know I was going to get us out and on the roof." He said the most frightening moment was when he was being lifted by the helicopter up in the air towards roaring helicopter blades. They never flew on any aircraft before in their life. He related how the helicopter's basket kept swinging and turning. And he's laughing about it. And I'm almost in tears.

Here they are telling me this story in a very nonchalant, casual, calm and soothing way. I'm listening and I told them, "You gave me more strength in the telling of that story than I'm giving to you, and I'm supposed to be your comforter as a Red Cross volunteer." When I came back the next day, they were evacuated from the Red Cross shelter in Lafayette and sent to Houston. One of my biggest disappointments since the evacuation was not being able to locate and follow up with this beautiful couple. God bless them for giving this ill-prepared "Red Crosser" comfort.

* * *

New Orleans municipal elections were initially set for February 4, 2006, but were postponed on account of state officials' contentions that the elections could not be

held so soon after the devastation of Hurricanes Katrina and Rita. This postpone-
ment prompted a lawsuit by New Orleans citizens who argued that an indefinite
delay in the electoral process deprived citizens of their voting rights, in violation of
the Constitution and the Voting Rights Act. Consequently, the state set April 22,
2006, as the date for the primary election.

On February 23–24, 2006, and March 27, 2006, Judge Lemelle heard oral argument by plaintiffs' counsel in *Wallace v. Chertoff*, Civil Action No. 05-5519 (which
was consolidated with *Tisserand v. Blanco*, Civil Action No. 05-6487, and *ACORN
v. Blanco*, Civil Action No. 06-611), arguing for a delay in the New Orleans municipal elections scheduled for April 22, 2006.

At the March 27, 2006 hearing, plaintiffs' counsel, among other things, challenged the state election law requiring first-time voters who registered by mail to
vote in person at a registrar's office or polling place. Counsel contended that requiring evacuees to return to Louisiana to vote imposed on them an unreasonable financial burden, comparable to a poll tax. Plaintiffs' counsel also challenged the state
election law that provided for in-state satellite voting in cities in the state of
Louisiana. Counsel contended that satellite voting should also be made available in
cities outside of Louisiana.

On March 27, 2006, Judge Lemelle denied the Wallace Plaintiffs Motion to
Reconsider, Alter, or Amend. "Recognizing the importance of the issues raised and
the need for further monitoring," Judge Lemelle also ordered that the parties' counsel meet within that week and submit, among other things, "an update on voter
information/outreach initiatives, absentee balloting, polling locations and staffing of
same, election day monitoring" so that he could address any remaining issues. See
Order of Mar. 27, 2006.

New Orleans's first post-Katrina municipal elections were held on April 22, 2006,
as scheduled. A runoff election was held on May 20, 2006. Incumbent Mayor Ray
Nagin was reelected after defeating Louisiana Lieutenant Governor Mitch Landrieu.

* * *

SM: Let me turn to the litigation that came before you in your court. I've looked at
some of the court filings and the record. The NAACP Legal Defense Fund and others have argued that the state of Louisiana's emergency election scheme was going to
disenfranchise a lot of displaced New Orleanians, and disproportionately African
Americans. The plaintiffs were concerned that first-time voters would have to come
back to Louisiana to vote and that they could not vote by absentee ballot. The plaintiffs were also concerned that evacuees would be disenfranchised because there was-
n't satellite voting outside of the state of Louisiana. The plaintiffs wanted Atlanta,
Houston—some of the places where a lot of people had been dislocated—to also
have satellite voting sites. Consequently, plaintiffs asked for a delay in the primary
election. Can you tell me about your decision not to permit that delay?

JUDGE LEMELLE: I will try giving you the procedural background, how it got
started and, to some extent, how it ended.

The first of three lawsuits was brought by a group contending that Governor
Blanco's call for an election in September [2006] was going to lead to disenfranchisement of minority voters, and by not calling the election before then would have

some impact on minority voters. So it was quite the opposite of what was eventually argued when the NAACP LDF was allowed by me to join the case.

After several hearings, there were two other lawsuits that brought somewhat similar, yet dissimilar, claims in terms of what relief they were seeking. The cases were heard on an expedited basis.

After evidentiary hearings, I, ruled that if the state did not take action to call the election on the date that the original plaintiffs wanted, I might be forced to do so—on the basis that it would lead to disenfranchisement of African American voters. In ruling, I strongly urged the state legislature to consider some other methods of voting, because after Katrina a lot of things—not just in terms of elections, but in terms of everything that affected normal life activities—changed and were being done in different ways. So I directed the parties to meet with the Legislature and the Governor's representatives. I wanted everyone to get together to see if they could come up with a plan—resolve some of the differences. If they couldn't, they would come back to me.

A plan was eventually presented. Initially, the plan was rejected by the Legislature, causing the Legislative Black Caucus to walk out of the session. When I heard news of the rejection, then obviously, I was disappointed. I expressed that disappointment to parties' counsel and told them to try again. Shortly thereafter, the Legislature reconsidered the proposals. All the proposals passed that were being sought, except for two. The two that did not get implemented, or were not considered, were the one that was going to set up satellite voting outside of the state and the other concern[ing] first-time voters. The major changes that passed involved certain provisions for first-time voters and satellite voting outside of Orleans Parish in other parts of the state.

Ultimately, another hearing was called by me to consider what was done and not done and whether or not there was evidence of voter disenfranchisement. None was found. I signed a judgment recently on all three cases, and those cases are now closed. I'm told that the election returns showed the percentage of black turnout was comparable to what it was pre-Katrina in mayoral elections. I don't know if that's correct or not. It's possible that another lawsuit involving similar claims might urge similar issues for a different election. Courts of law can only rule based upon the record evidence and applicable law, not upon outside-the-record influences. I based my rulings on evidence presented to me in court, and that evidence led to the relief I ultimately approved.

SM: I want to go to some of the media coverage at the time you made your decision. There were a number of civil rights organizations that argued that requiring first-time voters to vote in person was equivalent to a poll tax. Did that argument resonate with you at all, given the historical racial discrimination in Louisiana and other states in the South?

JUDGE LEMELLE: I know the history of the Voting Rights Act. The firm that I was with—all of us in the firm—at one time or another, had a civil rights case. We worked on voting rights cases; we were involved in a number of election lawsuits as lawyers for black litigants.

I think that the lawyers in this case, particularly the lawyers for the original set of plaintiffs, who are local evacuees themselves in Lafayette and Houston—did a fine job.

Subsequently, when we allowed the NAACP Legal Defense team to get involved, they did, as I expected, excellent work on behalf of their clients. Their legal arguments were all on point. Did it influence me? It had influence. But again, we can argue legal points of view and argue what the Voting Rights Act stands for, but you still have that burden of proof, to show with factual evidence that this is a disenfranchisement in some manner or another. And if [disenfranchisement is] proven, the relief has to be balanced and focused in light of several factors. There was insufficient evidence to fashion any more relief than what was already obtained.

SM: Why not delay? If you could have permitted more people to be enfranchised, then why not push back the election to make that possible?

JUDGE LEMELLE: There was a group of plaintiffs who wanted the election called earlier, and I didn't allow that; it would have been too early. There was also a belief that holding it too late would also lead to disenfranchisement. Voters were entitled to vote, and the elections were supposed to be held in February [2006]. They couldn't be held in February [2006]—that was too early. Do you delay it until possibly September? Would that effectively lead to disfranchisement because of a likely impact upon voter interest or turnout? Should that matter? It's an important election—mayor, city council, assessor—elections were designated under the city charter for a certain date. To delay it any further than necessary would be problematic.

New Orleanians wanted to come home; they wanted normalcy in their lives—at least that road to normalcy, if not that exactly. And to do otherwise, to me, would deflate their expectations. The most powerful tool of minorities in any society, in a democratic society like ours, is the power of the vote. Every vote counts—regardless of wealth, color, or any other factor—to determine the direction of society through elected representatives of the people. That's power. The Voting Rights Act, while in some respects imperfect, is designed to help protect that power. The people of our city had a need, a right, to feel like they have a right to determine where this city's headed. There were some angry voters out there. People, for one reason or another, justified or not, felt official efforts to promote normalcy and rebuilding were ineffective or untimely. I cannot speculate whether my decision—coupled with outside demonstrations related to it—contributed in positive ways to voter turnout. I like to think it did.

My object or role in deciding a voting rights case is not what the turnout is, because that's after the fact. It's to deal with it before, and to make certain that legal rights are protected. There was some speculation that black voters did not trust absentee or mail-in balloting. But more black voters voted absentee and by mailing in this election than ever before, because they were out of town, displaced. The same motivations that cause a white person to vote absentee will be the same motivations that cause a black person to vote absentee. Even more so here.

SM: Did you ever think, though, because the NAACP Legal Defense Fund, and ACORN, and Jesse Jackson and Marc Morial [Urban League President and former mayor of New Orleans]—because major civil rights organizations and leaders were taking a position counter to yours that your position was not the correct one?

JUDGE LEMELLE: I respect anyone's right to disagree. However, I would not be a good judge if I based my judicial rulings on matters outside the trial record.

In January 2005, Iraqi expatriates were able to participate in Iraq's first democratic elections held in fifty years, by casting their ballots at satellite polling places located in five cities throughout the United States. Had the attorneys come into court and given me some evidence, some proof, something to show for instance, how did the government set up the system of voting for displaced Iraqi nationals, I could have considered such as evidence—they and such evidence were not offered.

I asked them, "Well, how do you do that? How is that done? What did it take?" If I had gotten some proof along those lines, then it could have affected the extent of available relief, but I never got those answers to my questions. I cannot base my decision in any case on what I read about in the press, see on TV, or hear on the radio.

My sister-in-law participated in the march [challenging the state's emergency election scheme]. She was back in New Orleans. She felt as if my decision was going to make people come here and vote because they were going to be ticked off, they were going to be mad. Later, when she saw the electoral turnout, she said, "Maybe you did the right thing after all." I hope she's right. I called it based on the facts presented in court.

SM: There was an article that came out in the *New York Times* that said something to the effect that you had chastized social activists for using people who were suffering.[1] Do you recall anything like that?

JUDGE LEMELLE: One of the lawyers during a hearing argued what some protesters were saying. I asked in response, "Where are those civil rights leaders? They're not in my courtroom today. I've asked for this before. If they have testimony on something bring them to court. I can't go beyond the record. Where is that proof?"

I respect Jesse Jackson, I respect Marc Morial. Particularly Jesse Jackson, for his early role in the hard and very dangerous period of the civil rights struggle. We're the beneficiaries of his efforts. Certainly I would have been impressed by having Jesse Jackson in my courtroom. Just his presence, and listening to what he had to say in my courtroom, would have been a momentous occasion. But I couldn't base my ruling on what he had to say to the press. So it wasn't chastising them; it was more to the lawyers saying, "Are you going to call them as witnesses, counselor?" My intent was not to chastise anyone. I didn't have a right to. The press, perhaps, took it that way.

I do question anyone's underestimation of the human resolve, the will of the people to overcome Katrina's wrath, by participating in the electoral process. Black voters, like white voters, have a right to participate in the electoral process. And to say that simply because you are going to have to now vote absentee or by mail is not in itself a violation of the Voting Rights Act. You have to show how that is going to negatively impact black voters. My people are displaced. The evidence showed that they were going to vote—for a number of reasons. I had one witness who was a major civil rights leader in New Orleans and displaced to Georgia. He said that come hell or high water he was going to vote, no matter what. That impressed me. He disagreed with my ultimate findings. He thought there should be satellite voting outside the state. I respectfully disagree with the unproven notion that the absence of some, as a legal matter, would lead to voter disenfranchisement.

SM: Do you think there are lessons here that have been learned?

JUDGE LEMELLE: Quite a bit. In terms of voter election laws, the public learned more about the electoral process. I authorized, for instance, election day monitors to further assist and educate voters and voting day commissioners. New Orleanians are rebuilding and are on that bumpy road to normalcy. Voters in New Orleans East, Gentilly, Lakeview, and Lower Ninth Ward, among others, will probably have to return in the future to mega precincts in order to vote. I voted in Lafayette in both elections. I was thankful to see and hear about voter commissioners and registrars throughout Louisiana who were very hospitable, courteous, and professional to displaced voters.

SM: If there's data that shows that some of the election procedures had a disproportionate impact on African American voters, then changes could take place for the September election. For example, satellite voting polls outside of Louisiana and permitting first-time voters to vote by mail. Maybe those kinds of changes could occur in time for the September election.

JUDGE LEMELLE: And you've got some other variables too. A big unknown right now is, "Of the New Orleanians who evacuated, who's returning back? Are they returning back?" We did a lot of soul searching to determine whether or not we're going to rebuild or return to New Orleans. I wish I could convince everybody who left to come back, but I realize they have to make their own personal decisions after assessing many factors that affect individual concerns.

SM: As the only African American judge in the Eastern District,. . .you are [probably] seen as a leader and a role model. Do you think that people understand what was behind your decision to return?

JUDGE LEMELLE: I'm one little cog in the wheel amongst several cogs. Am I flattered by the thought that somehow I may inspire people to action? Of course. I'm taking a risk like everyone else does in rebuilding. The future is uncertain, with no guarantees.

SM: Did you ever think about not coming back?

JUDGE LEMELLE: No, never did; that never entered my mind.

SM: Why?

JUDGE LEMELLE: Suzette, when I came back the first time—and as devastated as my home was—I didn't want to leave. I didn't have power. I didn't have water or gas. I didn't have anything, but I wanted to stay on the second floor; just me, I wouldn't put my family through that. But there was mold everywhere and people said, "Oh no, you're going to get sick." My family would have suffered more out of concern for me had I done that. So I didn't stay. But every time I go back home, it's very difficult for me to leave. . .very difficult. Concern over the suffering conditions of homes and neighbors drive me to return. Remembrances—Lake Willow constantly pulls me back and one day I will stay.

What really hurts is when I hear people who are not coming back, like my neighbor, a physician, at the peak of his career, in a home he and his wife put everything into. Yeah, [*sighing*] those are the hurting things—to lose neighbors, friends, family, et cetera. We learn our lessons and one of the things I guess you learn in this is who are your true friends. And you learn a lot about human nature as a result of it.

SM: And about yourself.

JUDGE LEMELLE: Oh! A lot about myself. A lot of things I didn't even want to know! It scares me! [*He laughs.*] Like I told you when I didn't want to leave the house. I look back on that and I said, "Hmm, what would have happened to me had I stayed?" Then again, I was there yesterday. I go as often as I can.

Life in the "Big Easy" has everyone Katrina-weary, yet steadfast in seeing the city's rebirth.

NOTE

1. "Judge Orders New Orleans to Proceed With Election," *New York Times*, March 28, 2006.

PART II

CULTURE, TRADITION, AND IDENTITY

NEW ORLEANS'S AFRICAN AMERICAN MUSICAL TRADITIONS: THE SPIRIT AND SOUL OF A CITY

MICHAEL WHITE

It has become fashionable to label August 2005's devastating Hurricane Katrina as "the greatest natural disaster in American history," when considering the combined loss of life, massive destruction, economic impact, and prolonged displacement of citizens. What is less often discussed is that from the perspective of New Orleans's African American population, Katrina's aftermath has potentially launched the greatest cultural disaster in this country's history. The United States is a comparatively young nation that lacks the presence of many original vernacular traditions and the rich cultural history that can be found in Africa, Europe, and Asia. Despite this, the city of New Orleans embodies the variety, longevity, level of community involvement, and social relevance of its several one-hundred-plus-year-old African American folk traditions.

Though not generally understood beyond its entertainment value, the best known New Orleans cultural tradition is jazz. First taking shape around the late 1890s, jazz's lively rhythms and largely improvised instrumental melodies facilitated its transition from its birthplace throughout the world. By the 1920s, this musical genre had begun to influence most popular music and had become the ideological basis for the spirited "Jazz Age." It was also well on its way to being universally viewed as "America's only truly original artistic contribution." Other local customs—such as brass bands, social club parades, jazz funerals, and Mardi Gras Indian processions—have, until recent decades, remained almost hidden in the heart of New Orleans's African American community—socially, physically, and psychologically removed from most of white New Orleans and a large number of culturally isolated blacks.

The African American population of New Orleans shares with the rest of black America a varied and distinct existence formed by centuries of blending of African, European, and other ethnicities. Many black visitors from other American cities, Africa, the Caribbean, and Latin America notice New Orleans's uniqueness, yet they also express a strong spiritual and ancestral identification with the city as feeling like "home." New Orleans's physical distance from other large urban areas, its varied social makeup, and its cultural history have combined to form one of the most rich, diverse, and distinctive subcultures in American history. The local black population contributed greatly to New Orleans's unique nature, spirit, and identity. Though not often recognized or acknowledged as such—even among natives—jazz represents a synthesis and transformation of several cultures, with a dominant West African influence, that laid the fertile background from which emerged the plethora of New Orleans's rich cultural traditions. Many who view the New Orleans jazz tradition rarely recognize that jazz and other local music are all aspects of an endemic community lifestyle and philosophy that also manifests in a unique local approach to neighborhoods, family, language, cooking, dancing, humor, celebration, worship, and other aspects of life.

MUSIC IN NEW ORLEANS AND THE AFRICAN AMERICAN POPULATION

Since its founding and early colonization by the French in 1718, New Orleans was destined to be different from other American cities. Its location along the Mississippi River near the Gulf of Mexico contributed to its use as a major port—and made possible long-term exposure to Caribbean and other Latin American cultural influences. From the beginning, hardship and tragedy resulting from a geographically harsh and vulnerable environment and a series of devastating natural disasters led to a common special appreciation for life. Many holidays and feast days observed in the predominantly Catholic city also contributed to a longstanding obsession among New Orleanians with excessive pleasure seeking through food, alcohol, music, dancing, and celebration. These customary "necessities" often took precedence over "less serious issues," such as punctuality, business, and progress.

During the nineteenth century, New Orleans experienced a wide range of musical activity—from opera and classical forms, to military marching bands, dance music, religious songs and various ethnic folk music styles. The large, diverse population of slaves and free blacks was often exposed to and participated in an unusually wide range of musical styles. Throughout much of the nineteenth century, a tradition of authentic West African drumming, chanting, dancing, and celebrating was performed by scores of slaves and later free blacks on Sunday afternoons at an open public field known as Congo Square. African-style musical practices existed in New Orleans far beyond the legally sanctioned public spectacles of Congo Square in secret voodoo ceremonies, in private locations and along levees, bayous, and Lake Pontchartrain. While these ancestral traditions evolved and disappeared underground by the twentieth century, their existence is strongly felt in the spirit of exciting rhythms, creative dancing, and collective community celebration that colors and serves as the foundation for most indigenous and local versions of music, ranging from jazz, gospel, and rhythm and blues, to funk and contemporary hip-hop styles.

The cultural diversity within "black New Orleans" is also unique in America, consisting of distinctive strains of African tribes and significant blendings of Haitians, Americans Indians, French, Spanish, English, and other ethnic groups— all of whom contributed to the rich cultural heritage of pre-Katrina's native population. Miscegenation, a practice that stems from the French practice of long-term illicit liaisons with black women, brought about an almost unheard-of third social tier: the Creoles of Color. For many generations, these mulattos were a highly visible and vibrant force in New Orleans culture. They were a privileged class, comfortably wedged between black and white society. Among the largely free Creoles, many were well schooled in European classical music, with several developing active careers as performers, composers, teachers, and devotees.

The tense post-Reconstruction social climate had important effects on black New Orleans culture. New Orleans had been a city in which blacks were accustomed to a certain degree of freedom, comfort, and mobility. The optimism resulting from winning legal battles, successful civil rights protests, and having black elected officials and government promises began to disappear under a new wave of white supremacy, violence, and racial oppression. Creoles of Color were officially and legally defined as "Negroes," resulting in the loss of their privileged status, economic power, and other rights. An air of confusion, anger, and defiance arose among a collective black population whose quest for freedom and equality would not be easily compromised.

ORIGINS OF JAZZ

While national black leaders like W. E. B. Du Bois and Booker T. Washington sought ways to organize, combat racism, and improve the black condition in American society, black New Orleans's most significant response to these issues was a locally contained cultural revolution. As many legal struggles were being stifled— as by the landmark *Plessy v. Ferguson* case, which originated among black New Orleans activists—social and economic conditions worsened. The defiant spirit of the times and the traditional celebratory character of New Orleans synthesized to create a new musical form—jazz—which expressed, through sound and related movement (dancing), the hopes, aspirations, emotions, and needs of black New Orleans. This new music (not called "jazz" until years later) was a collective creation drawn from various musical, social, cultural, ancestral, ethnic, and religious elements that formed the total black New Orleans experience. There were a number of practical and logical reasons for the origins of jazz.

Along with the forced legal merger between blacks and Creoles of Color came a cultural one, which led to a great degree of teaching, exchange and blending of distinct musical traditions. The Creoles' familiarity with European classical instruments, techniques, and repertoire synthesized with the local African American folk tradition of work songs, spirituals, blues, ragtime, dance music, ethnic folk songs, rhythmic interplay and authentic African-style celebrations to create a new form of expression that would have far-reaching social and cultural implications and would also change the direction of western music.

Regardless of any economic or social conditions, the local obsession with music, dancing, and good times never waned among most black and whites in late nineteenth-century New Orleans. African American musicians had many opportunities to perform in a variety of contexts for all social classes. A tradition of skilled black and Creole reading musicians, who played in refined dance orchestras and European-style marching bands, had been a source of pride and respect. This practice grew steadily after the Civil War.

Like other Southern areas with large black populations, New Orleans also had various types of African American folk music in the form of work songs, street cries, spirituals, and dance music. After the Civil War and into the twentieth century, thousands of poor black immigrants from neighboring states brought musical traditions, such as the rural blues, which were absorbed into the rich gumbo pot of local sound and expression.

It was in this climate of social turbulence and vibrant musical activity during the late 1890s that legendary cornetist Charles "Buddy" Bolden and others were among the first to employ a looser, freer, more exciting, personal, improvised approach to playing ragtime, blues, hymns, marches, and popular dance music. Music had been an attractive option for African Americans to earn extra income and achieve a greater degree of mobility. The local demand for music grew to exceed the number of trained reading players and available musical scores. As a result, bands were often smaller, and by necessity many groups consisted of nonreading musicians who memorized, adapted or improvised their parts. Bolden was among the earliest to experiment with various music styles and to apply vocal effects, feeling, and rhythms of black blues and hymn singing to horn playing. Soon, this hotter, more exciting style became popular among both musicians and dancers in public parks, dance halls and community parades. Though never universally accepted, jazz groups gradually replaced many of the refined reading society orchestras and marching bands that had previously dominated. Early jazz did not attempt to ignore or destroy the rules of European classical music (which governed most popular music of the day); it extended and reinterpreted conventional musical principles by blending them with an African-influenced folk approach to tone, melody, and rhythm. Underlying jazz performance was an almost existential philosophy of music and life, which offered the public a living, visible, and audible democratic model of the freedom and equality being sought by African Americans in social, economic, and political arenas throughout the nation.

More than a limited repertoire of specific songs, early New Orleans jazz was an approach that was applicable to music of almost any style. It was characterized by collective improvisation and steady driving rhythms—created by various combinations of a horn section (trumpet or cornet, clarinet and trombone) and a rhythm section (banjo, bass, drums and piano). Under a steady, danceable rhythmic pulse, the trumpet played a melodic lead as the other horns "answered" in call-and-response-type musical conversations. Much more than the stereotype of just "playing what you feel," there was a consciously constructed ensemble approach consisting of freedom with a general role or direction for each instrument. In this context, equal importance was given to the development of both individual and collective aspects. Possessing an individual personal tone and expression was as important an objective as was using one's sound for the sake of creating a more swinging, unified and identifiable group. Most early New Orleans jazz emphasized this instrumental ensemble approach and limited the use of vocals and improvised solos.

By the early 1900s, jazz had become a functional musical expression of the core local black community (and had also begun to be adopted by some whites). It was seen and heard in many neighborhoods at almost any time, accompanying a wide variety of functions, including dances, picnics, boat rides, sporting events, bars, weddings, and parties. Many families used jazz as a form of diversion by forming bands with each member playing a different instrument. An active musical family tradition developed, which contributed greatly to the core of professional and seminal New Orleans jazz players. This custom continues as a vital aspect of local culture and music across several generations and musical styles. The range of early jazzmen went from self-taught nonreaders to well-trained reading musicians. Several prominent African American music teachers had provided varying degrees of schooling—both privately and in groups—to a large number of young future jazz players.

BRASS BANDS AND COMMUNITY PARADES

One of the most important aspects of the New Orleans jazz tradition—in terms of its social significance and potential for widespread community interaction—is that it has also functioned at the center of a vibrant street culture: open to all ages, classes, genders, and ethnic groups. A common occurrence in the early 1900s was seeing jazz bands playing atop advertising wagons that rolled through various neighborhoods. When two wagons met, the bands would engage in a competitive "cutting contest," which drew large crowds of "judges." The most widespread and socially relevant function of jazz among New Orleans's African American population was the tradition of street parades, which continued throughout the twentieth century and into 2005's Hurricane Katrina. By 1900, dozens of benevolent societies and "social aid and pleasure clubs" had become a well-needed black staple that addressed a variety of social, political, and economic concerns and provided much-cherished entertainment. It was these organizations that became the principal sponsors of jazz-related activities in the African American community, the most popular of which were dances, parades, and funerals.

The tradition of black social club parades was very different in style, purpose, and form from the better-known, mainly white Mardi Gras parades of New Orleans. These annual social club functions passed through black neighborhoods on Sundays and lasted several hours. The community-oriented processions consisted of three main components: social club members, brass bands, and the crowd. Social clubs may have had several divisions of elaborately adorned members. Often, a considerable amount of money was spent on special outfits purchased for one day's use. Many times, club members wore extremely bold color combinations or extreme variations of the same color. As they walked and danced through the streets, club members carried beautifully adorned umbrellas, canes, baskets, fans, or handkerchiefs, all of which were used for dancing.

Forming a visual contrast to club divisions were one or more hired brass bands, which were traditionally adorned in dark band uniforms or black pants, white shirts and white band caps. A typical traditional New Orleans brass band is a ten- to twelve-piece "marching" (in reality, just walking) jazz ensemble, consisting of three trumpets, two trombones, a tuba, a clarinet, a bass drum, a snare drum and two lower brass instruments (later replaced by saxophones). As with smaller jazz combos, collective improvisation and a loosely executed role for each instrument are the essence of their performance style. The bands walked between club divisions to a static military beat when not playing their standard repertoire of jazzed-up marches, hymns, blues, and other lively songs. The louder, more raucous brass band sound was characterized by a distinctive, very danceable underlying beat—drawn from West African rhythms. This medium tempo pulse and its variations were prominently played on the bass drum—in counterpoint to a steady basic tuba line. This "second line beat" served as the motivating force behind spirited mutual inspiration and interaction between musicians and dancers. The distinctive brass band parade rhythm is the foundation for and a dominant influence on not only New Orleans jazz, but also other local musical styles.

The last but equally important part of a black social club parade was the "second line"—a crowd of up to thousands of anonymous followers who seemed to appear from nowhere with the first hypnotic three-beat song introduction of the bass drum. Much more than casual observers, this multitude of souls followed along the side and in back of the parade—dancing, cheering, and collecting others throughout its duration. The "second line" dance of the club members and crowd was a free-form West African–influenced set of movements that accompanied and paralleled jazz—in terms of its characteristic collective and individually improvised variations. "Second liners" often used a basic shuffle step with feet close to the ground, but an endless variety of jumping, crawling, shaking, spinning, and other motions were also common. Despite significant changes in brass band sound and appearance after the late 1970s, the overall look and spirit of social club parades continued into August 2005 much as they had for over one hundred years. Finding oneself in the midst of such a spectacle could cause one to feel like one was being hypnotically swept along in a massive spiritual "human tsunami." It was impossible to see or hear more than a fraction of the millions of subtle ripples of motion, sound, and ecstasy that regularly occurred during these processions. A casual glance in any direction would reveal frenzied dancers everywhere: on the tops of cars, swinging from light posts, "hovering" over the roofs of building, riding through the crowd on the backs of others, moving in and out of impromptu circle formations, or gliding belly down in the street—all the while keeping perfect time and pace with the parade.

In this sea of various shades of black skin there were young faces and old faces; soft faces and hard faces; faces on bicycles and in cars, doorways and windows. Normal sensory perception was altered, as time, space, weather, age, class, and physical ailments seemed to disappear in a steamy tidal wave of hot music and dancing. Walking canes and crutches that rarely left the ground were suddenly hoisted high, like the club members' umbrellas. Elderly faces, wrinkled, tired, and helpless from years of struggle—now glowed with youthful joy. Babies, too young to walk, gleefully wiggled in their mothers' arms—also in tempo with the music. Even house pets occasionally came along and got caught up in the frenzy. Major disappointment often came at parade's end, when the participants "awoke" from the trance to discover themselves on another side of town and facing the difficult task of having to "walk" back the same distance they had just danced.

To some participants these activities were nothing more than fun—a pleasant diversion from daily routines. For many more conscious African Americans of New Orleans, a redefinition of existence—pointing toward the unjust social and political structure—was a fundamental underlying theme of early jazz and social club parades. One typical example is the popular standard European march song (with its common challenging array of strict rules and varying key changes, several sections, and harmonic and melodic shifts). In New Orleans both musicians (and dancers) regularly transformed this originally static form into a freer, more rhythmic, relaxed, personal, lively and enjoyable set of musical possibilities.

Thus, at the time of their origin, these parades offered the black community a euphoric transformation into a temporary democratic world characterized by free, open participation and self-expression through sound, movement, and symbolic visual statements. A variety of passions and attitudes that may have been repressed in normal daily life could be released in the security of these processions. Impositions and limitations of "second class" social status could be replaced by a democratic existence in which one could be or become things not generally open to blacks in the "normal" world: competitive, victorious, defiant, equal, unique, hostile, humorous, aloof, beautiful, brilliant, wild, sensual, and even majestic.

The social club parade often became a massive, almost religious-like gathering of African Americans, which, under the guise of entertainment, could also be a form of protest: a show of strength and unity and a defiant march toward freedom and democracy in a society where such assemblies would normally be discouraged or illegal.

There were also other implications and symbolism: from the loud, colorful club outfits (in contrast with the more conservative band uniforms), through the unrestricted freedom of the "second line" dancing, which often included dancing with or on top of any objects or structures that one came upon, to the occasional songs, which had breaks (brief pauses) that were filled in by shouts from the crowd. What was so blatantly (and simply) demonstrated time and again by various aspects of social club parades was that all forms of existence are acceptable and can function together. To a larger extent the concepts of possibility and transformation in music and dance had implications pointing toward a more just restructuring of society, politics, and life in general. Over the course of a several-month-long social club parade season (and throughout the twentieth century), a large segment of the community—people of all ages and backgrounds—had the opportunity to reap the benefits of participating in these unique customs: psychological and social uplift, release and coping devices, acknowledgement and acceptance, and a shared spiritual connection to community and ancestral traditions.

Funerals with music, labeled "jazz funerals" in recent decades, were a time-honored tradition and variation of social club parades. While the most-publicized twentieth-century jazz funerals were held in honor of deceased musicians, the majority of these ceremonies were given for members of benevolent societies and social aid and pleasure clubs. The funeral ceremony juxtaposed two different perspectives on death in the context of an open community celebration. After the religious service, a brass band (adorned in black suits) lined up outside the doors of the church or funeral home and played a slow, solemn dirge or hymn as the casket was placed into the hearse. Then began a procession led by an elaborately adorned grand marshal, followed by the band, social club members (also in black), the hearse, and finally the family (in funeral cars). As the procession proceeded, accompanied by a muted snare drum beat and more sad music, a crowd gathered and walked respectfully alongside. Slow mournful songs, like "Flee As A Bird" and "Nearer, My God, to Thee," highlighted the sadness and sense of loss felt by family and friends. The solemnity of this part of the service was reflected in the music, which was characterized by preacher-like trumpet melodies, "amen" responses from other horns, shrill high note clarinet wails, and a gloomy death march pulse throughout. As the procession neared the cemetery or had proceeded for several blocks, the band lined both sides of the street and played one last dirge as the remaining procession passed through. This final public sendoff was called "cutting the body loose."

After the burial, or when the hearse and family were a respectable distance away (en route to the cemetery), then began a joyous celebration accompanied by lively up-tempo songs like "Oh, Didn't He Ramble" and "When The Saints Go Marchin' In" and "second line" dancing by the crowd. The ensuing procession lasted for several blocks—sometimes miles—in this final recollection of the deceased's good time on earth. More importantly, this jubilant expression reflected a symbolic happiness because the dead person has attained the "ultimate freedom" from earthly burdens and has transitioned to a glorious union with the Creator. This is a New Orleans representation of the biblical reference to "rejoicing at death."

Religious community members—who sometimes avoided the frenzied, "sinful" social club events—were exposed to and participated in local music–related customs through a longstanding (but now virtually extinct) tradition of church parades. Held early on Sunday mornings to celebrate anniversaries and religious holidays, these Protestant (usually Baptist) processions had divisions of church members attired in white dresses and black suits. Brass bands played jazzed-up, lively versions of traditional hymns like "Lord, Lord, Lord" and "Bye and Bye" as church members and ministers strutted gracefully through otherwise quiet community streets. Here, the demonstration of strength, unity, and desire for freedom and social justice was acted out from the more solemn religious perspective of paying homage to and petitioning the Creator.

Another unique New Orleans African American cultural tradition that developed during the turbulent post-Reconstruction period of the 1880s and continued into the 2000s was the Mardi Gras Indians. Self described as "gangs," the Mardi Gras Indian tribes consisted of elaborately costumed blacks (predominantly males) who paraded through neighborhood streets on Mardi Gras and St. Joseph's Day. The outward purpose of these groups was to pay homage to the Native American spirit of resistance and to recognize cultural and ancestral ties between the two races. In appearance, music, and procession style, the Mardi Gras Indians were actually expanding the local tradition of transformed West African–style celebrations seen at Congo Square and other locations. This was yet another poplar music-related custom that involved large segments of the black community through massive open attendance and participation. Most Indians were also from New Orleans's lower-economic-level neighborhoods. Much time, money, and effort were spent preparing their boldly colored, breathtakingly beautiful costumes—all original creations that reflected traditional folk characters or individual statements. Several months of collective work from family and friends were necessary to produce each year's cherished new suits made of feathers, beads, sequins, and other materials. Within the dozens of tribes—which had names like the Creole Wild West and the Golden Eagles—there were coveted positions, like the "big chief," "wild man," and "spy boy"—each carrying different levels of responsibility and degrees of honor and respect.

In addition to preparing new costumes, other year-round activities, such as musical rehearsals and parading on special days, all contributed to a transformed existence in which pride, strength, respect, and nobility masked harsh everyday realities and difficult socio-economic conditions. So serious was the Mardi Gras Indian persona that deadly confrontations often resulted from random encounters of rival tribes on carnival day. Fortunately, modern day "battles" were of a friendlier nature, taking the form of competition for the most beautiful costume or the most skilled dancing. As they paraded through black neighborhoods, the Indians acted out a series of rituals through chanting, song, and dance. Special traditional Indian songs were sung in English and in unknown or secret dialects. Singing was accompanied by a "small band" of nonuniformed marchers, who used drums, tambourines and other percussion instruments to play long stretches of increasingly intense African-style polyrhythms. Indian dances were a series of West African–inspired movements and gestures, yet very different from "second line" dancing. Like social club parades and funerals, Mardi Gras Indian processions attracted large crowds of followers, who usually walked, clapped, and cheered along the entire route. These parades traditionally proceeded defiantly without legal permits over unplanned and improvised paths.

Though traditional New Orleans–style jazz was the only musical form to actually originate in the Crescent City, the local musical heritage is also marked by stylistic diversity. A unique approach to singing, piano playing, drumming, and horn playing for the basis of a "New Orleans sound" has been transmitted into other more contemporary musical styles. This community has also produced distinctive local versions of other musical genres—such as rhythm & blues, funk, and hip-hop. These styles often reflect inspiration and influence from the vernacular music of brass bands, jazz, and Mardi Gras Indian traditions, especially in terms of rhythm, melodic phrasing, and conveying the local community spirit. Fats Domino, the Dixie Cups, the Neville Brothers, Allen Toussaint, Irma Thomas, Juvenile, Master P, and Lil Wayne are few of the local artists from the wellspring of New Orleans talent to garner national attention. A small modern jazz movement, fostered by saxophonist/producer Harold Batiste in the 1950s and 1960s, gained momentum after the 1980s through the efforts of performers/teachers like Ellis Marsalis, Alvin Batiste, and Edward "Kidd" Jordan. These men taught several of the current international stars of contemporary jazz—including Wynton and Branford Marsalis, Terrance Blanchard, Donald Harrison, and Nicholas Payton. Though not having as widespread community involvement and social significance as traditional jazz, social club parades and Mardi Gras Indians, other local musical genres are the products of the same environment of supportive family and community traditions in a culture where musical activities, creativity, and community celebrations are a way of life.

No uniquely New Orleans style of religious music came to prominence; however, common Southern gospel forms were also influenced by the sound and spirit of local traditions in the city's many vibrant Protestant church choirs. A strong family church tradition has produced many singers and musicians who performed only in religious settings or have had successful crossover careers in other genres. It is no coincidence that the rich environment of jazz and street parades of her hometown helped to nurture a young Mahalia Jackson (1912–72), who would go on to become "the world's greatest gospel singer."

When considering the potential impact and implication of Hurricane Katrina on New Orleans's culture and traditions, one must recognize that the city and all of its customs (which evolved remarkably little throughout three-quarters of the twentieth century), had been undergoing a series of dramatic changes during the thirty years prior to the storm. Before Katrina, New Orleans music and culture found themselves in a state of transition in which commercialism and other changes had infused and transformed local traditions to the point that both authentic and generic styles existed together. In a popular, commercially oriented environment, the idea of "New Orleans music" came to mean many things to many people: local rhythm and blues, black traditional New Orleans jazz, modern jazz, Bourbon Street, brass bands, Dixieland, Cajun and zydeco music, Mardi Gras Indians, white rock groups, etc. Factors that affect lives across America—like computers, cell phones, cable television, video games, and other technological developments—have also affected local cultural traditions. Younger generations often approach concepts like tradition, race, culture, music, etc., from an understandably different perspective than their elders. Their thoughts, values, orientation, needs, views, and experiences are of a more modern world; they face different realities. Some younger African Americans do sense a uniqueness about being from New Orleans, but many are more in tune with the majority of American youth—and think more generically than regionally. Education, family background, and class differences affect the likelihood of participation in local traditions. Though some continue customs practiced by their ancestors for generations, others bypass those traditions and seek a "better life" and take advantage of "greater" opportunities.

Contemporary inner-city social problems have had more than a "healthy" existence in the black New Orleans neighborhoods from which local traditions originated during the thirty years prior to Hurricane Katrina. In some ways, it is miraculous that one-hundred-year-old-plus folk traditions could remain alive and vital in a community in which drugs, murder, poor health, illegitimacy, poverty, unemployment, illiteracy, urban decay, and other problems regularly rank among the worst in the nation. Like several other American cities, New Orleans saw a massive level of urban decay during the post–civil rights years; this was caused by diverse factors such as "white flight" to the suburbs, ineffective political leadership, corruption, indifference, failure to keep up public facilities and institutions, the Louisiana oil bust, incompetence, racism, and an seemingly unstoppable influx of illegal drugs and weapons into the black community.

Race relations in New Orleans were, as always, a confusing mixture of genuine interaction and brotherhood, open and covert hostility, and a disturbing mutual tolerance or acceptance of the longstanding socioeconomic class structure and racial boundaries. Despite a 67-percent majority black population, significant progress in higher education and other areas, a rising black middle class, and the election of black mayors and other public officials beginning in the late 1970s, African

Americans in New Orleans continued to retain a high level of poverty and a low level of business and economic power. A small number of prominent white families tightly retained control over much of the social, economic, and industrial condition of New Orleans. The imbalance of power is reflected by a static, poor, black labor force destined for menial service-industry jobs. Apathy, a breakdown of social and family values, and a poorly maintained public education system are factors that have continued to maintain the "status quo" and influence the direction of local traditions. A once-thriving and competitive school music program (often an indirect feeder to brass bands and other musical forms) has suffered greatly due to program and budget cuts, as well as an overall decay of the local public school system.

Despite New Orleans's continuing social problems and a decline of activities between the 1930s and early 1960s, all of the city's black cultural traditions saw a period of transformation and renewal that brought them to the height of their popularity during the decades leading up to Hurricane Katrina. Beginning in the 1940s, a worldwide "revival" of interest in traditional New Orleans-style jazz led to the first authentic brass band recordings in the late 1950s. Though far from national best sellers, collections of this underground "traditional soul music" of black New Orleans (and subsequently the rituals and organizations around it) made it accessible to the outside world for the first time. One repercussive effect of the "revival" began in the early 1960s, when the brass bands, in particular Dejan's Olympia Brass Band, were promoted as symbols of traditional jazz and New Orleans itself in the city's growing tourist industry. The environment of hotels, conventions, tourists, and businessmen—vastly different from that of social clubs, "second liners," and neighborhood streets—resulted in a commercialization of brass bands that greatly altered their size, appearance, sound, and demeanor. Ironically, the outside success and visibility of a few brass bands—which came to include performances at major festivals, international travel, television and movie appearances, documentary films and high profile political engagements, helped to renew and inspire interest in brass bands within the black community.

By the late 1970s, Mardi Gras Indians (and to a lesser degree social aid and pleasure clubs) began to make occasional appearances in the outside commercial arena. New economic possibilities, an ever-present local celebratory nature, and the homologous and static condition of many inner-city residents all contributed to a renaissance of community cultural traditions that continued into the early 2000s. Newer generations and current social conditions resulted in several changes in these traditions, which yielded both positive and negative effects. The most drastic, far-reaching change in local black musical traditions was the evolution and explosive growth of brass bands that began in the late 1970's and has gained momentum since the 1980s. In addition to increased work opportunities for brass bands in the commercial and tourist industries, several other factors contributed to a new sound, look, and attitude in what can be called the "brass band revolution." The deaths and retirements of older players, Sunday jazz brunches, increased travel by established groups, the rising cost of professional union musicians, a static traditional jazz repertoire, the heyday of a vibrant popular high school band tradition, the initiation of dozens of younger players into veteran trumpeter Ernest "Doc" Paulin's popular nonunion brass band, and the efforts of legendary guitarist Danny Barker, who founded the first children's brass band (Fairview) to help young people learn about their jazz heritage, were all important factors in causing this evolution.

Many changes in society were reflected in the new wave of brass bands. Soon the traditional marches and repertoire, collective improvisation, uniforms, and some instrumentation were being replaced by T-shirts, street clothes, smaller groups (eight pieces or less), riffing horn sections, faster tempos, and bebop solos. The new sound was still "New Orleans," but dominant overtones were not as much from "traditional jazz" as they were from a combination of rhythm and blues, modern jazz, and Mardi Gras Indian music. First the Dirty Dozen, then the Rebirth, and then a continuous flow of dozens of new, younger bands virtually replaced the few remaining traditional brass bands—both in the black community and the commercial world. By the early 1990s it had become as common for several competitive younger groups to gain international success as recording artists and night club acts, as well as to play stage shows and concerts, as it was for them to play for social club parades and funerals in the local black community. Some "retired" from community processions altogether in favor of more lucrative outside work. In 1986 the Rebirth Brass Band released a vocal/instrumental recording; "Do What Ya Wanna," which became the first-ever brass band local radio hit and an enduring favorite at sports events, parties, weddings, Mardi Gras, and a host of other New Orleans festivities. In a sense, this song's title, simple form, and hypnotic tuba riff signaled the spirit and direction of a new era of brass bands and community processions.

The younger brass bands presented a steady repertoire of original songs and themes from various sources, which were converted into their modern style. Despite complaints from some older club members and brass band veterans that the newer groups' music was too fast and frantic for the traditionally relaxed, medium tempo "second line" dancing, the competitive modern groups held a firm reign over most community street processions. The fresher contemporary sound, extremely casual appearance, and defiant attitude of change satisfied the youthful needs of self-expression and rebellious identity for new generations of social club members, "second liners," and musicians throughout the decades leading to Katrina. Partly as a response to both community and commercial interests, a few semi-traditional brass bands were formed. These groups wore more conventional attire, but used instrumentation and a style that hovered between modern and classic brass bands.

Another major change in New Orleans's black cultural traditions during the years prior to 2005 was a significant increase in the membership and number of new social clubs and Mardi Gras Indian tribes on the streets. Many of the newer clubs seemed to have been organized primarily for "pleasure"—mainly parading—as most of the "social aid" activities and rituals of older clubs had disappeared. Some modern club parades were marred by overtly vulgar gestures, language, and behavior on the part of second liners, club members, and bands (which now also "sing" or "rap" in the streets). A new generation of Indians and tribes had initiated a number of changes that threatened to alter the purpose and meaning of that tradition. Some of the newer Indians seemed more interested in the *idea* of parading, marching, and having fun, as well as in the attention gotten from the community, than in the original meaning, purpose, symbolism, and nobility of the tradition. A few had begun to take shortcuts in the ritual of suitmaking by gluing instead of hand-sewing pieces of their costumes. Some had avoided the valued recognition of individual creativity and group effort altogether by paying others to completely make their suits. Other staples of Mardi Gras Indian tradition, like the symbolic march from "uptown" to a "downtown" meeting at a location called the "battlefield," had been altered by some who removed their costumes, threw them in trucks and rode to the other side of town.

No longer the exclusively black functions of earlier days, it had become common to see several white faces (often more bohemian transplants than white natives) dancing among the throng of "second liners" who follow social club and Indian parades. Random violence (nonracial)—ranging from robbery and assault to drug dealing and murder—had been an occasional, unwelcome, and unfortunate presence at some community processions since their earliest days. A small number of criminal-minded individuals used the massive crowd cover and confusion to exact revenge or seize opportunities to commit illegal activities. These random individuals marred the celebration for the hundreds of participants who regularly came out to share and enjoy the events. At various periods, parade-related violence resulted in legislation and police actions that limited parade lengths and routes.

Another major change among New Orleans's black music–related traditions is seen in contemporary "jazz funerals" (often irreverently called "second lines" today). More frequent and popular than ever before, many funerals are dominated by a carnival or circus-like atmosphere consisting of family, friends, and band members wearing commemorative T-shirts or casual street clothes; fast music throughout; "second liners" who eat, drink, party, and behave disrespectfully; video and television cameras that disregard order and influence behavior; vendors selling drinks along the way; and several disturbing newer "rituals" like pouring beer over the casket and stopping the procession to stand the corpse up in a house to give final "tributes." To many older community members and musicians, the time honored jazz funeral has largely degenerated into a mockery of what was once a spiritual, glorious, and majestic tradition. On some rare occasions, more traditional style "jazz funerals"—with conventional brass bands wearing black suits, slow dirges, and ceremonial dignity—can still be seen in community neighborhoods. A tragic sign of the times is that many funerals of recent decades are more likely to have been given for young victims of violent crime than for deceased older social club members or musicians.

While the argument can be made that local social traditions and music have evolved to reflect the lifestyle and needs of today's black community, necessary change has too often resulted in an overall decline of values, quality, integrity, symbolism, unity, professionalism, pride, purpose, and meaning—the essential, uplifting, community-strengthening characteristics upon which all of these traditions originated and were maintained. One of the more positive changes—perhaps as a more-controlled expression of "New Orleansness"–is that the tradition of "second line" dancing has found renewed popularity. Many local, mainly black weddings, parties and other celebrations (among all social classes) often have small brass bands at the end to bring these events to a frenzied climax. Several small "underground" neighborhood clubs (of varying degrees of longevity) also used modern brass bands that played long hours to jam-packed crowds of locals and visitors.

Prior to Hurricane Katrina, nationally successful local rhythm and blues, funk, modern jazz, and hip hop artists—though popular—made only sporadic appearances at home due to an overabundance of talented regularly performing musicians and the lack of good pay, which is a reality for most New Orleans musicians of any genre. Many talented black musicians and singers performed regularly (sometimes exclusively) in the city's vibrant Protestant church circuit of popular (mainly contemporary) gospel choirs, services, and musicals.

The spirit of open acceptance, tolerance, flexibility, and a lax approach to rules—qualities that were initially essential to the creation and development of local musical traditions—also contributed to the formation of a "pseudo–New Orleans culture" in the years leading up to Hurricane Katrina. For much of their history, black New Orleans cultural traditions—except for the prototypical traditional jazz combos—existed almost in a vacuum; they were detached from many local whites and much of the black middle class and largely ignored by mainstream media, business interests, educational institutions, and city officials. When the already-popular tourism became the city's main support industry, local government and business leaders were reminded that music and food were the main interests of visitors. Many popular Bourbon Street nightclubs featured fast, loud, slick, generic "Dixieland"— a commercialized, denatured (white) imitation of black New Orleans jazz that, though using similar instrumentation and repertoire, sounds and feels radically different. This rather bland and comic version of traditional jazz is often played by whites and some blacks who were less familiar and concerned with the more authentic black New Orleans jazz style. New Orleans soon saw a steady influx of mostly white American and foreign-born musicians who quickly established careers (sometimes very successfully) on the local scene.

For many natives and transplants living in New Orleans pre-Katrina, the idea of "New Orleans music" had become more important than the actual music itself. An often-shifting and fabricated image of local musical traditions frequently served mainly as a means of financial gain. In a climate of little accountability or responsibility, many musicians, writers, media figures, businessmen, club owners, teachers, and politicians—often indifferent to or ignorant of the history and meaning of authentic black musical traditions—had become self-appointed arbitrators and dictators of "New Orleans culture." As a result, much of what was presented as "authentic" and "traditional" New Orleans music, as well as some highly publicized "preservation" efforts, had little actual essence or relevance to perpetuating the city's musical heritage. One example of this deception is how since the 1980s, movies and commercial interest have misled much of the world into believing that central Louisiana Cajun and zydeco are indigenous to the Crescent City. In truth, these musical styles are as foreign to native New Orleanians and local culture as are Tibetan wedding songs. One fact that is hardly recognized today is that since the deaths of remaining early-generation players during the 1970s, 1980s, and 1990s, authentic black traditional New Orleans–style jazz is almost extinct. More often than not, most "traditional jazz" played by African Americans was a limited commercial repertoire played in a dominant mainstream rhythm and blues or modern jazz style that had little of the distinctive "New Orleans sound" and could be heard anywhere. A very small number of conscious players and descendants of jazz families have continued the more authentic style and sound of traditional New Orleans jazz, both in terms of performing a more "classic" repertoire and in using traditional jazz principles as the basis of new songs and creative expression. Far from a unified concept or direction of "New Orleans music" or "traditional jazz," the local music scene was an evolving hodgepodge of different sounds and styles that had varying degrees of relevance to authentic black New Orleans traditions or their survival.

One year after Hurricane Katrina caused flooding of 80 percent of New Orleans, leading to massive loss of life and property, the effect on the city's unique longstanding black musical traditions and culture has potentially been devastating. More than half of the city's total population—including a core of musicians, Mardi Gras Indians, and social club members—have lost loved ones, lost their homes, suffered severe losses of property, and remain displaced. Many neighborhoods that were alive with the sounds of brass bands and Mardi Gras Indian chants lie in ruins. Many New Orleanians have established lives in other cities. Others have returned and are struggling with rebuilding. Many others wish to return home but cannot, due to a lack of jobs, affordable housing, and basic services. Hundreds of thousands of New Orleans natives, both at home and abroad, cope with a variety of new and continuing Katrina-related hardships: housing, schools, health care, depression, stress, suicide, illness, death, unemployment, insurance companies, accidents, government agencies, displacement, loss of neighborhoods, and separation of families.

It is difficult to put a price tag on or predict the total cultural losses that Katrina has brought to the black New Orleans area. While the more immediate threats to the amount of participants in, frequency of, and existence of social club parades, jazz funerals, and Mardi Gras Indian processions are obvious and measurable, other losses are immeasurable. This writer lost a valuable archive and collection of materials on New Orleans, Louisiana, and African American and African history, culture, and music; it consisted of thousands of rare, out-of-print, vintage, and original books, recordings, videos, and sheet music transcriptions, as well as new compositions, interviews, research, and photographs, plus vintage instruments and musical memorabilia, the sum of which rivaled and exceeded the collections of many local public libraries, university archives, and museums. More than a hidden, private stash of cultural treasures, this material was being actively used to promote and perpetuate authentic black New Orleans traditions in a number of ways through teaching, exhibits, special concerts, consulting, band development, new compositions using traditional styles, lectures, publications, et cetera. Many of these items were slated to become books, to be loaned to museums, to be the subjects of films, et cetera— which would have made a great contribution to the history, nature, and understanding of local music-related traditions from a rarely documented African American perspective. Several other musicians, photographers, writers, and filmmakers lost priceless and irreplaceable amounts of equipment and local cultural materials.

Many New Orleans musicians have been scattered across the United States. Bands have been broken up. There is an increased difficulty in finding and hiring local musicians in New Orleans. Much of the work that supported musicians in New Orleans, including parades, clubs, funerals, conventions, jazz brunches, and house parties, is greatly diminished or has not yet returned. Several neighborhood bars have again begun to regularly feature small modern-style brass bands. One fortunate consequence of the worldwide attention placed on the disaster has been a temporarily renewed interest in New Orleans music—resulting in various international tours, television appearances, tribute concerts, and commemorative recordings—which has employed and reunited dozens of local musicians in various genres.

Despite the devastation, displacement, and uncertainty that Katrina has brought, the irrepressible flame of ancestral traditions has not been completely extinguished. In addition to major commercial events like Mardi Gras and the New Orleans Jazz & Heritage Festival, the spring of 2006 also saw the return of some local community traditions. A large, commemorative, traditional-style jazz funeral for Katrina victims and a massive all-club "second line" parade seemed to give the message that neither time nor nature would obliterate the spirit of the New Orleans black community. Some Indian tribes paraded on Mardi Gras. A number of other funerals and social club parades also took place. These positive signs of return and renewal mask a serious concern in the community that the city's pre-Katrina masses of lower-income African Americans are now in the way of "progress." A common fear is that a conglomeration of businessmen, land developers, and politicians seeks to capitalize on the city's post-hurricane climate of mass destruction, confusion, and displacement. Some believe that several means are being employed to discourage residents from returning and to create a chaotic environment in which the takeover of black neighborhoods would seem a viable and welcome solution. Some view nationally broadcast signs of renewal as part of the deception. They feel that contrived and limited aspects of recovery are intentionally overblown in an effort to reduce widespread concern—by giving the false impression that most of New Orleans is back to normal. Many fear that current proposals for rebuilding and redevelopment would turn the city into a caricature of itself: a fabricated "pseudo-New Orleans" in which genuine culture and community would be replaced by a Disneyland or Las Vegas-type atmosphere designed for mass tourist and commercial consumption.

Those who express concern about the survival of the local black community and its musical traditions cite steady, quiet, pre-Katrina urban regentrification; the current slow progress of rebuilding efforts; questionable political actions on all levels; and recent tensions between parading community organizations and city government (and police) as signs of impending trouble. Just months before Katrina, popular big chief Allison "Tutti" Montana collapsed and died in the city council chambers while protesting police harassment of Indians during a St. Joseph's Day procession. In the spring of 2006, the city council used recent shootings at community parades as the motivation behind levying a prohibitive increase of nearly 400 percent in the parade tax for social clubs. Though the Black Men of Labor club began the September 2006 social club parade season on the streets with a full membership of thirty-six, several other clubs cannot afford the fee and will not parade. It is hard to predict what long-term effects the combined actions of prolonged displacement, the post-Katrina climate, and city government will have on the survival and direction of local black traditions.

While some may see the hurricane as a welcome opportunity for ethnic and demographic restructuring, the disaster has also brought about the potential for a major renaissance of local black music-related traditions and their more positive aspects. The survival and future of local traditions depends on several factors—some beyond the control of man. First, no matter what is or isn't done to restore people, culture, and landscape, ultimately New Orleans's current geographic reality is a risky and uncertain future to be determined by the random whims of nature. Toxicity of the environment, the threat of more frequent and severe hurricanes, an inadequate and weakened levee system, steady city-wide land sinkage, a powerful Mississippi River fighting to run a more natural course, and a continuously eroding Louisiana coastline, are all realities that many simply will not face. For various reasons, many citizens have decided to live off of hope, prayer, and the belief that Katrina was a once-in-a-lifetime event. Some point out that millions worldwide choose to live in areas that have constant threats from tornadoes, earthquakes, tsunamis, and other natural disasters. For local traditions to survive, New Orleans needs a concerted effort by leaders in city government, business, and industry to decide how to bring people back or allow those who wish to come back home to do so. More than accepting and expecting a return to the deplorable socio-economic conditions of pre-Katrina, citizens would need better and adequate housing, jobs, pay, education, sanitation, health care, and basic services. Effective and realistic ways need to be sought to attack problems like drugs, crime, and black-on-black violence, which—although they are major national problems—have recently returned to New Orleans in epidemic proportions.

Ultimately, the will and spirit of the people will help determine how, if at all, local neighborhoods and vernacular musical traditions will return. Thousands of citizens are struggling with thoughts of returning to a damaged, compromised city with an uncertain future. Many indicate that they have begun lives and found better opportunities and conditions elsewhere. Among those that have returned, confusion, anger, and frustration resulting from the slow rebuilding process and continued discomfort caused by present conditions have not yet revealed how much of the neighborhoods' family traditions, collective community spirit, and interest in previous musical practices will return. New Orleans is a wounded city with a wounded community. The healing process will take much time.

One thing is certain: the next several years will be an important defining period in the cultural history of New Orleans. There is a tremendous opportunity for native descendants and concerned citizens not to allow their history and culture to be washed away. It is an opportunity for all people to come together for the common good. It is a chance to do away with the widespread and divisive clannish mentality that for too long has limited growth and progress in many areas. It is a chance to use New Orleans's most significant contribution—jazz—as a central component of rebuilding and improving the city. In order for that to happen, a serious, honest, introspective self-examination needs to be directed by both citizens and civic leaders concerning the state of local jazz culture. A massive effort needs to be undertaken to better educate the general public, and especially business, political, and educational leaders, about the little-known history, meaning, style, values, importance, and socioeconomic potential of jazz.

With that in mind, many common myths and attitudes need to be brought forth and examined, especially by those in important decision-making positions. The selection of decision makers in the cultural arena needs to be carefully done and based on more reliable criteria than favoritism and popularity. The community as a whole needs to understand that while it is good to feature a variety of musical styles, the majority of visitors expect to find the only musical style that originated in New

Orleans—traditional New Orleans jazz. A major emphasis should be placed on pre-senting and promoting more authentic and artistically relevant traditional music from the wide spectrum of classic jazz, "revival" styles, and creative new traditional music, in addition to the usual commercial fare.

A serious look needs to be taken at how other cities—with less indigenous cul-ture, authentic traditions and talent—have built tremendously successful music-based economies. A delicate balance between authenticity and commercial interests should be of central importance. A number of harmful, longstanding myths about local jazz need to be exposed and explored in an effort to more honestly and realis-tically make assessments. We can no longer afford to let fame, self-promotion, and deception become easy ways for those who pimp and fake involvement in the authentic jazz tradition. It has been common among most previous decision makers to be oblivious to facts like these: Most music teachers and many musicians are not experts or trained in jazz. Many harbor negative attitudes toward it, especially tradi-tional jazz. Most don't realize that different jazz styles are like different languages, and a knowledge of and proficient performance base in one style does not usually translate into expertise in other jazz forms. Many do not understand that under the umbrella of what is labeled "traditional jazz" in New Orleans are several different styles and approaches that have varying degrees of relevance to the authentic New Orleans jazz sound or tradition.

At the risk of upsetting favoritism, political correctness, and indifference, which have governed many local culture-based selections and decisions of the past, a more responsible ethic marked by authenticity, knowledge, and accountability needs to be implemented to educate, develop, and promote genuine local talent in the area of traditional jazz music. When considering the future of the descendants of local black musical traditions, one must question the possible effectiveness of nonnatives and specialists in other genres and their motives for taking positions of dictating local cultural policies and dominating some performance activities. One influential for-eign musician, who made a successful career off of black traditional jazz musicians and serves on a policy-making board, candidly expressed his feelings this way: "I don't give a damn about no culture. I just came here for the music. Hell, if I wanted culture, I'd have stayed in Europe."

The original style of New Orleans jazz as handed down from the generation of Buddy Bolden is often misunderstood and unjustly labeled as "dated period music" that has little musical or cultural value in today's world. In reality, both the style of jazz and its principles can be used for new creative expression. It can also serve as a powerful metaphor in the rebuilding and future of New Orleans. Very often, good authentic, traditional jazz still proves itself to be an exciting, danceable, and spiritu-ally satisfying music to countless varied audiences worldwide. It should be viewed and treasured as the "classical music" of New Orleans. Jazz is a collective style based on a combination of teamwork and individual creative expression. It transforms con-ventional reality into a series of new ideas and possibilities. Jazz can be used in a number of ways to address social and economic issues.

Traditional New Orleans jazz needs to be taught in public and private schools by qualified individuals. For the youth, jazz can be an inroad to learning everything from local history and racial harmony to a team-oriented set of values like those associated with sports. Traditional jazz music can offer more easy access to early travel and employment opportunities than the more competitive and less open popular music fields. In many parts of the world, more authentic black New Orleans style jazz remains as a popular "cult" music performed in clubs and churches and at festivals. Those destined for other styles of music could gain the advantage of acquiring extra and special melodic and rhythmic instincts by having some experience with local traditions. Not everyone will become jazz musicians, but exposure to the field can lead to other employment possibilities in a much-needed local music business support system: entertainment attorneys, music writers and journalists, teachers, nightclub operators, recording technicians and manufacturers, musical instrument sales and repair specialists, et cetera. At the very least, the history and value system of jazz could offer young people positive ancestral connections and a sense of pride, self-worth, and accomplishment. It could also serve as an inspiration for achieving individuality and mobility in many areas of life. In this regard, jazz could become one possible deterrent to youthful unemployment, crime, and despair.

New Orleans has a new opportunity to become a standout in the tourist industry by setting a higher standard of offerings based on authentic local traditions. A cultural district or center with museums, performance venues, and retail outlets, largely jazz related, could be established. Regular promotional mini-festivals of jazz and interactive events could be sponsored and promoted by local government. Ways should also be found to reestablish and improve neighborhoods conducive to cultural traditions.

More African Americans must learn to value and recognize jazz and other aspects of their unique cultural heritage. Recognition of the post-Katrina threat to the survival of local music-related traditions has led to some promising new developments. For the first time, members of several younger groups have expressed interest in learning about the history, style, and repertoire of early jazz and brass bands. They have engaged in a number of workshops and concerts with members of more traditional-style groups. Both younger and older musicians involved in these collaborations demonstrate a renewed and heightened recognition of the value of heritage and tradition. They are coming together with the intention of learning, preserving, and passing on authentic New Orleans music to future generations, as well as infusing it with new life. In a positive and timely fashion, two African American universities, Xavier and Dillard (both badly damaged by Hurricane Katrina), have resumed a series of recently established programs that highlight local musical and cultural traditions. In addition to helping to document, preserve, and promote jazz and local vernacular customs, these efforts also serve as a way to bring together various groups from the community.

While racism cannot be completely eliminated, realistic solutions to deal with issues of race and economic structure should be sought. More whites must share economic opportunities and not be threatened by the idea of positive black self-promotion. Asian, Hispanic and other cultures use and celebrate their cultural heritage in positive ways without the perception of being hostile, militant, or anti-white. Jazz promotes a democratic construct characterized by acceptance, unity, cooperation, and a celebration of differences. Another optimistic development in Katrina's aftermath has been the appearance of several dedicated white and black individuals, groups, churches, and organizations who are spending considerable time, money and effort to aid and assist all elements of the New Orleans cultural community. This positive demonstration of humanity should serve as inspiration for the community as a whole. Everyone needs to recognize the real possibilities of losing jazz and the cultural traditions around it, without which New Orleans simply would not be New Orleans.

Many important questions concerning the survival of one of America's most culturally rich and diverse cities will remain unanswered for years to come. Will an injured New Orleans culture become a victim of "progress," greed, ignorance, and Hurricane Katrina? Can the city overcome its many problems and rebuild and restore its traditional neighborhoods, community bonds, and local spirit? What shape and direction will the future Crescent City take? Will it become a community that is authentic, fabricated, or ethnically restructured? Will New Orleans lose its "soul"? As several hundred thousand citizens struggle for survival in a limbo of continuing problems, unanswered questions, and uncertain futures, one can only wonder if the sad, sweet dirges or happy, dancing melodies of a recent authentic, traditional jazz funeral foreshadowed certain death or a glorious resurrection.

CHAPTER 8

HERO, EULOGIST, TRICKSTER, AND CRITIC: RITUAL AND CRISIS IN POST-KATRINA MARDI GRAS

CHELSEY LOUISE KIVLAND

> The way everybody is looking at Mardi Gras 2006 here, it's sort of the communities planting the flag in the ground saying, "We're back!"
>
> —Stephen Perry, President, New Orleans Convention & Visitors Bureau

The year 2006 marks the 150th anniversary of Mistik Krewe of Comus's parade, in which hooded club members, born of New Orleans's most elite families, paraded on horseback down crowded streets, founding Mardi Gras of New Orleans. Mardi Gras 2006, where exclusive clubs now parade on papier-mâché floats while throwing cheap charms to a crowd shouting "Give *me* beads!" also marks nearly six months since Hurricane Katrina claimed thousands of lives and displaced many more in its wake. Performing an intricate web of social unions and exclusions through the cultural costume of carnival, Mardi Gras 2006 manifested familiar tensions between tradition and change, where somewhere between the quotation of its past and the foreboding of its future lay the (re-)invention of the city of New Orleans, its relationality and politics, communities and power, which comprise, in their embodied performance, the city's "agora." The becoming of this agora—as it emerged at the intersection of the ritual of Mardi Gras; the orchestrated performance of Carnival rites; and the crisis, the critical social change, of Hurricane Katrina—is the focus of this essay.

To foster appreciation of this complex intersection, I explore how different communities addressed the crisis through their ritual performances. The disparate forms through which the crisis was given presence or absence, voice or silence, within the ritual performances demonstrated a politics of representation for the reclamation of divergent histories of the event and, accordingly, of communities' practical orientations to (and through) them. Indeed, each performance demonstrated a community's struggle to reconcile the past with the ominous future. Yet, seen collectively, the ritual performances played against each other to compete for the authority to assert a particular form of reconciliation. While each historical representation assuaged the crisis's impact within a community, the interplay of the representations fortified the very tensions that the crisis unleashed.

Four pivotal performances during the post-Katrina Mardi Gras, observed during my participation as a spectator, will serve as the basis of my exploration.[1] They are: (i) the "Rex" parade, a predominantly white social club, or krewe, founded in 1872 and comprised of the city's elite businessmen, which stands as the culmination of the festival, where the "King of Carnival" is crowned; (ii) the "Zulu Social Aid and Pleasure Club" parade, a predominantly black working class club, founded in 1909, which is the most popular parade and precedes the Rex parade; (iii) the "T-shirt brigades,"[2] groups of ten or more black men who, wearing satirical T-shirts, march through the crowds along the parade routes; and (iv) the "unofficial parades," groups of young people of various racial backgrounds who stand atop flatbed trucks or gather on foot and perform in unofficial festival areas. These four performances constitute *communities of performance*, a categorization that designates the momentary solidification of actions and affect by performers and audience. Describing each community in terms of their concerted social action highlights how the designation of ascribed features (e.g., race, class, gender) obliges the performance of membership.

COMMUNITIES OF PERFORMANCE

To enter the scene of the post-Katrina Mardi Gras is to inhabit a complex space of performance, but to grasp its social significance one must locate within this space the play of potent symbols. My descriptions of the four performances illustrate the particular exchange of discourse, actions, and objects that gave rise to four mythic personalities: the "hero" in the Rex parade, the "eulogist" in the Zulu parade, the "trickster" in the T-shirt brigades, and the "critic" in the unofficial parades.

If the mythic personalities are read as performative representations of the communities, these personalities constitute, following Durkheim, the *totemic* incarnations of the communal consensuses generated through performance.[3] The mythic personality, as totem, is the tangible symbol that represents each community's collective feelings regarding the crisis by consolidating a "shared past" and a "rallying cry" for the future.[4] Constructed through the communal ritual performance, it is a living symbol in a double sense: its enactment is as much the sign of a mythic personage as its embodiment is the myth itself. This duality means that the embodied totem carries both the meaning of the representation (e.g., King) and the confines of the one who enacts it (e.g., the white man is king). The symbolic assemblage that makes the representation and the assemblage that makes up the social body construct the mythic personality. Hence, the performers—and also, crucially, those to whom they are performing directly (audience) and indirectly (beholders)—actively signified the symbols used to covey each community's relation to the crisis. Though both groups vitally participated in shaping the totems, the audience was "seeing with" the performers, entering the community of performance, whereas the beholders were "seeing against" the performance, distancing this community from their own.

Rex Parade: The Hero

The King of Carnival, Rex, parades as the culminating event of Mardi Gras Day with his "court" plucked from the city's elite white men and his queen selected from the city's "leading debutantes." Rex is adorned with a crown set atop a faux white wig and beard that accents a gold-trimmed white velour robe, while his young queen wears a gold dress with a large feathered crown and his court "maskers" wear hooded velour costumes of green, purple, and gold, the royal colors. Their premier float proclaims their theme: "Beaux Arts and Letters." Icons of Jazz and Blues remain largely absent, as the figures of white, male authors and artists hearken the embedded "civility" of Rex's full name: "The School of Design." The next float's banner announces "Renew Orleans" alongside the parade's motto, "Pro Bono Publico."

Winding through city center, the parade shields itself from the eyesores of Hurricane Katrina. Their throws meet the hands of two tiers of spectators: elite families who, waving signs of their affiliations in the parades, collect the signature throws—fat stuffed-animal goats that symbolize the boeuf gras—from constructed platforms atop rented street space; and a mixture of local and tourist communities, which, segregated by race and class, vie for "good spots" on the street space between the stands. When the last float, which bears the insignia "Dignity and Strength," passes, the parade pauses before Gallier Hall for a moment of silence to honor, as King Logan made clear, "what [not who] has been lost and to recognize those who have given so much to help the city survive and rebuild." After Rex raises a toast to the black mayor of the predominantly black city, C. Ray Nagin, the maskers signal the end of the parade to scarcely visible authorities and are escorted by police to their gala housed at the local Marriot alongside hundreds of evacuees. Against the exodus of pedestrians, a fierce panorama of black graffiti on boarded-up storefronts surfaces across the empty trolley tracks: "U LOOT, WE SHOOT." And soon all that is left is the thick accumulation of waste in the eerie silence of the sparsely populated city.

The hero, White argues, when cast as the protagonist of a historical narrative, is the romantic avenger who embodies "transcendence over the world of experience," a victory over the forces of nature.[5] This transcendence is made possible by bravery in the face of a danger and a gallant design to overcome it. Similarly, the final declaration to the spectators, "Dignity and Strength," constitutes the heroic transcendence of the Rex parade in the face of crisis: to *renew* their ritual and their city. As an embodied declaration, this semantic pairing is a claim to the manner of strength and strength of manner that will usurp the social unsettlement and physical damages of Hurricane Katrina.

The parade's enactment of a kinship between the thematic statement of civility and the honors it bestows expressed the performers' claim to a dignified social status. The cultural exaltation of royal culture concealed the parade's racially segregationist practices by eliding the language of race for that of civility. Ornamental costumes and court practices hearkened back to the cultural authority, as opposed to class status, of their imagined colonial descendents, while the selection of authors in "Beaux Arts and Letters" solidified this authority. Hence, the parade presented evidence that the elite's ability to rebuild the city is owed to their superior culture rather than economic, racial hegemony. This intimation echoed the original parade, which, debuting a decade before emancipation, embraced the theme "Milton's Paradise Lost," a civilly disguised attempt to wrest control of the city from the "demon actors" (i.e., slaves) soon to "upset our way of life."[6] Seen against the backdrop of the aggressive graffiti, the awakening of this heritage highlights how the hurricane ignited colonial appeals to civility as a master trope with which to define racial groups and justify their social positions within the city. Finally, the spatial demarcations revealed an inextricable bond between the economic and cultural exchanges used to define the community of performance, the performers, and the audience. Opposed to the obstructed street space, the elevated street stands translate the performance into a private event that can only be fully appreciated from this view above the chaos. Through signaling affiliation, the parade implicated certain viewers into the ancestral line of the Rex court and its just rewards, whether cultural pride, beads, or a good view. In this way, the parade suggested an exclusive community built on the performance of a shared history.

Furthermore, by separating their chivalrous performance from the eyesores of Hurricane Katrina, Rex and court substitute the damaged reality of New Orleans with an image of their common power "to *show* what *we* can *do*," metonymically referencing New Orleans's power as a city. Exemplifying this common power, their moment of silence, taken at Gallier Hall, where Jefferson Davis lies in state, and dedicated to lost property and rebuilding efforts, revealed the maskers' ability, as heroes, to rebuild the city as before. The maskers also exhibited their strength through their control of the city. In the face of nearly disbanded civil services, the parade laid bare the mayor's civic acknowledgement of and acquiescence to the commercial elite. Not only was the Rex parade able to have a police escort despite the untended, crowded streets, but they were also able to hold their performance of waste despite the city's crippled sanitation infrastructure. This latter act, which transformed their performance of excess into the accumulation (not expenditure) of waste, boldly emphasized the present neglect of the dire situation in the city.[7]

Dignity and strength gave form to the task of recovery by substantiating the declared wish of the Rex parade "to 'Renew New Orleans's for the 'public good.'" However, while the members of Rex demonstrated their imperviousness to the graver consequences of the hurricane, the embedded contradictions of their wish revealed themselves. For renewal, as a remaking of what came before, vividly dissimulated and dismissed the needs and desires of the very public they claimed to aid. All three of the other performances would uniquely highlight this contradiction.

Zulu Parade: The Eulogist

Begun as a lending organization, Zulu Social Aid and Pleasure Club parades before Rex along a notably different path. Donning thick blackface with white outlines around the mouth and eyes and outfitted in "primitive African attire" of grass skirts, zebra print belts, and large afros, Zulu thrusts its theme banner draped across the lead float—"Zulu Leading the Way Back Home"—to the crowds awaiting Rex at city center. "Authentic Zulu warriors" who traveled from Africa and performed before the parade obstruct the lead float, asserting the equivocal hope of the parade's claim: Toward which home? The third float confirms this question, as it reads, "9th Ward, Gentilly," a memorial for the neighborhoods lost. Close behind, last year's king and queen stand, with their golden sashes but without crowns, before the word "Reunion." They wait "till all our family is home" before a "real reign" can begin.

With a final wave to city officials, Zulu begins its second route. Ordered to dismount their floats in this "ghostly territory," the maskers mark the first official steps of revelers beyond city center. Accompanied by marching bands, they tread through their neighborhoods in the fashion of a "second line," the New Orleans–style jazz funeral. Here is a vast crowd of predominantly black families, who stand atop the dirt of now vacant plots of land with arms reaching anxiously for Zulu's—and indeed Mardi Gras'—most "prized throws" of hand-painted coconuts. The throws portray images ranging from the standard "Z" insignia to pictures of people in acts of prayer and carpentry. Contrary to the Rex parade, Zulu pauses for silence abreast the Convention Center, where crowd and maskers have gathered to "cut the body loose." Following this tribute to the individuals who awaited accommodation and rescue in a crowed pool of bodies, waste, and darkness, the music takes an upbeat turn, dancing resumes, and the party really "comes home."

There are two intertwined responsibilities of a eulogist: to honor the deceased and in so doing to restore the spirits of the living. By tying together the acts of mourning and festivity, of "cutting loose" and "coming home," the performance of Zulu personified the mythic eulogist. The intricate interplay of this duality, shifting constantly through the parade, facilitated for the community of performance the *wish* of *reunion* that the maskers proclaimed.

This initial duality of emotion occurred in the parade's first scenes with the thematic declaration of Zulu's initiated homecoming and the tribute to lost neighborhoods. The juxtaposition captured the "need to honor the missing and honor those who must keep going." The absence of new royalty continued this theme with its demonstration that *full* restoration of the Zulu community necessitates the presence of those displaced. Echoing practices of ritual regicide, the dethroned old king and queen appeared here as effigies holding the tiers of the hierarchy in place until successors could be named and the devotees returned.[8] The most forceful demonstration, however, occurred in the parade's enactment of a second line through the damaged areas. This act undoubtedly communicated the mournful burial of all that has been lost: the homes, livelihoods, and lives. Yet, in the full tradition of second line, the joyful music and dancing accented the funeral as a burial of the crisis itself, where it no longer would haunt the present. Such exorcism within the scene of abandoned neighborhoods was remarkably vivid in the maskers' memorial throws, which articulated the sequence of custom, cleansing, and change constitutive of the traditional funeral. Indeed, festivities fully surged only after the final parting gesture of the funeral, "cutting the body loose," an occurrence that exemplified the role of the eulogist: to simultaneously honor the dead and revitalize the living.

Through uses of racially coded symbols, such as stereotypical costumes of African warriors, the parade designated the homebound community. A young black man, when asked about the seemingly offensive use of such stereotypical allusions to African "savagery" after the crisis's exposure of the city's unconscionable racialized oppression, clarified the complex message of this costume: "You know, they're making fun of white people." The reappropriation of the stereotype was plainly visible in the performance: the *real* Zulu warriors, referencing the ancestral strength of the black community, figuratively positioned a cultural home, one that outlasts the devastation of property in their territorial home. Suggesting a pride in the cultural separation of the races, the glorification of this stereotype performatively challenged the degradation of black culture in the city, especially salient in the conflation of poor, black people with criminals in the rumors of violence following the storm.[9] For this reason, the intensified allusion to Africa in 2006 not only indexed its historical use as an unthreatening mask that avoids the scrutiny of elites by accommodating their stereotype, but also communicated the fragility of this stereotype by exhibiting its equivocal meanings.[10] Indeed, Zulu performed the ineptitude of those—the maskers of "whiteness" beneath the blackface—who promulgate such stereotypes, challenging the authority of Rex parade's claim that the city's rebirth lay in their hands. Acting as a strategic dual emasculation, the performance's hyperbolic caricatures of race attenuated the elite's backlash and their power, the interpretation of which depends, as the young man clarified, on who is watching.

T-SHIRT BRIGADES: THE TRICKSTER

Though most audience members dress in costumes, one occurrence renders itself especially salient in its mass and its deviations from the parade audience containing it. From the end of the main thoroughfare of the parades, an oncoming group of ten or more black men walking in uneven, but discernible lines with a steady, measured beat approaches the crowds of the mainline parades. The coordinated "black mob," as one beholder calls it, moves against the stationary, mostly white crowds of tourists and locals, drawing the attention of all those they pass over and around. Long after the disruption passes, the striking commentary of the T-shirts that each marcher wears remains. Their white T-shirts boldly display the words "Willy Nagin and the Chocolate City, Semi-Sweet and a Little Nuts" surrounded by the mayor digitally rendered in the costume of "Willy Wonka," complete with cane, top hat, and three-piece suit. This satire thrown in the face of the crowds seeking to forget the crisis welcomes the convulsions of shock, laughter, and shouts; "That's just wrong!" is heard long after they have turned the corner. Awaiting the parades, the crowds are passed by two more brigades. One wears black shirts with a white machine gun encircled by the words "Battle of New Orleans 2005," and another the picture of a plasma TV to the side of which is an old-fashioned scroll that reads, "I escaped Hurricane Katrina. . .with nothing but a Land Rover, New Plasma TV. . . ." Aside from condemnations, one repeatedly hears the outcry, "Where did they get those?"

Identifying the role of the trickster figure of Esu-Elegbara in Yoruba mythology, Gates states, "tricksters are mediators and their mediation is tricks."[11] He argues, drawing on Bakhtin's concept of "double-voicedness," that tricksters join disparate discourses through two technologies of trickery, parody and the hidden polemic.

Bakhtin suggests that in parody the "author employs the speech of another, but introduces into that speech an intention which is directly opposed to the original one."[12] The author of the hidden polemic, in contrast, "brings a polemical attack against another speech act, another assertion, on the same topic" without adopting another's speech.[13] In their double-voiced speech, the T-shirt brigades mediated the contradictory discourses of tragedy and comedy to produce a satire of the vices and follies of Hurricane Katrina. Their trick was, in the "Willy Nagin" shirt, accomplished through parody, but a new trick emerged in the other shirts, "I escaped Hurricane Katrina" and "Battle of New Orleans," through the combination of parody and hidden polemic; a possibility generated by mass mediated coverage of the crisis.

The "Willy Nagin" shirt stole the speech of the mayor, who infamously compared in a public address the masses of black bodies wading through floodwaters in the days following the breach of the levees to a "chocolate city." By parodying this comment, the tricksters altered the dire state of the victims to a joke, protecting the dignity of those so labeled. This act also gained the force of condemnation by questioning the black mayor's affiliation with the black community. The shirts mocked the authority of the mayor, as a high-status black man, to speak for the majority of poor, black residents in the city, bringing attention to not only the racist but also classist connotations of his comment. The other two shirts deployed trickery straddling parody and hidden polemic by taking full advantage of unauthorized speech. Following White's apt description of rumor, this "ghost speech," as speech belonging to everyone and no one, subverted the accusatory rumors applied to the stranded victims of Hurricane Katrina by asserting that their deeds were valorous acts.[14] Because the national media promulgated such accusations without verification, heightening the event's sensationalism, the act, described as a "polemic of valor," subverted the accusations of a mass of potential authors. The beholders' fierce condemnations reacted against the trickster's embodied suggestion that the shirt's message was directed at all of them, as each potentially voiced the original accusation.

To *resignify* the tragic with the comedic in a satirical appropriation of the plight and accusation of the victims of the crisis was the wish of the trickster. Their control of this appropriation, accomplished by their independent, occult production of the shirts and by the symbolic value of their bodies, empowered the tricksters against those satirized. Since the maskers were exclusively black, their acts were not general comments crafted by all for the comedic release of all. Rather, the accusations targeted those not included in the original accusations, those not socially defined as black and poor, as "looters" or "wading bodies." Confronting the predominantly white beholders, the trickster laid the final trick in defining the performance community through their ability to pull off the joke, as a white man commented, "I could never get away with that."

Unlike the other parades, the trickster combined the symbols of ritual and crisis by questioning tourism as a means to restore the city. T-shirts are the characteristic souvenir of tourists, and Mardi Gras is the quintessential event for which tourists come. Using the T-shirt to deliver a crass joke—one exceeding "good fun"— mocked the city as a venue for tourism. More pointedly, the T-shirts ridiculed the use of the crisis as an opportunity for a value-added Mardi Gras experience, shown in bus tours of the destruction or the cottage industry of trinkets inscribed with "Katrina slogans" (e.g., "I survived Katrina"). The trickster's act of ridicule, though suggestive of the problematic relation between the hurricane and Mardi Gras, was given the valence of critique in the unofficial parades.

UNOFFICIAL PARADES: THE CRITIC

A walk to the parade outskirts meets the "hidden carnival." The floats of main parades have passed, the bead cries die, and loud political chants—"What do you want?" "Levees!"— crescendo. As the performers appear, their distinctiveness declares itself. Groups of revelers of various racial backgrounds wear outfits redolent of the mainline parades, but with a tinge of mockery, for in place of crowns and velour are cone-shaped heads and frilly satin suits. They stand atop flatbed trucks between posters emblazoned with messages that mock the festival of Mardi Gras by relating it to Hurricane Katrina. "Beads for Sandbags?" calls out from a disordered mass of bodies, posters, and frills. Openly contesting the parade tradition, the floats do not pass, but confront the street beholders who are walking toward the main parades. Their throws of political messages inscribed on strips of scrap or toilet paper mostly land in the palms of other revelers on the same float or one nearby.

Another set of revelers interweave between these flatbeds; on their bodies hang creative amalgamations of Hurricane Katrina and Mardi Gras. Two white male rollerbladers protested the work of the Federal Emergency Management Agency (FEMA) by wearing a bird costume made of blue tarp that urges, "FUCK FEMA"; six black men display a sandwich board with a scene depicting the stranded victims of the storm and the phrase, "It's No Party Here"; and two white men hold plywood tagged with a city map, the outer ring of which reads, "Tourists Welcome!" Close together, these performers march through the flatbeds, paying no attention to others in the streets.

A critic is one who pronounces judgment—not official evaluation, but public commentary—on others' actions or statements. Standing on the margin of the festivities, the unofficial parades made contact with revelers en route, yet were distanced from the legal sanctions of the ordered parades (e.g., street permits and authorized throws). Both sets of unofficial parades, the flatbed groups and the street performers, performed an exclusive exchange of symbols that defined their critical community against other revelers by protesting the tropes of Mardi Gras and Hurricane Katrina. The flatbed groups, by throwing actual waste (e.g., toilet paper) and political chants, and the street performers, by bodying forth denigrations, announced an evaluation of *this* Mardi Gras and the action to follow from *this* evaluation. Such gestures urged the beholders to evaluate the festival as an untimely performance of waste and to focus efforts on the physical and social damages from Hurricane Katrina.

When costumed in the carnivalesque, the critic emerges, Bakhtin suggests, within a momentary suspension of established custom through the medium of masquerade. Carnival allows the "lowly peasant" to mask the elite and play with their privileged customs, temporarily inverting the social order.[15] "Carnival," he states, "celebrates temporary liberation from the prevailing truth of the established order; it marks the suspension of all hierarchical rank, privileges, norms, and prohibitions."[16] In so doing, carnival opens a domain of "freedom" where communication becomes "directed at everyone," severing hegemonic relations by furnishing the people's "ancestral body."[17] However, this assertion has been shown to exaggerate the extent to which Carnival produces effective critics of established social order. As Turner notes, though the carnivalesque performs reversals of social roles, it is dependent on a set of rules and regulations that ultimately disabuses such reversals.[18] In this reading, the uninhibited reversal within Carnival presents ambiguity and contradiction within social relationships, but within "a structure of order" that ultimately projects the instability of this *temporary* space of play.[19] Carnival's acts of masquerade are redressed by the rules that allow them to exist; the performed fragility of the "ancestral body" makes desirable the social privileges and prohibitions that keep this body controlled.

In New Orleans's Carnival, though excessive libation and promiscuity take place on Bourbon Street, the Bakhtinian ideal of carnival has been virtually silenced by authoritative regulation. Since 1857, when "Krewe of Comus" decided to "control the disgraceful actions of the ruffians," Mardi Gras has not been a release from regimented life.[20] Demonstrated through officially ordered parades and forceful relegation of the public audience, Mardi Gras is a tightly controlled affair. Yet this regulation never completely silences the expression of social tensions; rather, it imparts to the critiques a fleeting and measured existence, which transforms play into an institution upholding city's social "rank and order."

An explanation of why the role of the critic intensified in the post-Katrina Mardi Gras emerges from the social *instability* and *ambiguity* that Hurricane Katrina exposed. The catastrophic effects of the event mirrored the ritual of Mardi Gras, for it put established norms of social life in flux. But crucially, the crisis disobeyed the ritualized rules of destruction and construction, leaving "undressed" the betweenness of order and disorder. By challenging the ability of the ritual to complete its redressive sequence in the face of the city's social liminality, the crisis "denatured" the symbolic structure of the ritual. It opened a new space where the symbols of ritual and crisis could be arranged in novel ways to critique social relations. Indeed, as two signs on flatbeds stated: "Rex Drowned" and "Guess who is Rex now?"

By taking advantage of this liminal state, the performances of the unofficial parades exposed a critique of the social relations that produced the very catastrophe of Hurricane Katrina, including the ineptitude of governmental organizations (e.g., FEMA), lack of state and federal support, the media's exaggerated emphasis on crime, and the neglect of revelers. This polemic was fronted with the aid of the carnivalesque, which referenced costumes of beads and capes and behaviors of partying and tourism, and used the ritual as a medium to highlight the catastrophic effects of poverty on the victims and the still untended exigencies of the city. Such use served to denigrate Mardi Gras as, again, an exclusive, inopportune, wasteful party not benefiting those who need help now. Viewed in combination, the performances of the critics expressed a wish of *resistance* against the concentrated forces of power, hoping "to *really* enliven Hurricane Katrina as a means for change. . .for after this, we want no more business as usual." In this way, the critics rejected the idea that the crisis could be overcome through a return to the status quo of social relations in the city, whether represented by renewal or reunion.

ANALYTIC OF CONSENSUS

I argue that the performed relations between crisis and ritual of the post-Katrina Mardi Gras depend on continual tension between the provisional establishment of agreement on the center ground of symbolic significances and the continual shifting and transforming of this center per the evolving settings where symbols are used. The specific dynamics of this constitutive tension of social action establishes what might be called the "signifying practice" of *consensus*.[21] The analytic of consensus is a tool to identify how communities of performance materialized a totem through the intersection of past representations and future imaginings and how they established a provisional accord on the sense and meaning of this totem. The first dimension marks the subject of consensus: the dialectic of historical materiality at play in the performances' construction of a specific kind of totem, a mythic personality, which harnesses shared history into a condensed rallying cry, a "wish" that orients action toward the crisis.[22] This dialectical totem is implicated in the act of performance and emerges through the interrelated acts of restoring the past, as well as meaningfully completing this restoration by implicating it in the active imagination of a possible future. The second dimension articulates the practice of consensus: the active process of attaining a provisional state of accord on the affective and semantic messages of the symbols deployed and assembled in performance. The formation of embodied agreement on the sense and meaning, the significance, of the totem is attained through each community's coordinated engagement with a symbolic dialogue, as well as their juxtaposition against other communities' performances.

Subject of Consensus: Dialectical Totems. Drawing together Benjamin's notion of the "dialectical image"[23] with the idea of the totemic mythic personality further clarifies how the mythic personalities, the hero, eulogist, trickster, and critic, embodied an orientation to the crisis that is retrospectively meaningful and prospectively useful for a community of performance as situated against other communities.[24] Situated at a historical juncture, the dialectical image, Benjamin states, is neither that the past gives light to the present nor that the present gives light to the past; instead, the dialectical image is where the past meets the future in a present constellation.[25] Because the performances enacted totems that uniquely represented a vision of the social wounds endured by New Orleanians and the efforts needed to address them, each totem materialized a representation that uniquely recovered the historical materiality of the crisis, surfacing certain artifacts of its impact and occluding others in order to mount a collective effort to address those wounds. Such recovery, however, necessitated the calling forth of past traditions, the "archaic images [*ur-images*]," as Benjamin terms them, of cultural practice (e.g., Mardi Gras, second line) to either front, blend, or hide the ruins of crisis. This play of ritual and crisis *ruins*— the archived tropes of both events—realized in a tension between revelation and concealment within each totem indicates a particular relationship with the future: a *wish*. Hence, the *wish of ruins* materializes the dialectic of past representations and future imaginings that established a more confined set of symbols that together gave form to a totem on which minimal consensus could be attained.[26]

First, the hero emerged from the ruins not only of longstanding traditions of royal costumes, courtly rituals, and cultural themes performed during the parade but also of the longstanding economic superiority and political clout of the social bodies performing. Where the former instantiated the "civilized" boundaries of the elite class, the latter allowed this class to showcase its power during its most unabashed performance of the year: Mardi Gras. Noted in the misrecognition of the dire state of city services and the absence of social concerns in their moment of silence, the hero's exaltation of tradition and ignorance of crisis oriented the performance to the archaic past before the fissures of Hurricane Katrina. Bringing forth, as one audience member stated, "our traditions, the traditions of the families in this city," demonstrated the hero's vision of the future as the perpetuation of the social order "we have built and will build again." The wish to renew the city, proclaimed by the hero, represented a dream to overcome the crisis as though its impact was but a minor unsettlement. Those who have "proven their resilience before," one masker noted, "will do so again." The "who" was those who have contributed to the city as it was and not, as many of the audience offered, "the criminals" who are now "thank goodness, gone" or the "whiners" who "cannot take matters into their own hands." Like the masks concealing the elite's faces, the hero's ignorance of the crisis concealed grave racialized, economic abuses while revealing an evaluation of them as attributable to personal (not public) responsibility.

Opposed to the hero, the eulogist emphasized the devastation caused by the crisis. But, like the hero, the eulogist attained the credence to amend the social suffering, beyond physical damages, through the confluence of ruined traditions. Integrating practices of mourning and festivity, a funeral procession and celebratory parade, the eulogist used the archaic images of established cultural practices to facilitate the reunion desperately needed after the crisis left many "not here with us." Evident in their choices of theme and the position and dedication of their moment of silence, as well as their insistence in performing with dethroned royalty and marching through the damaged areas, explicated the use of ritual to bring together the community of performance. This communal act of remembrance, articulated by an audience member as "our way of healing. . .[is] to move on with what we have left," merged past imagery of togetherness with the imagery of dissolution wrought by the crisis (i.e., the unsettling of Zulu's "family") in order to envision progress through the crisis. The wish of reunion, though oriented to the future, emerged from a symbolic blend that used archaic images of communal traditions to address the newly emergent images of crisis that challenged the community. This blend positioned a familiar pathway for the community that could, as long as the audience followed the cue, "lead them back home."

The orchestrated marching of the T-shirt brigades did not explicitly use the discursive force of any established traditions. However, the act of resignifying accusations in an effort to alter the meaning and direction of blame is well established in black cultural practice as a means to resist otherwise damaging significations.[27] Their embodied communication demonstrated how the "exchange of blame" arising from the crisis held the potential to negatively interpret the behaviors of black, poor victims who, in the wake of the storm, tried to cope with their destitute conditions. By appropriating the invidious accusations, the trickster's resignification of the ruins of accusations—helplessness, on the one hand, and unwarranted violence, on the other—demonstrated an effort to publicly and presently vindicate the community while retroactively blaming those supposed to have applied them, the mass of beholders. Their performance did not demonstrate an orientation to a glorious past of resilience or unity, nor a hopeful orientation to the future, but rather an orientation to the city's present racial tension. A resignification of behavior, their performance was perhaps an effort to "save face" for the accused and to deface the supposed accusers.

Finally, the unofficial parades, as the only performances that explicitly stressed the political abuse and neglect of the crisis, deployed the literal ruins of the crisis through the mask of Mardi Gras traditions. Calling out through chants or signs the characteristic imagery, objects, and social institutions of Hurricane Katrina, the critics highlighted the structural poverty and organizational failures exposed by the crisis. They did so by using the customary practices of floats, costumes, and throws of Mardi Gras in order to enable, in contrast to Rex, an emphasis on the crisis's impact and moreover, the social ills that caused it. The critic's refusal to silence the social forces (as opposed to the hurricane's natural force) at work in making the crisis oriented the community of performance to a future where the past, like its images, would be "turned upside down." Built on the subversion of ruins, their wish came to the fore through the performance's disavowal of the past and interest in change. To resist by questioning the past's concentrated forces of power was, at base, an iconoclastic hope to alter them, for the critics insisted that social change is part and parcel to remaking a viable New Orleans.

These four examples illustrate that these totemic mythic personalities materialized at the juncture of past and future as a representation of how the past can be rendered intelligible and how it can be made useful for the imminent future. Within each performance, the totem formed a different dialectical image through the use of a unique combination of ritual and crisis symbols, which communicated disparate orientations to not only the crisis, but the persistence of sociality in New Orleans. Performing consensus on the significance of these wishes of ruins—renewal, reunion, resignification, and resistance—within each community of performance, however, cannot be fully appreciated without attention to the coordinated and juxtaposed symbolic exchanges occurring within and between the communities.

Practice of Consensus: Sense and Meaning. Articulating minimal consensus within the performances demands an illustration of how each community established co-presence, engaged in a meaningful dialogue, and was differentiated from other coeval performances. Each of these conditions of emergence, however, cannot be apprehended sequentially; they are realized in their simultaneous and collaborative execution. The ensuing illustrations emphasize the layered communicative practices that associated emergent and traditional symbols with the consequence of forming relations and divisions among individuals. This emphasis will develop the concept of minimal consensus as an evolving practice, highlighting the extent to which this Mardi Gras actively performed the formation of common understandings of the relation between the ritual and the crisis and yet also reinforced contentious differences regarding these understandings.

"The very act of congregating," Durkheim states, "is a very powerful stimulant," for within a state of closeness emotions begin to "echo" each other.[28] This, he argues, leads inevitably to a "harmony and unison of movement," a rhythm where those present together, "co-present," become joined in the mutual expression of their affective excitement.[29] The intensified experience of co-presence occurred on the level of the collective through the inhabitation of the city during the post-Katrina Mardi Gras. Many revelers attested to the sensible intensification rendered through the heightened activity of the city, asserting *this* is the "first time the city feels alive since Katrina." Yet, beyond the citywide revelry, each of the communities of performance generated and occupied a unique space of co-presence.

Whereas the Zulu and Rex parades initiated a coordinated rhythm between performers through the measured beats of parades and throws, the unofficial parades did so through the dual acts of embodying stagnant, enclosed spaces where throws were exchanged and marching between these spaces in coordinated waves of movement. The T-shirt brigades, however, generated co-presence through orchestrating a highly rhythmic and tightly woven march against the muddled, stationary bodies of the mainline parade audiences. In this very act of coming into a space of closeness, all of the performers entered into an experience unlike their everyday encounters. Here, they entered a space charged with, what Durkheim calls, the "energized impulses of others' impressions"[30] where an intensity of experience emerged; as one unofficial parade participant stated, "It's so exciting here now. . .our rhythms, our bouncing off each other. . .the excitement is unstoppable."[31]

Building on Durkheim, I seek to draw attention to the unique forms of affective intensity that the performances produced through embodying their disparate totems. Within the entire ritual of Mardi Gras, the rhythmic echoes of bodily, verbalized, and inscribed gestures of the performances not only demonstrated the constitution of different spaces of intensity, but these intensities held different significance for the communities. This is because performance, in its production of intense co-sympathizing, is more than an exchange of symbols that designates semantic referents.[32] Because consensual practice highlights the affect of performance, its use, as an analytic, appreciates that an attainment of a momentary state of mutual accord on the meaning of symbols involves the practice of bringing individuals to a state where they share, and inextricably so, a sense of the symbols, a feeling that the symbols ignite. In so doing, consensual practice nuances how, and in what form, intensified co-presence fashions a unified social corpus, such as Durkheim's "collective effervescence" (the compression and exaltation of one collective[33]) or Turner's "communitas" ("the domain of equality where all are placed without distinction on an identical level of social evaluation"[34]).

As palpable constructions that emerge from symbolic exchanges within a state of collective intensity, the totems evolve *during* situated and distinctive interaction. The significance of the mythic personality hence reflects the needs and desires of individuals to settle on interpretations of events in order to collectively act on (or with) them. But, because the settlement is negotiated through communication, it resists attaining a homogeneous or fixed state of significance across an interacting collective. Rather, communication only temporarily and restrictively warrants consensus within the specific setting of interaction. Articulating how the different performances attained this *active* accord involves describing, in the tradition of Bakhtin, the structure and specificity of their communication or, more precisely, their dialogic interaction within the community and against other communities. Dialogic interaction demonstrates how the significance of each community's embodiment of a totemic mythic personality specifically evolves in an environment of performative unity and alterity.[35] As the very preconditions of performance, these coordinated exchanges of significant gestures carve up the nebulous region of intensified co-presence by giving *valence* to particular performances within the collective.

Dialogue, as the ever-present orientation to a "greater or lesser degree" of one's utterance toward some "other's utterances," Bakhtin suggests is foundational to all communication.[36] Appropriating the meaning in communication depends not only on the arrangement of a "neutral and impersonal language" but also on how the "gestural" symbols exist in dialogue within developing social situations. Situational use reflects a historical dialogue (engagement of past uses) and an interactional dialogue (engagement with others' uses), the intersection of which outlines the tension between tradition and change. The symbolic gestures communicated within the Mardi Gras performances of 2006 carry the traces of the conceptual and social changes they have made through their yearly renditions. This was made plain in the performances' uses of ritual symbols in relation to the emergent crisis symbols. Yet the meanings of symbols are also co-constructed by those with whom the community of performance is interacting within the present environment of the ritual. Bakhtin emphasizes that this co-construction locates a tension between the "coming together" of individuals on their similar uses of symbols and also the "coming apart" of individuals per their different uses of symbols.[37] Communicating meaning within interaction entails not the merging of all into one but rather the founding of a unique space of accord that is simultaneously with some individuals and against others.

Because the Katrina crisis impacted social relations within the city, creating fissures *within* social groups, the performances brought the community to a tenuous state of accord on the social implications of the event. The efforts to form a level of consensus were especially salient. Yet because the crisis also demonstrated strong divisions *across* social groups, it initiated the use of the ritual to front and comment on the rifts between groups. Such divisions were identified and created in the totemic features—from overcoming to mourning, joking to critiquing—and in the spaces of their deployment—from high to low, dry to wet ground. Merging dialogic accord and discord, minimal consensus settled through the construction of juxtaposed symbolic dialogues that took place on historical as well as two interactional planes, within the community of performance and against other communities, which furnished the evolving sense and meaning of each totem.

Within Rex, the hero aligned the community of performance around a feeling of hopeful cheer, incorporating the audience into the timeless parade rendition. Their insistence on having the parade met the willful enthusiasm of the audience members begging for throws, who called out "Thank you for parading!" on numerous occasions when throws landed in their hands. During the scene of the parade, audience members attested to both the pride that they have in their city (as Rex displayed its thematic icons of culture) and the assurance that New Orleans will recover. One audience member stated, "This parade reminds us of the history of our city. . . .It is a history that what happened cannot destroy. . . . We know that now with Rex here today." Moreover, the earnestness with which Rex orchestrated control of the city was echoed in the accounts of audience members who articulated how "this is not a party like what they show on TV. . .it is us bringing back our traditions." These traditions included the exchanges of throws, cheers, and also the honorary gestures of salutes and bows, which marked the embodied recognition of the elite power of the Rex parade.

As tied to the performed strength of the maskers, the performed loyalty of the Rex parade audience also undoubtedly rested on the creation of a space apart from the crisis. From the high, dry ground of city center, where the mark of Katrina is all but erased, the hero's exhortation to overcome the crisis through renewal of the city resonated easily across the audience. Such ease was also supported through the location of the moment of silence. Indexing an acclaimed Civil War general, this locale called forth among the community their actual past heroes, who have long been mobilized to instantiate conceit among the elite, as well as to make claims regarding the ownership of urban space and control of urban affairs.[38] The tenable status of this message, however, arose through the surfacing not only of the piles of waste and graffiti tags after the passage of the parade, but also in the references to the crisis played out in the three other performances.

Incorporating them into a funeral procession, which necessitated insider knowledge of the symbolic meaning of gestures, the eulogist of the Zulu parade brought its community to a feeling that rested between mourning and rejoicing. Following the cued dialogue of the procession, most noted in the synchronized raising of coconuts and the incitement of song and dance, the audience substantiated and solidified the intended affect of the eulogist. Where one audience member highlighted the "calm made through our parade," another noted how Zulu made "it now okay to celebrate." Joining these feelings within the very spatial and movement choreography of the parade, Zulu coordinated the community to merge in the act, as much as the belief, of reunion. In so doing, the Zulu parade invaded the parade to follow.

The Zulu maskers challenged the strength and dignity of the following Rex parade as its unwelcome reminder that the "hero's civilized city" rests on the misrecognition of the economic and cultural contributions made by the black, working class, that needs to be reunified for renewal to even begin. Moreover, the transgression of Zulu into the ruined neighborhoods and their tribute to them reminded the community of Rex that reunification was indeed necessary, for much needed to be rebuilt. One audience member developed this point: "We parade second but have more fans. . .makes no sense. But it's a good thing today 'cause we can show Rex that all those away must come home before anything is rebuilt." Yet, as the parades collided in Zulu's departure and Rex's beginning, Rex audience members replied to this assertion with the remarks that "we are not whiners like them" and "people should just start building their own communities." These totems, the hero and the eulogist, at this moment engaged in a symbolic tour de force where their wishes—renewal and reunion—battled for sequential precedent.

The T-shirt brigades, though mute, initiated a dialogue between the performers through their unified display of a single message on their bodies. The valence of their message, as a twisted assertion of blame, instilled itself against the shock of the beholders through their forceful, uniform movements through the crowds. Generating an obviously discernible rhythm, the tricksters' march circumscribed the space of the performance and also shaped the flow of the performed actions, insisting on the meaning of their wordless melody. The rhythmic march mirrored the satirical aim of the T-shirt, forming a crass parody of the mainline parades, which coded the brigades' intentions as derogatory. Moreover, the blank expressions across the tricksters' faces demonstrated an absence of shame in embodying the accusation against those confronting their bodies. Emerging at the intersection of their embodied movements and message, the tricksters' resignification of behavior depended on the ability to orient an affect within the community while producing an opposed affect among their beholders.

The tricksters, weaving between the Rex and Zulu parades and their audiences, impacted the symbols of these parades. The tricksters disturbed the flow and mood of the audiences of these parades, causing many to loudly condemn their actions. Second, they demonstrated to Zulu that reunification necessitates a struggle between communities to signify their social positions as much as a struggle within the community on the level of mutual mourning. Third, they fronted that which the Rex parade physically blocked and performatively ignored: the graffiti, the evidence of the moral panic that Hurricane Katrina spawned within the city. Last, by challenging the very definition of the community of the eulogist and hero, both based on racial ancestry (e.g., Zulu warriors, colonial royalty), the tricksters suggested that the community of performance could form through a mutual (accusatory) reaction against ersatz others.

Despite their division into two groups, the flatbed floats and the street performers, the unofficial parades formed a dialogue among themselves that oriented their engagement in mutual condemnation of the politics of the crisis. Their joint repetition of chants and pointed exchanges of throws, both of which excluded the participation of the beholders, fashioned the critics of each group as members of a collective project of protest. As such, their alliance was strengthened by the identification of a common enemy. One street performer confirmed this idea: "I came out today alone with my idea to say something to those watching, but. . .it is not just me here . . .now we are really making the partiers [of Rex] look." Additionally, the unofficial parades highlighted their outsider status by confronting these beholders from the low, wet ground of the city where the storm's destruction was most vivid. As opposed to Zulu, however, this act did not engender mourning but rather furthered the efforts of the critics to emplace evidence of their message by indexing and using the artifacts of the storm's surrounding debris. As such, the choice of place substantiated the dialogue of the critic, bolstering their acts of resistance.

Due to this oppositional stance, the critics, more than any other totem, purposively challenged the wishes of the other coeval performances. Emphasizing the parade of Rex, the critics' message took shape by, as the posters asserted, drowning "Rex," understood as the ideology, the naturalized exaltation of the cultural and political power, upon which the man who performs Rex stands. If Rex, however, can be taken to represent, as "King," Mardi Gras, then the critics can be seen as challenging as well the parade of Zulu, who, despite efforts to amend the festival in address of the crisis, articulated only its effects, leaving unacknowledged the causal factors that put at risk Rex's "family." By deploying its symbolic material to execute critique, the critics shook the very frame of Mardi Gras. Their actions impacted the festival air of the ritual, damaging the quests to "forget and party" heard across the entire scene. But, without the passages of the audiences of the mainline parades through their performances, the critics would have called out to an absent space. It was clear

that their chants climbed as the crowds en route thickened, for, as one street performer noted, "We are here to remind the 'bead whores' of what is really important
for the people of this city."

When analyzed collectively, each of the communities of performance can be seen
to have joined together in a moment of co-presence where they engaged in symbolic
dialogues that served to unify performers and audience while distancing this unity
from other coeval performances. The repertoire of practices that each of the performances choreographed, cued, and coordinated represented a symbolic dialogue
that aligned action in relation to the significance of the totem and thus, in relation
to the significance of the crisis, for then and for now. In this way, each performance
constituted an act of persuasion toward an ideology—an orienting framework for
action—of how to progress through the crisis socially. The hero demonstrated an
ideology of personal resilience, affirming the powers of individuals to "overcome the
crisis on their own" by exploiting their (assumed) independent economic capital to
rebuild the city; the eulogist demonstrated an ideology of moral unity, affirming the
power of their communal networks to heal the social wounds of the crisis; the trickster demonstrated an ideology of clandestine collective action, affirming the power
of orchestrated satire to reclaim their social pride from damaging accusations while
denigrating others; and the critic demonstrated an ideology of public protest, affirming the power of resistant action to mobilize the crisis, as well as its causes and implications, as a force for change.

Therefore, the performances notably revealed allegiances and contentions with
the dominant, contemporary economic-political model of neoliberalism, the ethos
of valuing and validating economic relations as guides for all social relations that is
represented by governmental structures and officials at federal, state, and local levels.[39] Explicitly fronting their ability to escape the crisis without suffering grave damages to property or welfare, the hero of the Rex parade left unacknowledged the
social networks that greatly advanced, if not facilitated, this ability. With their
emphasis on rebuilding, they demonstrated an allegiance to renewal through economic power, namely, the business infrastructure of the city. Contrarily, the eulogist
stressed, per tradition, the viability of social networks to mediate economic loss and
emotional suffering. Although not necessarily contradicting the neoliberal ethic of
nonstate support, this performative statement emphasized social networks, rather
than personal funds, as the alternative support mechanism. Specifically, it highlighted the need for redistribution in sustaining the city.

The tricksters reiterated this need, though they added a more clandestine valence.
Not only did the tricksters insist on exonerating criminal acts of redistribution (e.g.,
looting), they also hinted at the reluctance of the mayor to recognize the larger problem of economic oppression that provoked the metaphor of a "chocolate city." Most
fervently, the critics challenged the ethics of maintaining a racially hegemonic class
structure in the city. Through the juxtaposition of symbols in their performance,
these revelers asserted a relation between the elite's misrecognition of the crisis and
the state and federal neglect during the crisis. They emphasized the allegedly corrupt
partnerships between government officials (including Mayor Nagin, Governor
Blanco, and President Bush) and the business elite that exacerbated the crisis and
could furnish, as posters threatened, the "death of the city."

Though not discernibly revolutionary or anti-revolutionary, these ritual performances hence demonstrated "the politics in all ritual."[40] Specifically, they engaged the
motors of ritual and crisis in the political act of crafting an ideology with which to
evaluate current social relations and provoke their reproduction or change. The full
execution of this performative communication involved constructing and adopting
an embodied representation, the totem, able to signify the event of crisis in the service of imagining and thus engendering future social action. The totem's adoption
depended on the coordinated exchange of these symbols through semantic and prag-
matic communication during the scene of the performance. These dialogic interactions within communities of performances critically enabled each community's
exclusion from other communities. Explicating the complexity of the signifying
practice of consensus, these performances manifested not only the dialectical nego-
tiation of past representations and future imaginings to craft a significant representation of the crisis, but also the dependency of this significance on historically
situated yet presently interacted dialogues that disseminate the representation's sense
and meaning. The achievement of minimal consensus proved necessary for each
community to render the intersection of crisis and ritual sensible and meaningful
and to, as such, outline, ignite, and authorize social action accordingly.

PERFORMATIVE INTERSECTION OF RITUAL AND CRISIS

Because this Mardi Gras contained disparate reactions to the crisis, the ritual complicates the established tendency to understand crisis and ritual as inherently oppositional categories of social action.[41] Crisis, as an event that intensely demonstrates
the instability of social relations and ritual, as an event that constructs an intensified
vision of social relations, simultaneously exposes the structures of these relations and
engenders the active process of reorienting individuals. Therefore, though crisis and
ritual collide in the nature of their impact on social life (the former as "undoing," the
latter as "redoing"), their impacts curiously ape each other in that they both moti-
vate processes of making sense of sociality, the roles and relations, associations, and
divisions of persons within the agora. Rather than expressing the use of ritual to
redress the crisis, this intersection instead expresses the contentious, symbolic interplay of both events to provide a forum and motivation for asserting which "exposure" of social relations is valid and, moreover, socially useful as a representation of
the past and a motivation for action.

Hurricane Katrina and Mardi Gras collaborated in their mutual establishment of
a pool of potent symbols that individuals could creatively relate and deploy in their
efforts to publicly support or challenge the social aims of ritual and/or the social
effects of crisis. The performances' unique efforts uncovered the vast range of possibilities for the (mis)uses of ritual in order to form a degree of accord on the implications of the crisis and the (mis)uses of crisis to form a degree of accord on the social
implications of ritual. The performances can thus be seen to have established minimal
consensus not only on the significance of crisis (through using ritual to steer social
action in its wake) but also on the significance of ritual (through indexing the experience of the crisis *during* ritual enactments). This latter act, as seen most prominently

by the critics and tricksters, not only evaluated the ritual's social consequences in their general form but also indicated an evaluation of the ritual's specific role in promoting the crisis.

The post-Katrina Mardi Gras thus made plain the role of ritual performance, at times clever and discreet and at others frank and forceful, to address experiential conflict in order to orient a future relieved of its "menacing stance," as Turner states, "in the agora."[42] Yet, to the extent that this ritual also warranted the address of conflict, it reinvigorated it, for it facilitated a space for competing evaluations of crisis. These four performances opened a complex terrain of social interactions where communities, conjoined through performance, interpreted and influenced the social significance of the ritual and the crisis. This landscape of performance can be seen neither as ritual's use to redress crisis nor as crisis's use to change ritual. In conclusion, the intersection of ritual and crisis in the post-Katrina Mardi Gras vividly demonstrated the myriad ways that each event can be combined within historically specific contexts of interaction to tread an inextricable duality of tradition and change, between the reinforcement and condemnation of sociality within the agora.[43]

CODA

As the last float of Rex passed, an assertion cut across the sirens and footsteps: "By holding Mardi Gras, we are showing the rest of the world that New Orleans is not dead." Not long after, affirmative "seconds" followed, and soon all that could be heard was, "No, not dead; no, not dead; no, not dead. . ." repeating endlessly as the memorable stage of New Orleans's post-Katrina Mardi Gras emptied.

Each of the performances fervently instilled this echo of assertions in their efforts to awaken New Orleans. But if their performed addresses of the crisis foremost breathed life into the corpse of the city, the forms of life they revitalized were remarkably different. It is my hope that this essay has enriched our understanding of the dynamic and diverse intersection of ritual and crisis by increasing appreciation for the dual processes contained in performing—making by doing—consensual representations of the past in order to raise hope for a certain future. This duality is best represented through merging the process of constructing a totem that may momentarily stand within while standing for a community by, first, dialectically engaging symbolic ruins of ritual and crisis to orient a wish for the future, and, second, disseminating the sense and meaning of this totem through dialogic exchanges of these symbols. The symbolic assemblages of the hero, eulogist, trickster, and critic drew together the symbols of ritual and crisis to furnish a degree of consensus on how to (re-)invent social life in the traumatized yet emergent city, to, as this essay began, "plant the flag." In the most profound sense, the agora born of the post-Katrina Mardi Gras was a performative contest staged and waged for the very *life* of New Orleans.

NOTES

1. During my observations, I recorded daily field notes, which included descriptions of the performances and quotations from festival performers and audience members. All

of the quotations contained in this essay (unless otherwise marked) were gathered during informal conversations. I chose not to record names.

2. The "T-shirt brigades" is a term I have coined. I have been unable to discern a regularity of word choice among the festival participants.

3. E. Durkheim, *Elementary Forms of Religious Life* (New York: Free Press, 1995), 220–22, 229.

4. Ibid., 217–24.

5. Hayden White, *The Historical Imagination in Nineteenth-Century Europe* (Baltimore: John Hopkins, 1973), 8.

6. J. Curtis Waldo, *History of the Carnival in New Orleans, 1857–1882* (New Orleans: Ghaham & Son, 1882).

7. G. Bataille, *Consumption* (New York: Zone Books, 1991). Bataille observed that in a capitalistic economy where products accumulate at a rate surpassing their ability for consumption, individuals take relief in expending the excess unproductively. The accumulated waste of the Rex parade demonstrated quite boldly Mardi Gras's situatedness within an economy currently unable to support the elite's expenditure during the ritual, which put at risk the power of ritual act to effect the satisfaction of the elite.

8. E. E. Evans-Pritchard, *Witchcraft, Oracles, and Magic among the Azande* (London: Oxford University, 1976).

9. The most widely cited rumor was delivered by Mayor C. Ray Nagin, who reported during national programming of the *Oprah Winfrey Show*, that he had been "in that frickin' Superdome for five days watching dead bodies, watching hooligans killing people, raping people," but nearly all nationally distributed newspapers published numerous unverified reports of seven-year-old children being raped, nightly murders in the Superdome, roving bands of armed gangs in the Convention Center, or "dozens" of people being shoveled into the freezers of public shelters. It is necessary to add that after two weeks of flooding, city and state officials documented none of the rumors.

10. The use of blackface in the Zulu parade has a long history; its first use, as Roach (1996) notes, dates to the decade after *Plessy v. Ferguson.*

11. H. L. Gates, *The Signifying Monkey: A Theory of African American Literacy Criticism* (New York: Oxford University, 1988), 6.

12. M. M. Bakhtin, "Discourse Typology in Prose," trans. L. Matejka and K. Pomorska, in *Readings in Russian Poetics: Formalist and Structuralist Views* (Cambridge, MA: MIT Press, 1971), 186.

13. Ibid, 87.

14. L. White, *Speaking with Vampires: Rumor and History in Colonial Africa* (Berkeley: University of California, 2000).

15. M. M. Bakhtin, *Rabelais and his World*, trans. H. Iswolsky (Bloomington: Indiana University Press, 1984).

16. Ibid., 109.

17. Ibid., 11.

18. Turner, *Anthropology of Performance*, 131.

19. Ibid., 129–30.

20. J. Roach, *Cities of the Dead: Circum Atlantic Performance* (New York: Columbia University, 1996), 246.

21. J. Comaroff, *Body of Power, Spirit of Resistance* (Chicago: University of Chicago, 1895), 265. Following Comaroff, I seek to highlight social action as a communicative process denoting the "meaningful structure inherent in practice and the practical structure inherent in meaning." For the semantic complexity of the term "consensus" I am indebted to the etymological tracing of Raymond Williams' *Keywords: A Vocabulary of Culture and Society* (New York: Oxford University Press, 1976).

22. Durkheim, *Elementary Forms of Religious Life*, 220–221.

23. I focus my analysis on the contributions of Walter Benjamin in the posthumously published *Das Passagen-Werk* (Berlin: Suhrkamp, 1983), with notable support from the reading of this work by Susan Buck-Morss in *Dialectics of Seeing: Walter Benjamin and the Arcades Project* (Cambridge, MA: MIT Press, 1991).

24. Durkheim, *Elementary Forms of Religious Life*; W. Benjamin, *Das Passagen-Werk*.

25. Benjamin, *Passagen*, 579 (my translation).

26. Ibid., 124.

27. C. Mitchell-Kernan, "Signifying Loud-talking and Marking," in *Rappin' and Stylin' Out: Communication in Urban Black America* (Chicago: University of Illinois, 1972).

28. Durkheim, *Elementary Forms of Religious Life*, 217.

29. Ibid., 218.

30. Ibid., 217.

31. Ibid., 218–19.

32. D. Kapchan, "Performance," *The Journal of American Folklore* 108 (1995): 479–508.

33. Durkheim, *Elementary Forms of Religious Life*, 221.

34. Turner, *Anthropology of Performance*, 137.

35. M. M. Bakhtin, *Speech Genres and other Late Essays*, trans. V. W. McGee (Austin: University of Texas, 1986).

36. Bakhtin, *Speech Genres*, 92.

37. M. M. Bakhtin, "Author and Hero in Aesthetic Activity," trans. M. Holquist and V. Liapunov, in *Art and Answerability: Early Essays by M. M. Bakhtin* (Austin: University of Texas, 1990), 33.

38. The debate in 1993 regarding Mardi Gras Krewe integration and the removal of the Liberty Place Monument (which has inscribed in its granite base that the End of Reconstruction "recognized White Supremacy and gave us back our state" and stands in a central New Orleans park) provides a parallel example. See Roach, *Cities of the Dead*.

39. D. Harvey, *A Brief History of Neoliberalism* (London: Oxford University, 2005).

40. It is important to note that this insistence on viewing ritual as political follows the assertion of Kelly and Kaplan that key to understanding ritual is to understand the ritual in all politics and the politics in all ritual. See, J. Kelly and M. Kaplan, "History, Structure, and Ritual," *Annual Review of Anthropology* 19 (1990): 119–50.

41. Turner's influential model of the three-wave process of social dramas, which begins with a social breach, climaxes with crisis, and meets its redress with ritual, marks the solidification of an idea that has had longstanding in ritual analysis. See, for example, J. L. Comaroff and J. Comaroff, introduction to *Modernity and Its Malcontents: Ritual and Power in Postcolonial Africa* (Chicago: University of Chicago Press, 1993).

42. Turner, *Anthropology of Performance*, 34.

43. For a critical review of the literature on the intersection of history and ritual, in crisis contexts and noncrisis contexts, see J. Kelly and M. Kaplan "History, Structure, and Ritual."

CHAPTER 9

(Re)Imagining Ethnicity in the City of New Orleans: Katrina's Geographical Allegory

Stephanie Houston Grey

When Hurricane Katrina swept across the Gulf Coast in August 2005, the diaspora that it produced hollowed the city of New Orleans. In a recent issue of the *New York Times*, satellite photos revealed that the devastation is no longer evidenced by post-apocalyptic debris, but by the rows of now empty communities that have left much of the urban landscape bereft of vibrant human activity. As these ancient neighborhoods dig themselves out of the mud and silt, the displacement of hundreds of thousands of persons has created a new challenge for this iconographic city. This challenge stems from the endeavor to recapture an essence that many feared had been washed into the Gulf of Mexico. As the shrimp boats that now serve as debris gatherers extract Mardi Gras beads from the ocean, the cultural gulf that has been etched into the urban landscape raises questions about the future of this community in an age where the American city has itself become a mere echo of civic society and celebration. Reflecting upon the racial implications of this event, it is clear that the storm ignited a combustible mixture of social and economic issues surrounding race, class, and space. The following discussion explores the nexus between social geography and the political allegories used to define those that inhabit urban spaces. The city of New Orleans must now grapple with both its representation at a national level and the realities of alienation so vividly amplified by this catastrophe—a process that means examining the role that its own history has played in creating certain geographical containment fields that isolate those populations most deserving of assistance.

One common thread that runs through the accounts of those citizens who survived the storm emerges as a pronounced sense of betrayal. In the wake of Hurricane Katrina, the diaspora and the federal government's ambivalence have created a situation where citizens and their organs of public deliberation have been deeply fractured. Significant portions of the city's infrastructure, including large sections of St. Bernard and Jefferson Parishes, continue to suffer from the aftereffects of water damage and population displacement. The key to the successful reconstruction of New Orleans, with its unique and diverse cultural identity, revolves around the coordinated efforts of city leaders, contractors, and a citizenry that has largely lost faith in these governmental bodies. Recognizing the depth of this cultural rupture, the City Council of New Orleans adopted Resolution R-05-525 on October 12, 2005, establishing the Hurricane Recovery Advisory Committee. The charge of this entity includes rebuilding trust with the citizenry and business leaders, developing communication infrastructure, and speaking for the population of the city in a clear voice that reflects their concerns and anxieties. For a city to provide a genuine public space, it must address the gap between public and private identities so that urban designs and culture reflect the consciousness of those who inhabit them. One of the primary obstacles to realizing this laudable goal is the way that allegories of stigma and ethnic containment—social barriers that are often animated by racial dimensions—continue to inhibit this endeavor. Renewing a vibrant urban space means examining the grounds upon which the city's sense of authenticity has been established in the past and confronting the spatial ideologies that continue to inhibit the process of recovery.

Fredric Jameson notes that national allegories find their most vivid embodiment in the urban spaces where social divisions evolve into crushing material realities. The narratives that facilitate the citizen's identification with the *polis*, while critical to the process through which governing bodies reflect popular ethos, may be deflected by interests that diverge from those of the population. Rather than creating a public space where government and citizens participate in meaningful exchange, geographical allegories function to divide, isolate, and stigmatize. This chapter traces the way that these destructive allegories were mobilized against New Orleans's citizens of color during the Katrina event. This three-part analysis begins with the rationality of containment that marks many urban environments and the way that this logic "colored" the national response to the catastrophe. Then attention shifts to the history of New Orleans and exposes the factors that contributed to this national response by tracing the racial legacy of geographical allegory within the city—particularly the way that decadence has been linked to ethnic ambiguity in the lower Mississippi region. The final section proposes a template for appropriating these metaphors that link ethnic diversity to the unique physical and social ecology of the region and using this revised ethical vocabulary to facilitate a new consciousness that opens avenues of dialogue between the city's citizens and its representative bodies. At the core of this discussion is the assumption that urban environments are not simply determined by economic realities and do not merely reflect the natural teleology of population movement but are the products of communicative practices. Only by interrogating these models of population management can the city of New Orleans posit a new future where its metropolitan ethos facilitates social progress rather than retarding growth and development.

THE BIOETHICS OF CONTAINMENT

The federal response to Katrina demonstrated that the catastrophe was not to be addressed via outreach and rescue but through strategies of containment. With the exception of the United States Coast Guard, federal responders established a periphery around the city that blocked access to its main staging areas for at least five days. People sat in unbearable heat and on overpasses—scavenging for food and supplies while trying to keep sick relatives alive. This delay was the result not of a lack of access to the city itself but of an unwillingness to enter the urban zone for reasons that have never been fully explained. Some critics of the Bush administration suggest that this failure stemmed from the federal government's apathy toward these citizens and the region as a whole—that their responsibility ended after they simply stood in front of cameras and said that they cared. These explanations are, however, incomplete. During this time period, certain media outlets began to flood the airwaves with gruesome stories of rioting, murder, and particularly graphic accounts of child rape that were later revealed to be either fabrications or exaggerations. In their work on collective responses to urban crises, Stuart Hall et al. coined the term "moral panic" to describe situations where racial otherness is projected as a source of violence—leading to media hyperbole that feeds fear and hysteria, provoking a police response that is out of proportion to the actual threat that may or may not be posed. The moral panic generated by Katrina kept both police and military from entering a city that was in desperate need of relief, leading instead to the containment strategy. This sense of cultural anxiety regarding particular urban environments is not a new phenomenon, but it is nurtured and informed by metropolitan rationalities that place the isolation and management of ethnicity above other priorities.

During the later half of the twentieth century, the American city became a focal point for discussions of civic culture, control, and authenticity. In her book *The Death and Life of the American Cities*, Jane Jacobs suggested that the new design schemes being created by urban planning specialists produced a cityscape that inhibited genuine community by allowing inhabitants to circumvent their neighborhoods while moving through them. Rather than fostering interaction within these spaces, interaction that is critical to the life of any neighborhood, the modern city became a series of containment spaces. In his reflections on the city of Los Angeles, Mike Davis notes that ethnic isolation creates a series of social barriers that many cities reinforce through a unique confluence of strategies that include police enforcement of social and ethnic boundaries, control of press organs, and architectural innovations designed to isolate disenfranchised individuals from what little public life remains intact. He goes on to write that "the universal consequence of the crusade to secure the city is the destruction of any truly democratic space. The American city is being systematically turned inward. The public spaces of the new megastructures and supermalls have supplanted traditional streets and disciplined their spontaneity." Recent studies demonstrate that while many urban neighborhoods are more ethnically integrated than in the past, class distinctions remain intact due to the flow of commerce and social activities to the suburban periphery. The manipulation of modern urban infrastructure is designed to isolate both ethnic and economic groups from one another by creating a protective membrane around the privileged to shield them from those deemed less desirable. Thus, the federal response was not simply an aberration of administrative policy, but a logical extension of existing urban rationality.

What the media panic that followed in the wake of Katrina demonstrated is the extent to which these logics, while often obscured by tourist guides and cosmetic social policy, reside very close to the surface and can be activated very quickly with devastating consequences. Glenda Dickerstetsun states that the media "have given us an orgy of wall-to-wall coverage chronicling the catastrophe," likening it to a "block-buster film, which uses the misery of the developing world as an exotic backdrop for a story about the travails of white people." Here the citizens left behind to face the storm became a savage testimony to the dangers associated with color and the challenge that it posed to the rationality of law and order. This lexicon of violence and containment emerged within a media context where radio hosts were referring to the remaining citizens as parasites fleeing a dying animal. Fox reporter Bill O'Reilly labeled these citizens as thugs and drug addicts who were only attempting to leave the city because they could no longer get their fix. While rumors of ethnic savagery turned out to be fabrications, the racial codes used to define the looters activated the allegories of geographical decadence and transformed these survivors into a threat. Thus the chaos that emerged in the wake of the storm was characterized as an issue that originated within the indigenous culture rather than from these extraordinary circumstances. There can be few more vivid portraits of the impact that mediated environments and communication technology have on the management of space and population movement than in the days immediately following the storm. This continues to be reflected in the Katrina Smithsonian exhibit where the slow federal response is attributed to this violence rather than problematic social policies. Thus the savagery produced by the storm was naturalized by the composition of the population itself rather than the lack of supplies and medical aid.

Katrina reveals that the problem confronting New Orleans is not just an issue of rebuilding infrastructure, but it is one that is deeply rooted in issues of urban bioethics. In her now well-documented visit to the Astrodome, Barbara Bush explained to television cameras that "so many of the people here, you know, were underprivileged anyway, so this is working very well for them." This response demonstrates the lacuna that now exists between spheres of governance and the populations with whom they have either lost touch or against which they constitute their own interests. Henry Giroux refers to this ideology as the "biopolitics of disposability," an ethical stance that views individuals as either resources for advancing political interests or potential barriers to the social and economic development of empowered groups. He proceeds to draw an extended comparison between the visual images spawned by Katrina and the brutal murder of Emmit Till in 1955:

> Till's body allowed the racism that destroyed it to be made visible, to speak to the systematic character of American racial injustice. The bodies of the Katrina victims could not speak with the same directness to the state of American racist violence but they reveal and shatter the conservative fiction of a color-blind society. The bodies of the Katrina victims lard bare the racial and class fault lines that mark an increasingly damaged and withering democracy and revealed the emergence of a new kind of politics, one in which entire populations are now considered disposable, and unnecessary burden on state coffers, and consigned to fend for themselves.

While Giroux's comparison is designed to demonstrate the extent to which these national political boundaries have created undesirable bioethical ideologies, the metaphors of ethnicity, the South, and the river preexisted as extremely powerful components of a broader geographical allegory. The comparison of the drowning of New Orleans to the drowning of a victim of a brutal lynching reveals not a direct causal relationship but a topographical legacy that permeates both the local culture that has birthed the Crescent City and the stigma that it carries as the southernmost urban center on the lower Mississippi River.

GEOGRAPHICAL ALLEGORIES OF WASTE, POLLUTION, AND ETHNIC PERMEATION

The geographical allegories that mark New Orleans and the surrounding region are defined by a metaphorical interconnection between ethnicity and pollution. In the wake of the disaster, Mayor Ray Nagin asserted that New Orleans would remain a "chocolate city"—a comment for which he was later subjected to criticism. But as Michael Dyson points out, this characterization is not inaccurate. In many ways, this ethnic metaphor resides at the heart of the geographical allegory that demarcates the alluvial plains of the lower Mississippi.

Using this concept of allegory to foreground the containment strategies that define New Orleans, Mayor Nagin's metaphor represents a deeper symbolic under-current than is immediately recognizable when one ponders a map of the United States. As the Mississippi traverses the cities that run from North to South—from Chicago to St. Louis to Memphis and finally New Orleans—one sees a gradual shift in the allegories used to define these urban environments that belies a certain hier-archy. Residing as it does at the mouth of the Mississippi River, the city of New Orleans has been historically characterized as a site of expulsion, waste, and excess. When the unusual ethnic composition of the city is read through these geographical allegories, the impact upon both national and local policy can be devastating to the citizens of this region.

The ability to characterize American citizens as disposable resources stems directly from a topography of ethics and race that remains embedded within the American consciousness. Tara McPherson describes the visual representations of the South as displaying a lenticular logic—a "schema by which histories or images that are actually copresent get presented (structurally, ideologically) so that only one of the images can be seen at a time." In these depictions, the poverty and industrial pol-lution are often negated in pristine images of plantation ideals, allowing the social and economic problems associated with the region to remain more or less invisible to the national consciousness. This is compounded by the fact that the alluvial plain of the Mississippi is a sphere of literal expulsion where waste and pollution are forced from the nation's industrial core. For many historians and urban engineering schol-ars, this dumping associated with the chemical industry is justified by the ethnic composition of the area. Much as the nation expels its detritus and waste through the great river, New Orleans and its citizens are marked by their existence in this top-ographical boundary as extensions of this expulsion. Given the power of this allegory to isolate and stigmatize entire urban populations—when, for example, news com-mentators feed these stereotypes by asserting, without evidence, that an entire urban/ethnic population is fleeing a disaster due to its collective desire to acquire illegal drugs—the need to rearticulate this unique cityscape is a pressing one. One of the animating features of Southern Jim Crow policies was the assumption that African American populations were more prone than whites to vices such as cocaine addiction and therefore had to be contained via legislation. These images and commentary do not emerge from a political vacuum, but they remain dormant in the American consciousness until they can be activated in times of perceived threat or crisis.

As one moves south, narratives of progress and commerce are increasingly supplanted by metaphors of pleasure, waste, pollution, and decadence. The city of New Orleans has, in fact, turned these narratives into a source of income through its well-known tourist industry. While Ronald Pelias suggests that the Creole carnival has emerged as a site for lived critical practice where pleasure can become a source of empowerment, these exhibitions also mark these urban spaces as spheres of iniquity. Narratives of ethnic decadence would continue to mark the city's identity despite the fact that the carnival culture had itself been orchestrated by white business owners since the mid nineteenth century, when organizations such as the Mistick Krewe of Comos used the demonic language of Milton to fashion this city as a respite from the oppressive moral standards of the white middle class of the period. It is not surprising that the Hurricane has been likened to a Greek tragedy, complete with hubris and unheeded sibylline prophecies. Here in this great "ship of fools," the carnival continued until the eventual apocalypse that brings the participants before their eventual divine judgment. In her book *Southern Babylon*, Alecia Long notes that historically the sexuality associated with New Orleans was peculiar for the region because it manifested itself through the violation of ethnic boundaries. The response of city leaders at the turn of the twentieth century was not to regulate sexual behavior but to police racial integration. The creation of the Storyville district was an attempt not to eliminate sexual activity but to maintain the color line that separated various ethnic groups from one another. Thus the transgressions committed by the city—in the eyes of its critics in other parts of the nation—were not just those dealing with over-indulgence, but it failed to sufficiently police the ethnic boundaries that drive the ethnic rationality of other urban centers.

This obsession with ethnic division can be seen in Shirley Thompson's description of the case of Toucoutou, a woman of mixed ethnic background who sued one local citizen for referring to her as a "woman of color," a charge that threatened her marriage to a prominent white citizen. Her failure to achieve this status stemmed from a continuing struggle where the realities of ethnic ambiguity were coming into conflict with social policies that promoted racial purity in the face of licentious sexual practices. This ethnic ideology continues to animate the national debate surrounding the city of New Orleans as it remains a space marketed as a release from repressive ethical systems—a release facilitated by primal, ethnic color. Thus it was not sexual behavior itself that brought the wrath of God down upon the citizens of this city, but it was the inappropriate mixing of ethnic categories and the creation of the racial ambiguity that has come to define its amorphous borders. Mayor Ray Nagin would lend support to this allegory of retribution by publicly agreeing with Pat Robertson that the storm had been a divine judgment directed against the city for its iniquity. At the February 2006 Martin Luther King Day ceremony, he stated that "surely he [God] is mad at black America also. We're not taking care of our-selves, we're not taking care of our women, and we're not taking care of our chil-dren." This unfortunate choice of words lent support to the allegorical ideology that posited the city as a space where moral transgressions were characterized as the log-ical extension of the city's ethnic conglomeration.

The geographical location of the city, combined with its unique culture, has led to a presumption that the entire region exists in a natural state of degradation. In her analysis of the literary styles that followed Reconstruction, Jennifer Greeson argues that Southern culture was exported through images of a people "glorified by disas-ter." The presumption that the city and the region itself is doomed to suffer has, in many ways, permeated the local culture itself. Clyde Woods suggests that this has sparked a rebirth in the blues culture of the region, writing that this "indigenous knowledge system has been used repeatedly by multiple generations of working-class African Americans to organize communities of consciousness." Yet in some ways this indigenous epistemology views social isolation and economic deprivation as inevitable and inescapable conditions. This inability to adopt a restorative narrative is the product of a geographical allegory that transposes the ethnic fragmentation experienced within other urban cultures onto the survivors of Hurricane Katrina, rendering their urban space a withering appendage. Thus, the existing media insti-tutions and urban boundaries have reduced the disaster to a dramatic event that fur-ther isolates these communities.

RETHINKING THE ETHICS OF CONTAINMENT

This regulation of racial ambiguity animates these narratives that dictated that the city deserved to be punished for its transgressions. In a discussion of the social meta-physics of racial mixing, Naomi Pabst suggests that those caught in the middle "are regulated, disciplined, and punished by being cast as inauthentic," as "white" and/or "white-like." The problem that New Orleans presents is not an immediate physical threat, but it is a threat to the ethnic logic of racial division and containment. Teresa Zackodnik goes on to suggest that "the disruption posed by the mulatto's racial ille-gitimacy as neither white nor black, then, was neutralized to a certain extent by this concentration on issues of moral legitimacy." Thus frameworks of morality, or the decadence of uncertainty, become linked to the threat that accepted borders and eth-nic logics can be challenged and permeated. While the carnival is a site of release and sexual freedom, it also represents a site where the biopolitics of blood and alchemy threaten to destabilize existing racial hierarchies. Castigating New Orleans as a sphere of moral degradation stems directly from these attempts to maintain ethnic borders, thus isolating its urban populations of color.

The challenge for the city is to create a shift in the ethical categories and frame-works that have been used to define these urban spaces. If the Hurricane Recovery Advisory Committee (HRAC) is to be successful in reestablishing trust within the community and creating a thriving urban economy, the geographical allegories of pollution and decadence that directly breed apathetic public policies must be recog-nized and addressed. In their call for a renewed attention to urban bioethics as a unique field of study, V. Ruth Cecire, Jeffrey Blustein, and Alan Fleischman write that "what marks and, for some, stigmatizes urban existence is the magnitude and extent of poverty; the multiplicity, urgency, and severity of related social problems; and the uneasy coexistence of citizens and immigrants across a wide economic spectrum. The latter generates. . .tensions that arise from the challenge of creating a unifying moral consensus that will nourish the public good." This renewed attention to an ethical system that focuses upon the factors that can bridge communities remains one of the best possibilities for rebuilding the shattered ties between the citizens of the city and their local leaders. To directly address the allegorical degradation of their urban spaces, this commission should look for new, invigorating metaphors that draw upon the city's unique qualities to create a binding force that will lead it into the future. In an analysis of the Corps of Engineers' response to the disaster, Todd Shallat notes that "in the wake of Hurricane Katrina the agency confronts a conundrum beyond the scope of its dam-it, ditch-it tradition; how to let the world's third-ranking river approximate the rhythms of nature, and meander and spread its replenishing mud blanket across the Delta without disrupting navigation or risking a serious flood." While the natural topography of the area is often seen as an obstacle to development and economic prosperity, so are its urban populations. It is this assumption that the fluidity and ambiguity of the area constitutes a sign of its degradation that must be addressed and, if possible, revised.

Most urban spaces are constructed through a rationality of division where groups or classes of individuals are managed and separated via a series of containment fields to ensure that individuals can persist in isolation. In his most recent work on the new templates for understanding urban landscapes, Edward Soja notes that the binary logic of class no longer works because we now must deal with polymorphous identities that ebb and flow through unexpected gaps in social boundaries. The creative reformulation of the vacuum left by Katrina presents an opportunity to create a city that, rather than isolating humans as disposable debris, can be imagined using a new set of principles. An authentic space is one in which citizens move freely through space and acquire access to public spheres of deliberation and commerce—but do so in a way that allows for the mutual recognition of the essential humanity of the entire *polis*. Forty-five years ago, Jacobs suggested that the rationality of efficiency had killed the city because people moved in isolation from one another and in ignorance of the neighborhoods where local consciousness is produced. The challenge for the city of New Orleans stems from its ability to take its ambiguous, amorphous topographical qualities and use these to create a counternarrative to the allegories of expulsion and decadence that have been deployed to render its diverse and vibrant urban population a product of waste.

Historians of social engineering have long recognized that the city of New Orleans is peculiar, given that the integration of culture, history, geography, and nature combine to create a unique chemistry. The city itself is unstable and exists in a liminal zone that is constantly shifting and sinking as new land is added while other portions are eroded away. William Howarth suggests that swamps and wetlands have a peculiar ontology within western literature that impacts the manner in which they are represented within our culture. Often these places that are neither land nor water are viewed as "dangerous, useless, fearful, diseased, noxious."

Conversely, their topographical ambiguity can be reconstituted as a more positive discursive zone that is seen as places of "beauty, fertility, variety, utility, and fluidity." It is precisely this fecundity that can animate New Orleans as a unique urban space where the boundaries of modern urban rationality can be challenged. Here, the rich indigenous culture becomes a sign not of licentious moral behavior but of creativity and invention. Jazz music, with its focus upon "condensation, fragmentation, and the innovative use of found-material, articulated always with an eye toward prior tradition," creates a sphere where the very character of the city can emerge as a source of rich invention. While tourists are titillated by tales of corpses bubbling to the surface during rainstorms, the jazz funeral remains a unique tradition where the zone between life and death, much like that between water and land, is subject to scrutiny. The presence of the city is in its power not to reinforce the binary logics of division and containment but to embody a tradition that celebrates ambiguity as a moral strength rather than a source of weakness.

The central dilemma facing the city of New Orleans centers around city planners' ability to confront the urban vacuum that still exists at the heart of the city while creating an urban design that embodies its ethics of diversity. In his discussion of the allegories of colonialism that have come to define the city of Nairobi, Joseph Slaughter describes a city that provides the illusion of vibrant urban culture but is in many ways an absent city, where the forces of export lead to an economic emptiness at its core. In some ways, this phenomenon has become a hallmark of the American urban space as well: while providing for the transport of goods, it does not signal the presence of life. M. Christine Boyer suggests that this modern rationality creates urban environments that are unrecognizable and without historical location. Local communities were reorganized according to a logic of movement and commercial efficiency that erased their unique consciousness and produced the now-familiar modern "this place = any place" vertigo. As the public was further fragmented into a series of private interiors, the modern city evolved to accommodate this new ethic through the suburban trend labeled variously "spread cities" "edge cities" "disurbs" or "post-suburbs." The city of New Orleans stands at a unique point in history where its unique cultural flavor can emerge to contest this seemingly inexorable march of progress. Rather than a choice between the *present* core or the *absent* citizen, it has the possibility to become a place where its infrastructure and governance can evolve to reflect the carnival as a rejuvenating force. To avoid the fate of those spaces that have surrendered to become a series of edges, it can become an organic system where its citizens participate in the rebuilding process and, through their dialogue, create an authentic urban experience. The challenge that the recovery organizations face is substantial because the politics of space continue to marginalize citizens of color from the political and economic centers of commerce within which these ethics of division are generated.

CONCLUSION

For the public sphere to serve the needs of its populace, it must be ideally situated as a site for public engagement, deliberation, and trade; it must also be integrated with communication technology. For Richard Sennett, these issues of functionality are preceded by a concern over authenticity: a citizenry finding its sense of identity in increasingly privatized spaces leads to a balkanization of culture. This logic of urban containment has emerged in New Orleans from a series of problematic historical assumptions about the nature of the city and its unique consciousness. The geographical allegories of Southern degradation find their nexus in a city that has been denigrated as a zone of both environmental and moral pollution. Even many of its leaders have been implicated in this system of thought that has emerged over the course of the past two centuries. If the HRAC is to be successful, it must do more than simply hand out government contracts; it must develop cognizance of the specific geographical logics that led to the city's isolation during Katrina and address them accordingly at the national level. It must then address the extent to which these allegories have permeated the local consciousness and created powerful rifts between the citizens of color and their government. This paper has suggested that the local ethnic culture, rather than being viewed as an obstacle to recovery, can become a source of pride and reinvigoration. This is not utopian idealism but represents strategic reassessment of the city and its resources. The key to this process is to open channels of communication between the various communities that have been affected by the storm and work actively to reorganize the ethical character of the city at this transitional time.

The complexity of this process is not to be underestimated. The creation of a more authentic urban culture means that the citizens see themselves reflected in the mediated environments that constitute the city itself. One place to start is to use existing media technology both to project these voices within the urban environment and to articulate this reorganization at the national level. These voices of reconstruction do not emerge from the discourse of violence and victimhood, but they allow the unique and diverse identities that compose the city to project their own indigenous knowledge systems in such a way that they can have genuine influence upon the urban planning process. Rather than people being treated as debris or obstacles, they become partners and interlocutors in the rebuilding process. At present the dramatic impact of the storm has subsided within the national media, and the process of reconstruction is not itself immediately accessible to a news culture that craves spectacle and conflict. Rather than allowing these citizens to slip back into invisibility, this project suggests that communication scholars have valuable roles to play in identifying areas where social barriers can be permeated through existing technologies and new, more vigorous narratives that form the backbone of urban organization.

SELECTED BIBLIOGRAPHY

Allen, Barbara. "Cradle of a Revolution? The Industrial Transformation of Louisiana's Lower Mississippi." *Technology and Culture* 47 (2006): 112–19.

Boyer, M. Christine. *Dreaming the Rational City: The Myth of American City Planning.* Cambridge, MA: MIT Press, 1983.

Bullard, Robert. *Dumping in Dixie: Race, Class, and Environmental Quality.* Boulder, CO: Westview, 1990.

Brattain, Michelle. *The Politics of Whiteness: Race, Workers, and Culture in the Modern South.* Princeton, NJ: Princeton University Press, 2001.

Cecire, V. Ruth, Jeffrey Blustein, and Alan Fleischman. "Urban Bioethics." *Kennedy Institute of Ethics Journal* 10 (2000): 1–20.

Coclanis, Angelo, and Peter Coclanis. "Jazz Funeral: A Living Tradition." *Southern Culture* 11 (2005): 86–92.

Cohen, Michael. "Jim Crow's Drug War: Race, Coca-Cola, and the Southern Origins of Drug Prohibition." *Southern Cultures* 12 (2006): 55–79.

Constant, Edward. "Certainties of Very Low Probability." *Technology and Culture* 47 (2006): 249–52.

Colten, Craig. "The Rusting of the Chemical Corridor." *Technology and Culture* 47 (2006): 95–101.

Cuff, Dana. "The Figure of the Neighbor: Los Angeles Past and Future." *American Quarterly* 56 (2004): 559–582.

Davis, Mike. "Fortress Los Angeles: The Militarization of Public Space." In *Variations on a Theme Park: The New American City and the End of Public Space*, edited by Michael Sorkin, 154–80. New York: Hill and Wang, 1992.

Dickerstetsun, Glenda. "Katrina: Acting Black / Playing Blackness." *Theatre Journal* 57 (2005): 614–616

Dumenco, Simon, and Ann Marie Kerwin. "Bush's PR Problem? He believes it Equals Action." *Advertising Age* 76 (2005): 50.

Dyson, Michael Eric. *Come Hell or High Water: Hurricane Katrina and the Color of Disaster.* New York: Perseus, 2006.

Fischer, Claude, Gretchen Stockmeyer, Jon Stiles, and Michael Hout. "Distinguishing the Geographical Levels and Social Dimensions of U.S. Metropolitan Segregation, 1960–2000." *Demography* 41 (2004): 37–59.

Gilfoyle, Timothy. "White Cities, Linguistic Turns, and Disneylands: The New Paradigms of Urban History." *Reviews in American History* 26 (1998): 175–204.

Giroux, Henry. "Reading Hurricane Katrina: Race, Class, and the Biopolitics of Disposability." *College Literature* 33 (2006): 171–196.

Greeson, Jennifer. "Expropriating the Great South and Exporting Local Color: Global and Hemispheric Imaginaries of the First Reconstruction." *American Literary History* 18 (2006): 496–520.

Gumpert, Gary, and Susan Drucker. "Communication and Urban Life." *Intermedia* 33 (2006): 18–23.

Hall, Stuart, Chas Critcher, Tony Jefferson, John Clarke, and Brian Roberts. *Policing the Crisis: Mugging, the State, and Law and Order.* New York: Holmes and Meier, 1978.

Horne, Jed. *Breach of Faith: The Near Death of a Great American City.* New York: Random House, 2006.

Howarth, William. "Imagined Territory: The Writing of Wetlands." *New Literary History* 30 (1999): 509–539.

Jacobs, Jane. *The Death and Life of the American Cities.* New York: Random House, 1961.

Jameson, Fredric. *Postmodernism, or, the Cultural Logic of Late Capitalism.* Durham, NC: Duke University Press, 1991.

Kelman, Ari. *A River and Its City: The Nature of Landscape in New Orleans.* Berkeley: University of California Press, 2003.

Kodat, Catherine. "Conversing with Ourselves: Canon, Freedom, Jazz." *American Quarterly* 55 (2003): 1–28.

Kolb, Carolyn. "Crescent City, Post-Apocalypse." *Technology and Culture* 47 (2006): 108–11.

Lewis, Peirce. *New Orleans: The Making of an Urban Landscape*. Charlottesville: University of Virginia Press, 2003.

Long, Alecia. *The Great Southern Babylon: Sex, Race, and Respectability in New Orleans 1865–1920*. Baton Rouge: Louisiana State University Press, 2004.

McPherson, Tara. *Reconstructing Dixie: Race, Gender, and Nostalgia in the Imagined South*. Durham, NC: Duke University Press, 2003.

Pelias, Ronald. "A Personal History of Lust on Bourbon Street," *Text and Performance Quarterly* 26 (2006): 47–56.

Rambuss, Richard. "Spenser and Milton at Mardi Gras: English Literature, American Cultural Capital, and the Reformation of the New Orleans Carnival." *Boundary 2* 27 (2000): 45–72.

Reuss, Martin. "Searching for Sophocles on Bourbon Street." *Technology and Culture* 47 (2006): 349–56.

Sennett, Richard. *The Fall of Public Man*. New York: Random House, 1975.

Shallat, Todd. "Holding Louisiana." *Technology and Culture* 47 (2006): 102–07.

Shayt, David. "Artifacts of Disaster: Creating the Smithsonian's Katrina Collection." *Technology and Culture* 47 (2006): 357–368.

Slaughter, Joseph. "Master Plans: Designing (National) Allegories of Urban Space and Metropolitan Subjects for Postcolonial Kenya." *Research in African Studies* 35 (2004): 30–51.

Soja, Edward. *Postmetropolis: Critical Studies of Cities and Regions*. Cambridge, MA: Blackwell. 2000.

Thevenot, Brian. "Myth-Making in New Orleans." *American Journalism Review* 27 (2006): 16.

Thompson, Shirley. "Ah Toucoutou, ye conin vous: History and Memory in Creole New Orleans." *American Quarterly* 53 (2001): 232–66.

Trout David, and Derrick Bell. *After the Storm: Black Intellectuals Explore the Meaning of Hurricane Katrina*. New York: New Press, 2006.

Woods, Clyde. "Do You Know What it Means to Miss New Orleans? Katrina, Trap Economics, and the Rebirth of the Blues." *American Quarterly* 57 (2005): 1005–1018.

Zackodnik, Teresa. "Fixing the Color Line: The Mullatto. Southern Courts, and Racial Identity." *American Quarterly* 53 (2001): 420–51.

Žižek, Slavoj. "The Subject Supposed to Loot and Rape: Reality and Fantasy in New Orleans." *These Times*, October 20, 2005. http://www.inthesetimes. com/site/main/article/2361.

CHAPTER 10

THE REBUILDING OF A TOURIST INDUSTRY: IMMIGRANT LABOR EXPLOITATION IN THE POST-KATRINA RECONSTRUCTION OF NEW ORLEANS

LOREN REDWOOD

On August 29, 2005, the storm referred to as Hurricane Katrina, one of the most deadly and destructive in U.S. history, struck land, causing massive devastation in the states of Louisiana, Mississippi, and Alabama. Less than a month later, Hurricane Rita hit the same area, magnifying the damage to infrastructure, buildings, and services for basic survival. Now, more than a year after Katrina and Rita, the efforts at recovery continue. The massive needs for reconstruction and attempts at revitalization of economies in the region have served as catalysts to changes in the demographic landscape. Mainstream media coverage and academic attention to the injustices experienced by survivors of the hurricanes are continuing to expose tangible evidence of racial and class oppressions. There exists, however, an omission in the examination of the complexity of racial oppressions. I refer here to the large immigrant populations that now comprise the primary work force in reconstruction of the impacted region. This work force is made up of primarily Latina/o women and men who have been lured to the Deep South by unscrupulous labor recruiters to fill the demand for low-wage, exploitable labor. In the examination of racial oppression and class oppressions post-Katrina, the struggles and abuses of this highly marginalized population have most often received one of two responses: erasure or hostility.

A recently released report detailing ongoing racial division in New Orleans asserts, "In the aftermath of Katrina, the authors of media and political discourse wrote scrip about race war and job theft. They cast the actors as Black victims and Brown invaders, and told stories that distracted the public's focus from the institutional responsibility of government and private contractors to insure that all workers are treated with fairness. This 'bait and switch' has fueled the perception of racial conflict and competition. The conflict has been embraced by many, to the disadvantage of the excluded and exploited communities."

The presence of this work force of alternately invisible and vilified low-wage laborers, instrumental to the rebuilding of the region, is the product of the larger structures of capitalism and of U.S. governmental policy and intervention. Of key importance in the rebirth of the economy in Louisiana is the rebuilding of the tourism industry, the primary site being New Orleans. This chapter exposes the inextricable links between tourism and state-sanctioned exploitation of immigrant labor and expands upon current understandings of tourism as an enactment of neocolonial violence. An investigation of the most recent shifts in labor and immigration post–Hurricanes Katrina and Rita will follow, with particular attention given to labor exploitation in the rebuilding of a major U.S. tourist destination. Acts of resistance and protest by exploited immigrant workers are discussed. Also examined are the rising racial tensions between African American communities and Latino communities in the Deep South brought about by labor conditions, creating hostility between local poor and global poor. This examination concludes with a look at immigrant rights advocacy organizations in the Deep South, their growth, and the need for racial coalitions to address shared oppressions.

On August 25, 2006, the *USA Today* article "French Quarter Set to Roll" notified the American tourist that New Orleans was back and ready to receive visitors. The opening paragraph of the article stated, "On the river side of Jackson Square in the heart of the French Quarter, a dozen carriage mules stand in a line in the late-afternoon swelter, looking thoroughly bored from mane to hoof. Not clipping, not clopping, barely munching. Just standing there. Still. Mulishly underemployed." A few days later Mitch Landrieu, the Lieutenant Governor of Louisiana, expressed his appreciation of the article. Speaking of the need to rebuild New Orleans and the state of Louisiana following Hurricanes Katrina and Rita, Landrieu argued, "Tourism is critical to the rebirth of Louisiana and, as the state's second largest industry, it employs more than 100,000 people. This is why everyone can feel good about having fun while making a difference for those who make the Big Easy swing." However, little attention was given to a more important and ongoing story concerning eighty-two guest workers, women and men from South America and the Dominican Republic, who had joined a lawsuit filed by the Southern Poverty Law Center against Decatur Hotels LLC, the wealthiest luxury hotel owners in New Orleans and one of the largest hotel chains in Louisiana. The case, filed in the U.S. District Court for Louisiana, focused on the issue of labor exploitation of immigrant workers, which had become increasingly common in New Orleans and other parts of the Deep South in the rebuilding efforts that followed Hurricanes Katrina and Rita.

The suit alleged that Decatur Hotels LLC, violated the Fair Labor Standards Act when the company failed to reimburse workers for the exorbitant fees they paid to aggressive labor recruiters working as agents for the hotel chain. Guest workers, recruited to New Orleans by the hotel chain, were brought in to work in hotel support positions such as housekeepers, guest service attendants, and hotel maintenance workers. Federal regulations require employers to first attempt to recruit U.S. workers and to demonstrate proof of their efforts before requesting access to guest workers. The lawsuit highlighted the fact that U.S. workers were available and that the company's goal in hiring foreigners was to drive down wages. The number of immigrant workers currently employed by Decatur is estimated by the Southern Poverty Law Center to be approximately three hundred.

The laborers recruited by Decatur LLC under the H-2B guest worker program, paid labor recruiters between $3,500 and $5,000 to take them to New Orleans. After being lured to New Orleans, many of these guest workers faced enormous debt and were struggling to meet their most basic of needs while laboring in luxury hotels, including the renowned Astor Crowne Plaza. Workers reported that the recruiters promised full-time work and wages between $1000 and $1200 per month; however, most received low wages in the range of $6.02 and $7.79 per hour, were significantly underemployed, and had no overtime opportunities; additionally, workers had not yet been reimbursed for the huge expense associated with travel to the United States. For example, Luis Lopez, employed as a room service employee at the Astor Crowne Plaza, reported that he "spent most of the last two months without the work hours he was promised. . .[and that] his last paycheck was for just $18." Lopez further claimed that his wife, in caring for their three children in the Dominican Republic, was "getting desperate as debt collectors swarm in and she can't buy food." Lopez, housed in a Decatur-owned hotel for guest workers where he shared a room with three other men, paid $50 per week to the hotel for housing, the same hotel chain to which they were contracted to work.

These workers had few options. The nature of the federal guest worker agreements bound the workers to their employers and disallowed workers from leaving, even if the employer did not deliver on promises with regard to wages, work hours, or working conditions. Teresa Ortiz, a guest worker also employed by Decatur stated, "It's modern-day slavery. What are my options? I go home to Bolivia, poorer than when I got here and deeper in debt. Or I break the law to find another job."

The filing of the lawsuit was marked with a protest by the guest worker hotel employees staged outside the United States Federal Court building in New Orleans. Protest signs clearly reflected the injustices experienced by the workers. Workers and their allies held signs bearing the words "Dignidad" and "Dignity" while other guest workers wore enlarged copies of H-2B visas on ropes around their necks. Some held up enlarged signs of their visas, symbolizing the legal process that irreversibly bound them to their employer. In an effort to expose the extreme economic injustice, one man carried an enlarged copy of a check issued by Decatur Hotels Corporation for $18.08. Many of the guest workers also wore handcuffs to symbolize their bondage.

The study of tourism presents a significant site of inquiry into the ongoing understandings of empire. The growth of tourist industries globally, initiated by "First World North" countries (either by direct investment or as a result of structural adjustment programs) impacts countries of the "Third World South" on a massive scale. Sharon Bohn Gmelch argues that "tourism has enormous social implications globally. It represents the largest ever movement of people across national boarders, eclipsing migration and immigration, refugee flight, pilgrimage, business and educational travel." The enormous growth of this industry makes tourism an increasingly important site for investigation. Gmelch notes that "the travel and tourism industry generates $4.5 trillion (U.S.) in economic activity and provides over 2 million jobs globally." Gmelch further reports, "worldwide, tourism employs one in every twelve workers and accounts for 11 percent of global gross domestic product." This becomes a relevant concern to scholars when issues of labor in the tourist industry intersect with issues of subject position and imperial power. In this framework, tourists can be viewed as neocolonial actors with the labor of production for tourist sites performed by neocolonial subjects.

The study of tourism is inextricably linked to the study of transnational labor, immigration and migration, labor market theory, and racialization, and it therefore requires an analysis of Western Empire. The role of the United States through government funding and corporate investment includes both the internal colonization exemplified by tourist sites within the country, as well as international sites of occupation. Tourism is a continuing location of colonial enterprise and is an enactment of nation-state violence. E. San Juan Jr.'s articulation of the U.S. positioning in terms of postcolonialism and corporate and state violence provides a context for the consideration of tourism. Juan argues that postcolonialism, the ideology of globalized capitalism, "offers a metaphysics of legitimation for those groups who stand to benefit from the predatory economics of uneven development, namely, transnational corporations and their compradors, including their retinue of postcolonial rationalizers."

The nation-state manipulates the use of immigrant labor as an exploitable and disposable workforce, especially in the maintenance of the tourist industry, following an established historical pattern. The aftermath of the Katrina disaster has greatly influenced migration and demographic patterns in the Deep South, calling attention to the legal, civil, and human rights needs of a disenfranchised population consisting mostly of Latina/o immigrants. The federally funded rebuilding of this major tourist enterprise has strategically promoted the recruitment of a large superexploitable workforce comprised of both documented guest workers and undocumented immigrant laborers. This strategy included the Department of Homeland Security's deferment of I-9 employer inspections, which required employers to verify the citizenship status of their employees, and the Bush Administration's temporary suspension of the Davis-Bacon Act of 1931, under which employers must pay prevailing wage rates on federally financed construction projects. Additionally, the failure of the Environmental Protection Agency and Occupational Safety and Health Administration to uphold and reinforce safety regulations further demonstrates the inextricable links of the nation-state in the creation of superexploitable labor in the service of capitalist enterprise. The rebuilding of tourist infrastructure has proved to be a financial windfall for a select number of primary contracting corporations, most with connections to the federal government, who received no-bid contracts from the Federal Emergency Management Agency and now stand to collect enormous profits from the labor of immigrant workers. The opportunity for further exploitation is even more serious when coupled with the fact that the trail of contracting and subcontracting is quite difficult to trace. James Hale, a vice president of the Laborers

International Union of North America states "contracts let to the subcontractors are just plain invisible" and concludes that this environment creates "an open invitation for exploitation, fraud and abuse."

In addition, agencies such as The Hispanic Connection Inc., based in Baton Rouge, report a dramatic growth in the demand for recruitment of Latina/o laborers as H-2B guest workers needed for clean-up, construction, and service work in the aftermath of the hurricane. Guest worker recruitment agencies like The Hispanic Connection are in the business of recruiting women and men immigrants as H-2B workers for labor as janitorial and service workers in casinos and hotels and for domestic workers in private homes. Immigrant labor scholar Kristen Hill Maher asserts that "such businesses function less as employer agencies than as labor brokers, treating migrant workers as commodities for sale." Maher notes a frequency of fiduciary abuses present in the contracting relationship between the agency and the worker.[4] Maria Edwards, director of the Hispanic Connection, notes that her agency has been flooded with requests for laborers since the two hurricanes hit but is currently unable to meet the demand. She also observes that those employers she was unable to assist have gone to extensive lengths to find workers on their own.

Prospective immigrant workers are particular targets for labor abuses at times when attempting to participate in guest worker programs. In November of 2005, the publication *La Jornada* reported that businesspeople from the Gulf Coast, posing as representatives of three different companies, came to the Mexican city of Guanajuato to recruit laborers. These companies' representatives allegedly obtained permission to conduct business in Mexico from the Secretary of the Government and told prospective workers that through their connections to the Secretary of Exterior Relations, they were able to secure worker visas for those willing to come to the United States. Reportedly, many paid between $500 to $800 to these imposters, a fee that was alleged to be payment for a work visa and transportation to New Orleans; the representatives reportedly offered payment of $12 an hour for electricians, construction and carpentry; the day the workers were scheduled to travel to the United States, the recruiters absconded with the money. Unfortunately, this is not an isolated event but a recurring pattern of U.S. corporations who employ labor brokers. What is significant about this particular incident is the apparent complicity of the Mexican government, an act that further demonstrates the unequal power relations between the United States and Mexico and the reliance of the United States on Mexico for a large exploitable pool of labor.

A recently published study of Latina/o workers in New Orleans, conducted by U.C. Berkeley and Tulane University, details the conditions for both documented and undocumented laborers. Study findings indicate that "nearly half of the reconstruction workforce in New Orleans is Latino, of which 54 percent is undocumented." The findings of the study show that both groups are exposed to wage abuses, hazardous working conditions, lack of health care, inadequate housing, and police and Immigration and Naturalization Service harassment. In all cases, undocumented workers fare significantly worse than documented laborers. Undocumented workers are paid significantly less and are frequently the victims of labor theft, not being paid at all for their work. Employers, who many times do not provide workers with information regarding their identity, disappear or shut down worksites without warning, leaving workers with no recourse. H2–B guest workers, however, are not immune to wage abuses. Some guest workers relate that after being recruited to the Deep South, they have not worked for months, but they cannot look for other employment due to government restrictions. Still others report being virtually held captive by employers who confiscated their visas upon arrival to the job site and withheld wages. These workers report paying recruiters thousands of dollars to immigrate that has not been reimbursed.

The incidents of hazardous exposure to toxins are too numerous to assess. Thousand of workers report they live in the homes that they are dismantling and are exposed to mold, chemicals, sewage, and nonpotable water. In recent accounts by H2–B guest workers, laborers report being forced to live in the same hotels they are gutting. Others report being forced to live in squalid conditions with contaminated water sources and very little food. Undocumented workers also have little to no access to health care when sick or injured. Roberto Lovato has reported on numerous incidents of laborers working on military bases in Mississippi who "complained of suffering from diarrhea, sprained ankles, cuts and bruises, and other injuries" incurred on the worksite; they were given no medical attention "despite being close to medical facilities on the same bases they were cleaning and helping rebuild."

Additional investigations of worker abuses have uncovered incidents of immigrant laborers threatened with guns when they complain of not being paid. In one case, two workers report being forced to work at gunpoint by their employers, who were also New Orleans police officers, for no pay. United States Immigration and Customs Enforcement actions have soared in the last year with the deployment of more than "725 personnel to the Gulf, including approximately 400 special agents from the office of investigations, 200 officers from Federal Protective Services, and 100 offices from Detention and Removal Operations."

The urgent demand for labor in the Deep South as a result of hurricane damage and changes in labor restrictions that have allowed for a high level of exploitation is already causing friction between established disenfranchised groups and their new counterparts. While many residents are temporarily displaced and may never return, active recruitment with the promise of lucrative work opportunities in the region is luring nonresidents to the area from all parts of the United States, as well as Mexico and Central America. Frank Mercado, a construction worker from Puerto Rico currently employed in Mississippi, states, "We're brown people. We're just an apparatus." As a heavily disenfranchised group, these workers will likely join African American and black populations as another underclass of the Deep South.

The "rebirth" of a tourist mecca demands this pool of underclass laborers, to both rebuild infrastructure and provide service work labor. Raymond Mohl has documented the ways in which the processes of globalization have resulted in changes in the Southern economy. Mohl attributes the demand and growth of a transnational labor force in the South to "new economic investment poured into the region as American and foreign capital."

Eduardo Bonilla-Silva examines the changes in labor brought about by globalization and a "new world-systemic need for capital accumulations." Bonilla-Silva asserts that demand to maximize profits drives wages down and "has led to the incorporation of 'dark' foreigners as 'guest workers' and even as permanent workers. . .who are progressively becoming an underclass."

The presence of racial and ethnic tensions, which continues to grow in the Deep South, are the result of much larger systems of capitalism and a worldwide process of globalized labor exploitation. Edna Bonacich's theory of "ethnic antagonism" describes the consequences of globalized labor, which speaks to the split between working-class African Americans and the new workforce of immigrant laborers. The prevailing rhetoric that directs the focus on to issues of immigration only serves to obscure the true issue. This claim is further supported by Fred Krissman, who cites neoclassical economics as providing an "academic fig leaf, behind which those who decry immigration can hide their nativist sentiments even as they dictate ever more punitive actions against immigrants."

Saket Soni of the New Orleans Worker Justice Coalition addresses this conflict by referring to guest worker programs as "wedge policies that divide African Americans and immigrants." Guest workers brought in to the area for exploitable labor in the service industry create a "misperception of competition that positions labor issues as a wedge issue between communities of color." Soni notes that in the post-Katrina economic crisis faced by African American communities in New Orleans, unemployment is "7.2 percent. . .[and] survivors are locked out of the hotel industry even as they struggle to return home and regain their lives a year after Katrina."

In the examination of immigration policy and imported labor, the Bracero Program, active from 1948 to 1964, is important to the U.S. historical context of guest worker programs. As Mae Ngai points out, other cases of contract-labor were in operation during this time period; however, "the Bracero Program was by far the largest project, involving some 4.6 million workers." Ngai employs the concept of "imported colonialism," which the author explains has "produced new social relations based on the subordination of racialized foreign bodies who worked in the United States but who remained excluded from the polity by both law and by social custom."

The history of guest worker programs in the United States is one fraught with problems of abuse and exploitation. Cindy Hahamovitch exposes the role of the nation-state by framing guest worker programs as "state-brokered compromises designed to maintain high levels of migration while placating anti-immigrant movements." She concludes, "In the process they have drawn nations together in a new sort of dependency, in which the world's wealthy nations rely on foreigners to do their hardest and dirtiest work, and in which poorer nations depend on earnings abroad for their very survival." Concern for U.S. immigration policy as it applies to guest worker programs is echoed by the executive vice president of the Service Employees International Union (SEIU), Eliseo Medina. David Bacon notes that the SEIU is "one of the AFL-CIO's key policy makers on immigration." In response to the Bush Administration's proposed legislation regarding guest worker programs, Medina notes the hypocrisy of telling immigrants "you have no right to earn citizenship but tell[ing] corporations you have the right to exploit workers, both American and immigrant. . . .This proposal allows hardworking tax-paying immigrants to become a legitimate part of our economy, but it keeps them from fully participating in our democracy, making immigrants a permanent sub-class of our society. Such sub-classes are consistently vulnerable to a variety of abuses."

The labor abuses primarily experienced by an immigrant labor force in the affected regions go largely unchecked. According to the Gulf Coast Commission on

Reconstruction Equity, Mississippi and Louisiana, which fall into the category of "right to work" states, have no departments of labor at the state level, and there is "only one bilingual Department of Labor investigator to cover the states of Alabama and Mississippi." Thus, there are only two formal recourses available to the individual in response to labor abuses: a civil suit filed with the state where the work is performed, or a claim filed with the Federal Department of Labor. Neither action is likely to be taken by an undocumented immigrant.

In response to human rights abuses, the Deep South is currently experiencing a growth in immigrant rights advocacy by such organizations as The Southern Poverty Law Center, The Mississippi Immigrant Rights Association, and Oxfam American (Oxfam). Also responding to these abuses are faith-based organizations including the Catholic Legal Immigration Network Inc., Catholic Charities, and the Hispanic Apostolate of New Orleans (Catholic Legal Immigration Network). These organizations are providing another avenue for exploited immigrant laborers to pursue human and civil rights protections.

In conclusion, a visit to the French Quarter, including a stay at the Astor Crowne Plaza and a carriage ride from an "underemployed mule" obfuscates the reality of the superexploited laborers who are integral to the creation and continued operation of this major American tourist site. The study of tourism must acknowledge the relationship of transnational labor, immigration and migration, labor market theory, and racialization in its analysis of the Western Empire as colonial actor upon the global south. The multiple connections between tourism in the "First World North" and labor from "Third World South," the role of globalization and global financial institutions, the active encouragement of the U.S. government, and immigration policy, as well as the effects of environmental devastation on the health and safety of laborers recruited from the "Third World South," make this a relevant area of inquiry.

Another things obfuscated by the media and current scholarship is the presence of this workforce, which is either made invisible or vilified. What requires illumination here is the commonality of racial oppressions experienced by all communities of color now present in New Orleans and the Deep South generally. The rhetoric that is currently successfully dividing these communities must be confronted and interrogated. In the protest held by guest workers on the steps of the federal court house in New Orleans, Tracie Washington, Director of the NAACP Gulf Coast Advocacy Center and co-counsel in the case against Decatur Hotels, helps highlight this need when she joins hands with Latino and Latina American immigrants and calls for solidarity between communities. Washington notes that the New Orleans guest worker program is a "continuation of the racial exploitation that began with slavery in this country that is corporate-driven." Indeed, the rebuilding of New Orleans as a major U.S. tourist destination in the aftermath of Hurricanes Katrina and Rita requires a large pool of highly exploitable labor. In this context, building alliances across racial, ethnic, and class lines presents the opportunity for greater empowerment for all communities in the fight for dignity, human rights, and social and economic justice.

BIBLIOGRAPHY

Bacon, David. "The Political Economy of Immigration Reform: The Corporate Campaign for a U.S. Guest Worker Program." *Multinational Monitor* 25, no. 11 (November 2004): 9–13.

Barclay, Eliza. "As Locals Struggle, Migrants Find Work in New Orleans." *The San Francisco Chronicle*, October 12, 2005. http://www.sfgate.com/cgi-bin/article.cgi?file=/c/a/2005/10/12/MNG53F74SK1.DTL.

Bonacich, Edna. "A Theory of Ethnic Antagonism: The Split Labor Market." *American Sociological Review* 37, no. 5 (1972): 547–59.

Bonilla-Silva, Eduardo. "From Bi-racial to Tri-racial: Towards a New System of Racial Stratification in the USA." *Ethnic and Racial Studies* 27, no. 6 (November 2004): 931–50.

Brown-Danis, Judith, et al. "And Injustice for All: Workers' Lives in the Reconstruction of New Orleans." *Leadership Conference on Civil Rights*, July 6, 2006. http://newreconstruction.civilrights.org/details.cfm?id=46602.

Cass, Julia. "Guest Workers Sue New Orleans Hotel Chain: Immigrants Say Decatur Group Failed to Deliver on Promised Employment." *The Washington Post*, August 16, 2006.

"Center Exposes Exploitation of Immigrant Workers." *Southern Poverty Law Center*, August 16. 2006. http://splcenter.org/legal/news/article.jsp?aid =205&site_area=1.

Chandler, Bill, and Tina Susman. Interview with Amy Goodman. "Workers in New Orleans Denied Pay, Proper Housing and Threatened with Deportation." *Democracy Now! Online*, December 16, 2005. http://www.democracynow.org.

Diego-Rodríguez, Martín. "Timan en Guanajuato a más dos 2 mil posibles migrantes." *La Jornada*, November 10, 2005. http://www.jornada.unam.mx/imprimir.php?fecha=200 51110&nota= 022n2pol.php&seccion.

Fletcher, Laurel E., Phuong Pham., Eric Stover, Patrick Vinck. *Rebuilding After Katrina: A Population-Based Study of Labor and Human Rights in New Orleans*. New Orleans: Payson Center for International Development and Technology Transfer, 2006.

Gmelch, Sharon Bohn. "Why Tourism Matters." *Tourists and Tourism: A Reader*, ed. Sharon Bohn Gmelch. Long Grove, IL: Waveland, 2004.

Goodman, Julie. "Problems Growing Along Coast: Labor Department Looking into Disputes over Wages, Unsafe Work Conditions." *Clarion Ledger Online*, November 6, 2005. http://clarionledger.com/apps/pbcs.dll/article?AID=/20051106/NEWS0110/511060 344/1260.

"Guest Workers Charge Racial Exploitation, File Federal Suit Against Luxury Hotel Chain." *New Orleans Independent News Center*, August 17, 2006. http://neworleans.indymedia.org/news/2006/08/8452.php.

Gulf Coast Commission on Reconstruction Equity. *Good Work and Fair Contracts: Making Gulf Coast Reconstruction Work for Local Residents and Businesses*. Chicago: Interfaith Workers Justice, 2006.

Hahamovitch, Cindy. "Creating Perfect Immigrants: Guestworkers of the World in Historical Perspective." *Labor History* 44, no. 1 (2003): 69–94.

Krissman, Fred. "Sin Coyote Ni Patr"n: Why the "Migrant Network" Fails to Explain International Migration." *The International Migration Review* 39, no. 1 (2005): 4–45.

Landrieu, Mitch. Commentary. *USA Today*. August 31, 2006, p. A10.

Lavato, Roberto. "Gulf Coast Slaves." *Salon Online* November 15, 2005. http://dir.salon.com/story/news/feature/2005/11/15/halliburton_katrina/index.html.

Lydersen, Kari. "Immigrants Rebuilding Gulf Coast Suffer 'Third-World' Conditions." *The New Standard Online*, November 3, 2005. http://newstandardnews.net/content/index .cfm/items/2559.

Maher, Kristen Hill. "Good Women 'Ready to Go': Labor Brokers and the Transnational Maid Trade." *Labor Studies in Working-Class History of the Americans* 1, no. 1 (2004): 55–76.

Mohl, Raymond A. "Globalization, Latinization, and the Nuevo New South." *Journal of American Ethnic History* (Summer 2003).

Ngai, Mae M. *Impossible Subjects: Illegal Aliens and the Making of Modern America*. Princeton, NJ: Princeton University Press, 2004.

Oxfam America. "Disaster and Recovery Along the U.S. Gulf Coast: Oxfam America's Response to the Hurricanes." *Oxfam America Online*, February 2006. http://oxfam america.org/ (accessed March 16, 2006).

"Post-Katrina Tourism in Louisiana Packs Fun, State's Rebirth." *USA Today Online*, August 30, 2006. http://www.usatoday.com/travel/destination/2006–08–24–new-orleanstourism _x.htm?csp.

Roberts, Michelle. "Immigrants Sue New Orleans Hotelier." *USA Today*, August 18, 2006, p. B3.

Root, Jay, and Aaron C. Davis. "Undocumented Immigrants Flock to Jobs on Gulf Coast." *Knight Ridder Washington Bureau Online* October 12, 2005. http://realcities.com/ mld/krwashington/news/world/12885261.htm.

San Juan, E. Jr. *Racism and Cultural Studies: Critiques of Multiculturalist Ideology and the Politics of Difference*. Durham, NC: Duke University Press, 2002.

Shriver, Jerry. "French Quarter Set to Roll." *USA Today Online*, August 25, 2006. http://www.usatoday.com/travel/destination/2006–08–24–new-orleans-tourism_x.htm?csp.

Texeira, Erin. "Immigration is Dividing Blacks, Latinos." *St. Louis Dispatch*, April 9, 2006, p. B6.

"Update: Immigrant Rights Post-Katrina." *Catholic Legal Immigration Network*, August 23, 2006. http://www.cliniclegal.org (accessed August 27, 2006).

PART III

RACE AND REPRESSION

"DO YOU KNOW WHAT IT MEANS . . . ?": MAPPING EMOTION IN THE AFTERMATH OF KATRINA

MELISSA HARRIS-LACEWELL

> Do you know what it means to miss New Orleans
> And miss it each night and day?

—Louis Armstrong

> George Bush doesn't care about black people.

—Kanye West

On the first anniversary of the Katrina disaster, a documentary film about the events premiered on HBO. "When The Levees Broke: A Requiem in Four Acts" was produced by award-winning, African American filmmaker Spike Lee. Lee used the enduring images of human suffering in New Orleans and the compelling narratives of hurricane victims to give new meaning and poignancy to the tragedy. Although the film is political in its thesis and conclusions, it is fundamentally an emotional tale about the heart of the experience of Katrina for the people of New Orleans. HBO promoted the film as an "intimate, heart-rending portrait of New Orleans in the wake of the destruction that tells the heartbreaking personal stories of those who endured this harrowing ordeal and survived to tell the tale of misery, despair and triumph." In his discussion of the film and its importance, Lee makes a claim to the centrality of the emotional effects of the storm on its victims: "Post-Katrina, the obituary column in the *Times-Picayune* is 30 percent more. Suicides are up. . . .People are just buggin'. And there are no facilities to deal with the mental-health issue down there. This stuff is going to have reverberations for many years to come. When you have children who've seen their parents drown in front of them or parents who have seen their children drown in front of them, I mean, how do you deal with that?"

Building on this compelling articulation of the emotional effects of the storm, this chapter explores the interconnection of race, politics, and emotion in the aftermath of Hurricane Katrina. When the levees of New Orleans failed, it was not a colorblind disaster. The storm caused greater loss and displacement for black New Orleans residents (Cutter 2005; Gabe et al. 2005; Sherman and Shapiro 2005). The storm also provoked significantly different reactions from black and white Americans who viewed the unfolding disaster through the media (Bobo 2006; Sweeney 2006; Huddy and Feldman 2006). Not only did Americans of different races perceive vastly different realities about the events in New Orleans, but black and white Americans *felt* differently about what happened. The affective responses of African Americans were more pronounced than those of their white counterparts. These emotions are rooted in America's racial history and the resonance of that history in contemporary U.S. society. In the aftermath of Katrina, emotional devastation is a political response. Using data from several national surveys conducted in the weeks following September 11, 2001, and the weeks following Hurricanes Katrina and Rita in 2005, I map the differences in emotional responses among black and white Americans to both disasters. I then analyze this survey data to suggest that Americans' political and racial beliefs were significantly related to their psychological experiences in the weeks following Katrina. Finally, I argue that the emotional map of Katrina responses demonstrates the centrality of race over class in shaping how black Americans understood and experienced the disaster.

EVALUATING AMERICAN FEELINGS ABOUT DISASTER

Katrina wrought enormous devastation on those who lost property, sustained injury and suffered displacement as a direct result of the flooding in New Orleans. While compelling and important, it is hardly surprising that victims of a catastrophic natural disaster experienced profound emotional reactions. Symptoms of posttraumatic stress disorder (PTSD) are common among disaster survivors (Harvey and Bryant 1998; Kessler et al. 1995; North et al. 1999; Davidson et al. 1991; Green 1991). Far more remarkable are the occasions when those who are not directly victimized by a disaster experience negative emotional consequences simply as a result of their vicarious exposure to the vulnerability and suffering of others. The terrorist attacks of September 11, 2001, generated precisely this sort of collateral emotional damage for Americans who were not directly victimized by the events (Schuster et al. 2001).

Several national studies demonstrate that in the days and weeks following September 11, Americans experienced elevated stress and signs of probable posttraumatic stress disorder (Shuster et al. 2001; Schlenger et al. 2002). In the two months following the attack, 17 percent of respondents in a national sample reported posttraumatic stress symptoms. Although New York City was most directly affected by the attacks, Americans throughout the country shared in the anxiety and stress that the tragedy evoked. While there was a shared sense of distress, residents of New York experienced more negative emotions; follow-up studies revealed that New Yorkers were still experiencing emotional suffering while much of the country had begun to return to more normal psychological functioning six months following the attacks (Silver et al. 2002). Proximity to the disaster left New Yorkers more shaken at the outset and more distressed in the long term than initially empathetic fellow citizens who could more easily return to normal emotional states.

The National Opinion Research Center (NORC) at the University of Chicago conducted the National Tragedy Study between September 13 and September 27, 2001. NORC interviewed 2,126 Americans, with an oversample of residents in New York City and Washington, DC.[1] The National Tragedy Study replicated a series of questions used to gauge the emotional state of Americans in the days following the assassination of President Kennedy in 1963. The Kennedy Assassination Survey Symptom Checklist (KASSC) measures fifteen physical and emotional reactions to traumatic shock.[2] Table 11.1 reports the average score on the KASSC for black and white respondents in both the national and New York samples. New York residents (both black and white) have elevated KASSC scores compared with their counterparts in the general population. New Yorkers were more upset by the events of September 11, but many people throughout the country shared a sense of vulnerability and sadness after vicariously experiencing the attacks on Manhattan.

Table 11.1   Mean scores on Kennedy Assassination Survey Symptom Checklist

	Mean	SE		Mean	SE
National			*New York*		
White			White		
n=641	4.29	0.16	n=166	5.67	0.30
Black			Black		
n=95	3.85	0.41	n=81	4.59	0.44

Source: The National Tragedy Study 2001

Table adapted from Tom Smith, Kenneth Rasinski, and Marianna Toce, "America Rebounds: A National Study of Public Response to the September 11 Terrorist Attacks," *Report of the National Opinion Research Center* (Chicago: University of Chicago, October 25, 2001).

Importantly, when the KASSC score is modeled as a function of race, gender, age, education, income and employment and estimated with an ordinary least squares regression, the results demonstrate that in the days immediately following September 11, there was no statistically significant difference in KASSC scores among blacks and whites in either the national or the New York samples (Rasinski et al. 2002). In the case of September 11, proximity to the disaster was much more important than race in predicting initial emotional responses. Understandably, New Yorkers felt the tragedy most sharply, but all of America, black and white, mourned along with the city. NORC researchers concluded, "Nationally, the attack engendered anger, confusion and both defensive and altruistic behaviors. There is no indication that African Americans and Caucasians differed in their initial appraisal and behavioral coping. The overwhelming nature of the attacks appears to have cut across ethnic groups in response" (Smith et al. 2001, 6).

While September 11 provoked similar initial responses from blacks and whites, the aftermath of Hurricane Katrina revealed a wide emotional chasm between the races. A Pew Foundation survey[3] conducted immediately following Katrina asked respondents, "Have you yourself felt angry because of what's happened in areas affected by the hurricane?" Although a significant proportion of whites responded to the disaster with anger (46 percent), anger was much more prevalent among African Americans. Of black respondents, 70 percent reported that the events surrounding Katrina made them angry. Similarly, African American respondents to the Pew survey were much more likely (71 percent) than white respondents (55 percent) to report they "felt depressed because of what's happened in areas affected by the hurricane." These results suggest that Spike Lee's documentary film accurately captured a sense that black America experienced the aftermath of Katrina with intense emotion.

The Racial Attitudes and Katrina Disaster Study by the University of Chicago Center for the Study of Race, Politics and Culture[4] further explores the emotions that Americans experienced in response to Katrina. The Katrina Disaster Study, conducted approximately one month after the initial levee breaches in New Orleans, asked respondents, "In the past five weeks since the Hurricane Katrina disaster how often have you felt: so sad that nothing could cheer you up; nervous; restless or fidgety; hopeless; that everything was an effort; worthless; that difficulties were piling up so high you could not overcome them; and that you are unable to control the important things in your life." For each emotion, respondents could report that they felt this way very often, fairly often, not too often, hardly ever, or never. Figure 11.1 reports the percentage of blacks and whites who felt these emotions "very often" in the weeks after the hurricane.

There are clear and consistent differences between African American and white respondents. Approximately double the proportion of African Americans report the highest level of suffering from each of the negative emotions they were asked about in the survey. They are twice as likely to report being sad, nervous, restless, and hopeless. They are also twice as likely to feel overwhelmed, worthless, and as though everything takes more effort. Just over a month after Katrina, more than one in ten African Americans report feeling very often that they are unable to control the important things in their lives.

The aftermath of Hurricane Katrina provoked different patterns of emotional responses than the tragedy of September 11, 2001. There was little difference in how black and white Americans felt about the horror of the terrorist attacks. Predictably, New Yorkers felt worse in the days, weeks, and months following September 11, but these differences were true for New Yorkers of both races. The country mourned together in a sense of shared vulnerability. Although the aftermath of Katrina provoked strong emotional responses from all Americans, the impact of the storm seems to have been more deeply felt among African Americans. Black respondents were also more debilitated by their negative emotions. When asked if these emotions had interfered with life activities, 67 percent of whites reported that they had felt no

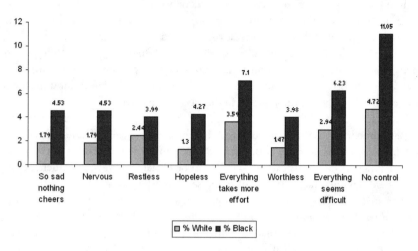

Figure 11.1   Percent who report feeling negative emotions "very often" in the weeks immediately following Hurricane Katrina

Source: Racial Attitudes and Katrina Disaster Study by the University of Chicago Center for the Study of Race, Politics and Culture

impairment of life activities. Only 58 percent of blacks believed they suffered no impairment. A full 42 percent of the black Americans reported that their negative emotions had interfered with their life activities.

Similar to the ways citizens of New York felt more distress following September 11, black Americans had more negative emotions after Katrina. While the response of New York residents is likely explained by their greater proximity to Ground Zero and perception of greater vulnerability to repeat attacks, these explanations cannot account for the heightened emotional distress of African Americans in the weeks following Katrina. Most African Americans have little reason to fear being struck by a hurricane with the potential to destroy their homes, neighborhoods and city. So why were black people more distressed than whites?

It could be that black people are generally more sad and anxious. There is some evidence from prior research suggesting that African Americans report chronically elevated levels of emotional distress compared to whites (Amato 1991; Breslau et al. 1998). This evidence, however, is decidedly mixed. In some studies black Americans report fewer symptoms of sadness. In other studies, the racial disparity in negative emotions can be explained by controlling for class. Because of conflicting findings across multiple studies, the evidence is neither clear nor compelling that black people express chronically higher levels of negative emotions. Analysis of data from the University of Chicago Katrina Disaster Study shows that when income, education, gender, and age are accounted for in a simple regression model, race still has a strong and significant correlation with reported mental distress. This suggests that the emotional disparity between blacks and whites following Katrina cannot be accounted for solely on the basis of enduring markers of social differences between the races.

Even if black respondents are not chronically more negative in their emotional reports, it is possible that the post-Katrina racial disparity is the result of African Americans regularly responding to national crises with more negative emotions than their white counterparts. Data on racial responses to September 11 show that this explanation is incorrect. In Table 11.1 there are no statistically significant differences between black and white respondents in the days immediately following 9/11. To the extent that there are differences in average emotional responses between blacks and whites immediately following September 11, those differences indicate that whites, not blacks were more distressed. The Katrina gap is not present in the case of national tragedy four years earlier, so it cannot be explained away as the mere extension of an ingrained racial pattern of response to disaster.

The analysis below explores a third hypothesis, suggesting that the reason for the racial disparity in emotional response to Katrina lies not with the essential psychology of black or white Americans but rather with historical and contemporary racial beliefs that shape the political and emotional lives of many black Americans, but of far fewer white Americans. September 11 was widely understood as an American tragedy that was national in scope, but Hurricane Katrina was perceived as a more narrow racialized disaster. It is therefore the politics of race that helps explain why whites and blacks feel so differently about these two events.

THE POLITICS OF RACIAL EMOTION

> Their government had forsaken them; they weren't citizens but castoffs, evacuees turned effortlessly, in language and life, into refugees.
>
> —Michael Eric Dyson

When newly emancipated black men entered the American polity at the end of the Civil War, it was with optimism about the possibility of becoming full partners in the American democratic experiment. Despite having suffered generations of forced labor, formerly enslaved persons allowed themselves to embrace the rights and responsibilities of citizenship. Having fought to preserve the Union, black men took on the franchise and ran for elected office. Black women and men opened businesses, founded schools, and formed organizations and associations to support their new status as wage earners, taxpayers, and citizens. Then in 1877 the Hayes-Tilden Compromise crushed the emerging dream of black citizenship. The country's political parties balanced their own power on the backs of black people, negotiating away the responsibility of the United States to protect its citizens. The federal government withdrew troops from the former Confederacy and initiated one hundred years of creative brutality as the South instituted Jim Crow, lynch mob rule, and disenfranchisement.

Still, with each major military conflict of the nineteenth and twentieth centuries, black men marched into battle believing that their service as citizens might translate into protection by the American state. Black people earned a fraction of all other workers, but they faithfully paid taxes despite having been shut out of the public schools and accommodations that these taxes financed. As taxpayers and soldiers, black men and women expected the full rights and protections of other Americans, but the country remained steadfast in its refusal to grant them even the basic rights of access to public education, public facilities, government aid, voting rights, or social equality.

Racial inequality has persisted into the twenty-first century. African Americans continue to earn a fraction of their white counterparts, suffer from worse physical health, enjoy fewer educational opportunities, are less well represented in politics and popular culture, and labor under persistent racial stereotypes. Despite significant and continuing racial inequality, one might argue that the last half-century has been the most hopeful time for black citizenship in American history. Having fought and died in the streets of the urban South, African Americans rolled back voting restrictions, asserted the right to fair and equal treatment under the law, and made real gains in education, access, and visibility. The months immediately preceding the Katrina disaster in 2005 witnessed several important milestones in the history of African American citizenship. The United States Senate issued an official and public apology for never passing anti-lynching legislation. The government exhumed the body of Emmett Till to bring closure to the murder that helped launch the civil rights movement. Prosecutors in the state of Mississippi reopened and successfully litigated the case of the murders of the three civil rights workers Chaney, Goodman, and Schwerner. These were positive steps in the history of black America, but when the levees broke in New Orleans, the destruction of the city was accompanied by a disillusion of emerging optimism about the contemporary state of black citizenship.

It is possible that this history of disappointed expectations influenced the emotional reactions of black people during Katrina. I initially began to hypothesize this connection when I traveled to New Orleans in November 2005, just seven weeks after the levee breach. While in New Orleans, I conducted dozens of interviews with survivors of the storm. I also attended several community meetings led by Mayor Ray Nagin and his Bring Back New Orleans Commission. The emotional devastation of those who had lived through the storm was palpable. Everyone I spoke with had been displaced following the city's mandatory evacuation; the majority had sustained unimaginable loss of personal property; many had survived the nightmarish conditions of the Superdome or Convention Center; some were still searching for missing family members and friends; and a few had confirmed that family members had died during the storm. As direct survivors of the storm, they manifested classic symptoms of posttraumatic shock disorder.[5] I was not surprised to find that these survivors were enduring painful and raw emotions, but I was stunned by the nearly universal agreement among African American survivors that their suffering was related to their status as second-class citizens.

An example of how Katrina's black survivors in New Orleans talked about their own experience emerged in the November 14, 2005, meeting of Mayor Nagin's Bring Back New Orleans Commission. The meeting was held in a large ballroom at the Sheraton hotel in downtown New Orleans, which had become a kind of headquarters for municipal action. Mayor Nagin presided over the meeting, and the official panel included representatives from several federal agencies, local utility companies, elected officials, and community leaders. After an update on the state of the city and the pace of recovery, members of the audience were allowed to come to a central microphone and address the panel.

The first four persons who spoke were white residents from the Garden District area who had a variety of concerns about power outages, mold, and general health concerns. The fifth person to speak was an African American man who owned his own trucking business. As a local contractor, he expressed outrage at the "storm followers" who were making "$18 per hour while I can only manage to get $15 per hour." In an impassioned plea to the mayor he shouted, "Listen. I got four hundred black men ready to work, and we are being talked to like dogs. This is our city and we are being treated like second-class citizens." In response to his statement, the audience broke into unrestrained applause. When the next African American man spoke, he pleaded for a moment "to pause and recognize the loss of thousands of people. The nation paused on 9-11, but not now. No one cares about our losses. I am a homeowner who is homeless. I am a taxpayer and a voter. I placed my trust in the elected officials to do what is right, but instead we got nothing. We are not refugees, we are Americans."

Subsequently, a black woman, who had stood while holding her sleeping toddler in her arms during the first seven speakers, continued the theme of government accountability to its citizens: "I was one of the people left behind. I was stuck on a roof in New Orleans East. I am a taxpayer and a registered voter. I am happy, but I am not rich. I have been shifted to five hotels all around the country. I am tired. I have never asked Louisiana for anything. I just want a place to call my own. I didn't need help before this. I was doing for myself and for my children. Mr. Mayor, all I want is a home for my children for Christmas."

After this woman, a black man who lost his home in the Ninth Ward and was displaced first to Denver and then to Dallas confronted the mayor, saying, "What is really going on? You are asking us to come back to work. I served this city for thirty-five years, and we are watching foreigners get paid to rebuild it while we are sitting on the curb. There is something going on, Mr. Mayor. I understand about the dollar bill situation, but I want to come back and function for my people, for New Orleans. It is wrong for us to be turned down. I was willing to stay an extra year to help my city. This is my home; these are my roots. We are not in Texas. We are here."

It is not surprising that individuals who suffered catastrophic losses would articulate impassioned emotional pleas. It is more provocative when so many of these requests are framed with an anxiety about citizenship. Not only did the people of New Orleans whom I spoke with express sadness and grief about their loss, they also provided explanations for that loss that were rooted in racialized understandings of themselves as disposable members of the American populace. They asserted their positions as homeowners, taxpayers, citizens, and government workers as they constructed an argument about their city and country's betrayal of them. These interviews proved to be a rich source for understanding the depth of the emotional trauma and the importance of political explanations for survivors. However, what remains remarkable about the aftermath of Katrina is that black Americans who were not directly impacted by the disaster also experienced significant elevations of negative emotions.

Survey data make clear that Americans not only felt bad as they witnessed Katrina; the vicarious experience of the disaster had immediate and dramatic political consequences. The Pew study reported that 67 percent of Americans believed that President Bush could have done more in his handling of the relief effort, and nearly 60 percent rated the response of the federal government as only fair or poor.[6] The Katrina disaster also caused many Americans to reconsider the nation's security, with 42 percent reporting that the events surrounding Hurricane Katrina made them feel less confident that the government can handle a major terror attack. In the aftermath of the hurricane, job approval ratings for President Bush plummeted,[7] and one year later they still had not rebounded to their pre-Katrina levels. Many observers point to the Katrina disaster as both a national tragedy and a political turning point, linking an emotionally difficult experience with a politically relevant change in public opinion.

The nation's emotional reactions to the Katrina disaster are important because they demonstrate the link between how Americans *think* about social and political realities and how they *feel* about national events. To test the hypothesis that Americans' affective responses to Katrina were primarily rooted in a particular understanding of America's racial history, the analysis below estimates a model of emotional distress among black and white Americans in the weeks following Katrina as a function of personal, political, and racial variables. This estimation is performed using data from the Racial Attitudes and Katrina Disaster Study by the University of Chicago Center for the Study of Race, Politics and Culture.

Researchers normally do not think of an individual's mental or emotional state as resulting primarily from world political events. Decades of epidemiological work have convincingly demonstrated that we can make predictions about the likelihood that an individual will feel depressed, angry, or fragile based on a number of personal characteristics and proximate life circumstances. Even in the days and weeks following Katrina we should expect most emotional variability between individuals to be directly related to durable and personal patterns that have been explicated in previous research. We should expect that poorer and less-educated Americans should generally feel more psychological distress than their more affluent counterparts. We should expect women to express more sadness than men, and for the very young and very old to express more sadness than those who are middle-aged. In light of these expectations, the equations estimated below control for education, income, sex, and age.[8]

Acknowledging that a significant proportion of the variation in individual emotional responses can be accounted for by these variables, the goal of this analysis is to determine whether there is an independent relationship between negative emotions and political variables after accounting for the demographic variables. To explore this question, the model below uses three categories of variables. The *partisan variables* used in the equation include partisan self-identification[9] and agreement with the statement "President George W. Bush represents the concerns of people like you." The second set of questions taps respondents' *attitudes toward America*. Respondents were asked to rank their level of agreement with the statement "I am proud to be an American" and with the statement, "America is the land of opportunity. If a person works hard in America he or she can accomplish almost anything." The survey respondents were also asked to assess which statement is truer: "America's economic system is fair to everyone" or "America's economic system is unfair to poor people."

The model tests the hypothesis that Americans' emotional responses to Hurricane Katrina were intimately linked to their beliefs about race and America's racial history. Several measures of racial attitudes are used to capture this idea. Blacks were asked if they believed that what happens to black people will affect their lives and whites were asked if they believe what happens to whites will affect their lives. These racially-linked fate attitudes are included in the model below. Respondents also indicated if they believed that blacks in America have achieved racial equality, will soon achieve racial equality, will not achieve equality in their lifetimes, or will never achieve racial equality. Responses to this question serve as a measure of racial pessimism.

Finally, respondents were asked a series of questions about their support for federal reparations for African Americans as compensation for a number of historic injustices. Blacks and whites were asked, (1) "Do you think the federal government should or should not pay money to African Americans whose ancestors were slaves as compensation for that slavery?" (2) "Do you think the federal government should or should not pay money to African Americans as compensation for the system of anti-black violence and legal segregation known as 'Jim Crow'?" (3) "Do you think that reparations should or should not be paid to survivors and their descendants of large, violent, twentieth-century anti-black riots such as those that occurred in Tulsa, Oklahoma and Rosewood, Florida?" Responses to these three items are combined in a single scale indicating overall support for reparations to black Americans.

The dependent variable is a scale derived from a factor of eight emotional response variables that asked respondents, "In the past five weeks since the Hurricane Katrina disaster how often have you felt: so sad that nothing could cheer you up; nervous; restless or fidgety; hopeless; that everything was an effort; worthless; that difficulties were piling up so high you could not overcome them; and that you are unable to control the important things in your life." The scale is computed as a factor and is constrained to a unit scale where 1 represents the highest presentation of symptoms on all indicators and 0 represents having none of these negative emotions.

Table 11.2 presents the estimated coefficients and standard errors from an ordinary least squares regression modeling the emotional distress scale from the University of Chicago Katrina Disaster Study as a function of these partisan variables, demographic variables, and national and racial attitudes. The model is estimated separately for white and black respondents to account for possible differences in the emotional processes operating for each group.

As expected, the model accounts for relatively little of the variation in emotional distress among blacks or whites. As predicted, most people feel bad or good based on personal and proximate causes. There is a substantial and statistically significant effect that gender and age bear on both whites and blacks. Men and middle-aged people experienced fewer symptoms of emotional distress in the weeks following Katrina. Somewhat surprisingly, education is not a statistically significant covariant for either group. Income is significant only for whites. Poorer whites are far more emotionally distressed in the weeks following Katrina than are their wealthier counterparts. For African Americans however, income makes no difference. Poorer and richer blacks were equally likely to feel sad and overwhelmed.

Table 11.2   Model of emotional distress among black and white Americans in the immediate aftermath of Hurricane Katrina

	Whites n=602 $r^2 = 0.16$ Coefficient	SE	Blacks n=345 $r^2 = 0.10$ Coefficient	SE
Partisan Variables				
Bush represents people like me	-0.05	0.04	0.06	0.06
Democrat	0.03	0.02	0.01	0.02
Republican	-0.03	0.02	0.02	0.06
Attitudes toward United States				
Proud to be American	0.06	0.05	0.05	0.06
America is unfair to the poor	-0.01	0.02	0.01	0.02
America is the land of opportunity	-0.15*	0.04	-0.10*	0.05
Racial Attitudes				
Racial linked fate	0.04*	0.01	0.05*	0.02
Blacks will not achieve equality	-0.01	0.01	-0.03	0.02
Reparations Support (3 item scale)	0.04*	0.02	0.03*	0.01
Demographic Variables				
Education	0.03	0.04	-0.03	0.06
Income	-0.14*	0.04	-0.07	0.05
Female	0.04*	0.01	0.04*	0.02
Age^2	-0.002*	0.0004	-0.001*	0.0006
Constant	0.62*	0.05	0.48*	0.07

Source: Racial Attitudes and Katrina Disaster Study by the University of Chicago Center for the Study of Race, Politics, and Culture

Coefficients estimated with ordinary least squares regression. The dependent variable is constrained to a unit scale. Coefficients can be read as the percent change in total emotional distress. For example, those whites who support all three forms of reparations are 4 percentage points more distressed in the weeks following Katrina than those whites who do not support any form of reparations, all else being equal.

* indicate significance at p<.05

   While demographic variables account for some of the variation in the emotional distress measure, statistically significant relationships between emotions and racial and political variables are present. Negative affective reports were not related to partisan attitudes for either blacks or whites. The emotions of Republicans and Democrats are indistinguishable from the emotions of those with no partisan preference. Also, those who believe that President Bush is representative of people like them had the same bad feelings post-Katrina as those who do not. Similarly, patriotism is unrelated to the emotional impact of the disaster. NORC researchers concluded that in the weeks following September 11, patriotism helped provide an emotional buffer against despair for many Americans. In the case of Katrina, patriotism has no discernable impact on emotional reports of whites or blacks.

   While these basic political beliefs had little influence on how Americans felt, racialized political variables were important. For both white and black respondents, there were three key factors associated with their emotional responses after Katrina. First, those who believed that America is a land of opportunity where individuals can accomplish anything were significantly less distressed following the hurricane. This belief in America's limitless potential served as a buffer for some who witnessed the events in New Orleans. Perhaps they reasoned that although terrible things were happening, America would provide opportunities for these citizens to be restored and reestablished. Among both blacks and whites, those who were optimistic about the nation's opportunities felt relatively better than those who were more pessimistic.

While American idealism mitigated negative emotions, belief in racially linked fate and recognition of America's historic racial injustices were related to more negative emotions after the hurricane. African Americans and whites who believe that their fate is linked to the fate of their race felt more distressed than those with less sense of linked fate. For African Americans, this finding seems straightforward. Just as New Yorkers felt more vulnerable than the rest of the country in the days following September 11, so too did black people, who perceived their life opportunities linked to that of other black people, feel more vulnerable in the aftermath of Katrina. Although probably not estimating that they were more likely to be affected by a natural disaster, blacks' perception of a linked fate likely heightened the sense that they were more vulnerable to the inadequate government response to human suffering.

More surprising, whites who sensed their fate linked to that of other whites also felt more distressed following Katrina. We might expect that they would feel less distress because most whites escaped the worst horrors of the New Orleans disaster. However, white respondents who have a sense of linked fate may be more racially empathetic because they are more aware of the operation of race in individual life outcomes. If this hypothesis is accurate, then these white respondents are people who recognize race in America and express that recognition through a belief that the outcomes of their racial group influences their experiences. Forman and Lewis assert that when Americans expressed shock about the racialized poverty exposed by Katrina, they were articulating a willful ignorance of race that characterizes contemporary racial understandings in America. Contemporary racism in the form of racial apathy is not the explicit desire to inflict racial harm but instead a willful expression of ignorance about racial inequality and its effects. It is not possible to fully test this hypothesis with the available data in this study, but simple bivariate correlations show that a sense of white linked fate is positively correlated with the belief that blacks will never achieve racial equality (0.11), positively correlated with the belief that America is unfair to poor people (0.08), and negatively correlated with the belief that America is the land of opportunity (0.04). White linked fate covaries with negative emotions in the aftermath of Katrina for those white people who recognize and acknowledge the continuing importance of race and inequality in American society.

Finally, those blacks and whites who support federal reparations for slavery, Jim Crow and twentieth-century race riots felt, statistically, significantly worse in the weeks following Katrina. Respondents were not asked to predict the likelihood that federal reparations would be forthcoming. We should not assume that the clear majority of black respondents who believe the federal government *should* provide reparations are optimistic that the federal government *will* provide reparations. Instead support for federal remuneration is more accurately understood as recognition of the lingering effect of racial injustice and the government's failure to acknowledge or make amends for that injustice. Both black and white Americans reported feeling greater levels of sadness and distress in the weeks following Katrina if they also believed that the federal government stilled owed black Americans for centuries of previous injustices.

CONCLUSION

To the real question, how does it feel to be a problem? I answer seldom a word. And yet, being a problem is a strange experience,—peculiar even for one who has never been anything else, save perhaps in babyhood and in Europe.

—W. E. B. Du Bois

Three tentative conclusions suggest themselves in light of this analysis. The first raises profound questions concerning race and class. Vulnerability to the storm's ravages fell disproportionately on those who lived at the intersection of poverty and blackness. Wealthier black people in New Orleans fared better in the aftermath of the storm than poorer blacks. While the disaster's direct effects had a class component, the emotional responses to the storm for black observers were structured by racial considerations, not class concerns. Income and education did not distinguish the emotional experience of Katrina for black people. Further, as Table 11.2 shows, questions concerning America's fairness toward poor people bore no direct correlation with negative emotions. Although some researchers and observers have worked to point attention toward issues of class and away from discussions of race in the aftermath of Katrina, it seems that race and not class is at the heart of the black affective experience of the disaster.

In his foundational text, *Behind the Mule: Race and Class in African-American Politics* (1994), political scientist Michael Dawson convincingly demonstrates that African Americans often use racial group interests as a proxy for determining their own interests. This black linked fate results in many African Americans expressing public opinions and supporting public policies that are not apparently in their class interests. Black linked fate emerges both from historical, race-based experiences of discrimination and contemporary realities that make class interests more complicated for African Americans. The black middle class is likely to be residentially segregated from their white counterparts, economically vulnerable, and linked through familial and social ties to the black poor. Black observers may have seen the black poor suffering in New Orleans, but they also saw themselves. The white middle class was more insulated from these negative emotions. Thus, while socioeconomic status was correlated with white emotional patterns in the weeks following Katrina, class was unimportant for understanding black American feelings.

The second conclusion we can draw from this research is that the disparity between blacks and whites is not because white people are essentially unable to empathize with blacks. The disparity grows out of differing meanings given to the events. If white people see the world the way that black people do, then they feel the way that black people do. The disparity emerges because only a tiny fraction of whites share African Americans' perceptual experiences. The responses in Table 11.3 underline the point that black and white Americans exist and operate in vastly different political realities. Blacks and whites score significantly differently on every political and racial variable importantly correlated with emotional distress. White people are much more optimistic about the opportunity for individual advancement in the American system. A clear majority (55 percent) strongly agree that America is the land of opportunity, while fewer than one third (31 percent) of black Americans join them in this optimistic assessment. While a clear majority of whites express a sense of linked fate with other whites (57 percent), this is dwarfed by the near unanimity among blacks that their race helps determine life chances. Finally, while a substantial majority of black Americans believe that the federal government should provide compensation to black Americans for historical injustices, only a tiny fraction of whites agree.

The small proportion of white Americans who perceive America's racial landscape in ways similar to the majority of black Americans also felt significant distress and sadness following Katrina. The enormous racial disparity is not attributable to the fact that whites are essentially incapable of racial empathy. Instead, the disparity

Table 11.3 Percent of white and black respondents who agree with indicators in model

	%White	%Black
Partisan Variables		
Bush represents people like me (yes, a lot)	22.6	3.4
Democrat	29.1	57.9
Republican	30.1	3.4
Independent or No Preference	38.6	37.8
Attitudes toward United States		
Proud to be American (strongly agree)	77.1	57.9
America is unfair to the poor	41.7	71.3
America is the land of opportunity (strongly agree)	54.9	30.8
Racial Attitudes		
Racial linked fate	57.2	79.2
Blacks will not achieve equality	27.4	67.2
Reparations Support		
Reparations for slavery	2.6	51.7
Reparations for Jim Crow	3.4	59.2
Reparations for twentieth-century race riots	4.9	63.7
Demographics		
Female	50	58
Income (mean on a 0–1 scale)	0.56	0.45

Source: Racial Attitudes and Katrina Disaster Study by the University of Chicago Center for the Study of Race, Politics, and Culture

exists because so few whites see the world through the lens that most black Americans use for understanding their world.

The final tentative conclusion to be drawn from this research is that America's racial history affects America's emotional present. September 11 proved to be a rallying point of American identity. Americans largely shared a sense of vulnerability and loss in the days, weeks, and months following 9-11. However, the national camaraderie after September 11 did not last. American's responses to the tragedy became increasingly complicated by race, region, and partisanship as the domestic and international responses to the attack emerged. The Hurricane Katrina aftermath did not provoke such a uniform affective emotional response. Black people found themselves relatively more isolated in their grief and fury. At the core of this differing response are the racial and political meanings that black people assigned to the Katrina disaster. Good citizens of conscience and kindness responded with generosity and concern for those who had been displaced and devastated. White doctors, nurses, preachers, lawyers, and everyday working people were moved to great acts of heroism, generosity, and benevolence. But still, in the nation as a whole there was a significant gap in the emotional experiences of the disaster. While most whites believed that the abandonment of New Orleans was the result of bureaucracy, inefficiency, lack of preparedness, or technical capacity, most blacks believed that race was the real issue.

The levee failure in New Orleans that trapped thousands of Americans and destroyed one of America's most distinctive cities was one of the few televised American tragedies. Until the events in New Orleans, only the assassination of President Kennedy in 1963 and the destruction of the World Trade Center buildings in 2001 allowed Americans to share in the trauma of their fellow citizens in real time. Unlike either of these earlier events, the aftermath of Hurricane Katrina was an unfolding drama that lasted for days. Rather than a single, terrible moment replayed by the media, the horror of New Orleans increased daily, produced new images of agony and death, and generated increasingly awful narratives of suffering. Americans witnessed hours of grisly footage. The nation was able to watch in real time as the faces of their fellow citizens were contorted in fear, pain, and hunger; unable to feed their children or comfort their parents or find their partners.

While the terrible consequences of Katrina were readily apparent to most, African Americans suffered unique horrors as they watched the aftermath of the storm. Hurricane Katrina was not colorblind in its effects, and Americans were not colorblind in their interpretations of the disaster. There is a vast racial disparity between how black and white Americans understood the lessons of the storm.

Empirical social science can open up the human experience of politics by prompting considerations about what it means to feel bad within the context of American politics. To be a citizen in a democracy is to be not only the ruled, but also the ruler. To be a citizen in a democratic republic is to have a voice in which you can name your experiences and pursue your good through public means. On September 11, the nation momentarily felt like vulnerable, attacked, but united citizens. This sense of vulnerability was less shared in the case of Katrina. Not only were the victims of the hurricane abandoned in their drowning city, but black Americans were abandoned in their grief as they once again confronted the fact of their second-class citizenship.

BIBLIOGRAPHY

Aneshensel, Carol, and Clea Sucoff. "The Neighborhood Context of Adolescent Mental Health." *Journal of Health and Social Behavior* 37, no. 4 (1996): 293–310.

Avis, Thomas-Lester. "A Senate Apology for History on Lynching Vote Condemns Past Failure to Act." *Washington Post*, June 14, 2005.

Ballie, Andrea. "Spike Lee Painstakingly Chronicles Katrina in Epic Doc 'When the Levees Brooke.'" *Canadian Press*, September 16, 2006.

Blau, Francine, and John Graham. "Black-White Differences in Wealth and Asset Composition." *Quarterly Journal of Economics* 105, no. 2 (1990): 321–39.

Bobo, Lawrence. "Katrina: Unmasking Race, Poverty, and Politics in the 21st Century." *Du Bois Review: Social Science Research on Race* 3, no. 1 (2006): 1–6.

Breslau, N., G. C. Davis, and P. Andreski. "Risk Factors for PTSD-related Traumatic Events: A Prospective Analysis." *American Journal of Psychiatry* 152 (1995): 529–35.

Breslau, N., R. Kessler, H. Chilcoat, L. Schultz, G. Davis, and P. Andreski. "Trauma and Post-traumatic Stress Disorder in the Community: The 1996 Detroit Area Survey of Trauma." *Archives of General Psychiatry* 55 (1998): 626–32.

Brunner, Borgna. "The Murders of James Chaney, Andrew Goodman, and Michael Schwerner: The Ringleader of the Murders, Edgar Ray Killen, Was Convicted on June 21, 2005, the 41st Anniversary of the Crimes." http://www.infoplease.com/spot/bhmjustice4 .html.

The Brookings Institution Metropolitan Policy Program. "New Orleans After the Storm: Lessons for the Past, a Plan for the Future." Publication of the Brookings Institution. Washington, DC, October 2005.

Chan, Emily. "Evidence-based Public Health Practices: Challenges for Health Needs Assessments in Disasters. *Hong Kong Medical Journal* 12, no. 4 (2006): 324–326.

Cockerman, William. "A Test of the Relationship between Race, Socioeconomic Status and Psychological Distress." *Social Science and Medicine* 31 (1990): 321–31.

Cooper, Frederick, Thomas Holt, and Rebecca Scott. *Beyond Slavery: Explorations of Race, Labor and Citizenship in Postemancipation Societies*. Chapel Hill: University of North Carolina Press, 2000.

Cutter, Susan, Bryan Boruff, and W. Lynn Shirley. "Social Vulnerability to Environmental Hazards." *Social Science Quarterly* 84, no. 1 (2003): 242–61.

Cutter, Susan. "The Geography of Social Vulnerability: Race, Class and Catastrophe." *Understanding Katrina: Perspectives from the Social Sciences* (2005).

Darden, Joe. "Black Residential Segregation since the 1948 *Shelley v. Kramemer* Decision." *Journal of Black Studies* 25, no. 6 (1995): 680–91.

Davidson, L. M., L. Weiss, M. O'Keefe, and A. Baum. "Acute Stressors and Chronic Stress at Three Mile Island." *Journal of Traumatic Stress* 4 (1991): 481–93.

Du Bois, W. E. B. 1903. *The Souls of Black Folk*. New York: Norton, 1999.

Dyson, Michael Eric. *Come Hell or High Water: Hurricane Katrina and the Color of Disaster*. New York: Basic Books, 2006.

Forman, Tyrone, and Amanda Lewis. "Racial Apathy and Hurricane Katrina: the Social Anatomy of Prejudice in the Post-Civil Rights Era. *Du Bois Review: Social Science Research on Race* 3, no. 1 (2005): 175–202.

Frymer, Paul, Dara Strolovitch, and Dorian T. Warren. "New Orleans is not the Exception: Re-politicizing the Study of Racial Inequality." *Du Bois Review: Social Science Research on Race* 3, no. 1 (2006): 37–58. http://understandingkatrina.ssrc.org/FrymerStrolovitchWarren.

Fullilove, Mindy Thompson. *Root Shock: How Tearing Up City Neighborhoods Hurts America and What We Can Do About It*. New York: Random House, 2005.

Green, B. L. "Evaluating the Effects of Disasters." *Psychological Assessment* 3 (1991): 538–46.

Harvey A. G., and R. A. Bryant. "The Relationship between Acute Stress Disorder and Posttraumatic Stress Disorder: A Prospective Evaluation of Motor Vehicle Accident Survivors." *Journal of Consulting Clinical Psychology* 66 (1998): 507–12.

HBO.com. Synopsis: When the Levees Broke; A Requiem in Four Acts. September 26, 2006.

Huddy, Leonie, and Feldman, Stanley. "Worlds Apart: Blacks and Whites React to Hurricane Katrina. *Du Bois Review: Social Science Research on Race* 3, no. 1 (2006): 97–114.

Kessler R., A. Sonnega, E. Bromet, M. Hughes, and C. B. Nelson. "Posttraumatic Stress Disorder in the National Comorbidity Survey." *Archive of General Psychiatry* 52 (1995): 1048–60.

Kessler, R. "Epidemiology of Women and Depression." *Journal of Affective Disorders* 74, no. 1 (March 2003): 5–13.

Litwack, Leon. 1979. *Been in the Storm so Long: The Aftermath of Slavery*. New York: Random House.

Logan, John. "Population Displacement and Post-Katrina Politics: The New Orleans Mayoral Race, 2006." June 1, 2006.

Massey, Douglass. "American Apartheid: Segregation and the Making of the Underclass." *American Journal of Sociology* 96, no. 2 (1990): 329–57.

McLeod, Jane and Shanahan, Michael. "Poverty, Parenting, and Children's Mental Health." *American Sociological Review* 58, no. 3 (1993): 351–66.

National Fair Housing Alliance, "No Home for the Holidays: Report on Housing Discrimination against Hurricane Katrina Survivors." December 20, 2005.

North, C. S., S. J. Nixon, S. Shariat et al. "Psychiatric Disorders among Survivors of the Oklahoma City Bombing. *Journal of the American Medical Association* 282 (1999): 755–62.

Nguyen, Manthi, ed. *Black Americans: A Statistical Sourcebook*. Palo Alto, CA: Information Publications, 2003.

Norris, F. "Epidemiology of Trauma: Frequency and Impact of Different Potentially Traumatic Events on Different Demographic Groups. *Journal of Consulting and Clinical Psychology* 60 (1992): 409–18.

Pew Research Center for the People and the Press. Hurricane Katrina Survey. September 6–7, 2005.

Rasinski, Kenneth, Alicia Matthews, Bethany Albertson, Tom Smith, and Jennifer Berkhold. *The 9/11 Terrorist Attacks: Ethnic Differences in Emotional Response and Recovery*. Chicago: University of Chicago National Opinion Research Center, 2002.

Schlenger, W. E., J. M. Caddell, L. Ebert, B. K. Jordan, K. M. Rourke, D. Wilson et al. "Psychological Reactions to Terrorist Attacks: Findings from the National Study of Americans' Reactions to September 11." *Journal of the American Medical Association* 288 (2002): 581–88.

Schuster M. A., B. D. Stein, L. Jaycox et al. "A National Survey of Stress Reactions after the September 11, 2001, Terrorist Attack." *Journal of the American Medical Association* 345 (2001): 1507–12.

Sheehan, Charles. "Till Is Exhumed: Experts Hope Science Can Unravel Case. *Chicago Tribune*, June 2, 2005.

Sherman, Arloc and Isaac Shapiro. "Essential Facts about the Victims of Hurricane Katrina." *Center on Budget and Policy Priorities*, September 19, 2005.

Silver, R. C., E. A. Holman, D. N. McIntosh, M. Poulin, and V. Gil-Rivas. "Nationwide Longitudinal Study of Psychological Responses to September 11." *Journal of the American Medical Association* 288 (2002): 1235–44.

Smith, T., K. Rasinski, and M. Toce. "America Rebounds: A National Study of Public Response to the September 11 Terrorist Attacks." *Report of the National Opinion Research Center.* Chicago: University of Chicago, 2001.

Sweeney, Kathryn. 2006. "The Blame Game: Racialized Responses to Hurricane Katrina." *Du Bois Review: Social Science Research on Race* 3, no. 1 (2005):161–174.

Weisman, Jonathan, and Michael Abramowitz. "Many See Storm as President's Undoing." *Washington Post*, August 26, 2006, p. A1.

NOTES

1. The National Tragedy Study was conducted between September 13 and September 27, 2001, by the National Opinion Research Center, University of Chicago. It was a telephone interview of adults (18 years old and older) living in households with telephones in the United States. The total sample size of 2,126 comprised a national sample of 1,013 households and additional samples in the New York City, Washington, DC, and Chicagoland areas. The overall response rate was 52 percent.

2. The Kennedy Assassination Study was conducted between November 26 and December 3, 1963, by the National Opinion Research Center, University of Chicago. The Kennedy Assassination Study Symptoms are derived from responses to the following: "I am going to read a list of things some people have said happened to them since they heard about the attack on the World Trade Center. Please tell me whether or not they happened to you: Didn't feel like eating. Smoked more than usual. Had headaches. Had an upset stomach. Cried. Had trouble sleeping. Felt nervous and tense. Felt like getting drunk. Felt more tired than usual. Felt dizzy at times. Lost my temper more than usual. Hands sweat and felt damp and clammy. Had rapid heartbeats. Felt sort of dazed and numb. Kept forgetting things."

3. Results for the survey are based on telephone interviews conducted under the direction of Princeton Survey Research Associates International among a nationwide sample of 1,000 adults, 18 years of age or older, including an oversample of African Americans, during the period September 6–7, 2005. The oversample of African Americans is designed to allow a sufficient number of interviews for reporting results of this demographic group. The national sample of telephone households was supplemented with an additional 103 interviews with African Americans whose households had been recently contacted for past Pew Research Center national surveys. Demographic weighting was used to ensure that the survey results reflect the correct racial and ethnic composition of national adults, based on U.S. Census information.

4. The Center for the Study of Race, Politics and Culture at the University of Chicago supported a national survey of Americans to gauge political and racial attitudes in the aftermath of Hurricane Katrina. Principal investigators were Michael Dawson, Melissa Harris-Lacewell, and Cathy Cohen. The data were collected by Knowledge Networks between October 28, 2005, and November 17, 2005. Knowledge Networks employed a Random Digit Dialing (RDD) telephone methodology to develop a representative sample of households for participation in its panel. Once a Knowledge Networks household was selected, members were contacted first by an express delivery mailing and then by telephone for enrollment in the *Knowledge Networks* panel. The panel structure enables clients to conduct surveys of low-incidence populations, such as African Americans, more efficiently and inexpensively than would otherwise be possible. Every participating Knowledge Networks household receives free hardware, free Internet access, free e-mail accounts, and ongoing technical support. Participants

receive a short multimedia survey about once a week. Surveys are delivered by e-mail on the same standardized hardware, through the television set. The data include responses from 1252 Americans. The racial composition of the respondents is as follows: White: 703, Black: 487, Hispanic: 52, Other: 10. Interviews are conducted in person by Melissa Harris-Lacewell in various locations in New Orleans November 11–18, 2005. Interviews include than 28 personal discussions with local residents and hours of transcripts from three community meetings about rebuilding efforts.

5. The name Post-traumatic Stress Disorder first appeared in 1980 in DSM-III, the American Psychiatric Association's Diagnostic and Statistical Manual of Mental Disorders Third Edition. The diagnosis was updated in 1994 in the latest edition, DSM-IV. The diagnostic criteria for Post-traumatic Stress Disorder (PTSD) are defined in DSM-IV as follows:

> The person experiences a traumatic event in which both of the following were present: the person experienced or witnessed or was confronted with an event or events that involved actual or threatened death or serious injury, or a threat to the physical integrity of self or others; and the person's response involved intense fear, helplessness, or horror. The traumatic event is persistently re-experienced in any of the following ways: recurrent and intrusive distressing recollections of the event, including images, thoughts or perceptions; recurrent distressing dreams of the event; acting or feeling as if the traumatic event were recurring (eg reliving the experience, illusions, hallucinations, and dissociative flashback episodes, including those on wakening or when intoxicated); intense psychological distress at exposure to internal or external cues that symbolize or resemble an aspect of the traumatic event; physiological reactivity on exposure to internal or external cues that symbolize or resemble an aspect of the traumatic event. Persistent avoidance of stimuli associated with the trauma and numbing of general responsiveness (not present before the trauma). Persistent symptoms of increased arousal (not present before the trauma) as indicated by at least two of the following: difficulty falling or staying asleep; irritability or outbursts of anger; difficulty concentrating; hypervigilance; exaggerated startle response.

6. News sources mattered for the conclusions that Americans drew about the efficiency of presidential response. Seventy-three percent of CNN watchers reported that the president could have done more, but only 50 percent of Fox News viewers agreed. Forty-six percent of those whose primary source of Katrina coverage was Fox News believed that the president had done all he could.

7. The Pew survey reports a 50 percent approval rating for President Bush in January 2005 and a 40 percent approval rating immediately following Katrina.

8. Education is coded as years of education. Income is measured as self-reported household income and coded on a unit scale where 0 represents the lowest income category and 1 represents the highest income category. Sex is coded as a dichotomous variable with female = 1. Age is coded as reported age at time of survey. The variable used in the equation is age squared to account for the hypothesized parabolic relationship between age and mental health.

9. Democrat and Republican self-identification are included in the model. They should be read against the excluded category of those who say that there are independent or have no preference.

CHAPTER 12

WITNESS: THE RACIALIZED GENDER IMPLICATIONS OF KATRINA*

KATHLEEN A. BERGIN**

INTRODUCTION

Rechelle Carter. Brianna Carter. Linda Watson. Errolyn Warden. Barbara Richards. These five women came to Houston following Hurricane Katrina on a bus from Violet, Louisiana, a small black community located on the banks of the Mississippi River in St. Bernard Parish east of New Orleans. I met them while volunteering at the Reliant Complex, where 25,000 hurricane evacuees lived for three weeks in the fall of 2005. There I learned their story.

Several neighborhood streets in Violet are sandwiched between Judge Perez Drive and St. Bernard Highway: Guerra Drive, Lucciardi Drive, Caluda Lane. They are also surrounded by water. Behind St. Bernard Highway is a levee that holds back water from the Mississippi River basin. On the opposite side of Violet, behind Judge Perez, is a floodgate that traps the water from Lake Borgne.

On August 29, residents of Violet breathed a sigh of relief when the local news announced that Hurricane Katrina had passed. There were a few downed tree

* Originally published in 31 Thurgood Marshall L. Rev. 531 (2006) with additional source attribution.
** Many friends and colleagues read earlier drafts of this chapter and kept me stable through my own experiences, however comparatively trivial, with Hurricanes Katrina and Rita. I hope they will understand if I thank them collectively and use this space exclusively to honor and dedicate this chapter to the brave women of Violet, Louisiana, whose unbreakable spirit I could only hope to emulate.

branches and a few puddles on the lawn, but the neighborhood was intact. Violet was not destroyed by the hurricane. But it was destroyed. After the storm was over, police cruisers and fire trucks rolled down the streets with bullhorns warning residents to take cover: "We're about to open the floodgate."[1]

Rechelle was cooking lunch for her family when she heard the ominous warning. Within minutes a violent surge of water burst through her house. Before she could react, the water was knee deep. It was waist high by the time she made it to the living room to save her daughter. She knew her elderly neighbors in the home across the street would never make it out alive.

Errolyn saw the warning about the floodgates on the television news ticker. The water came so fast she nearly drowned before she could reach the bedroom to warn her sister. Neither of them could swim, so they clung breathlessly to a floating dresser that had overturned in the surge. Water reached the ceiling within minutes but neither of them had the strength to punch a hole to the roof. They were lucky, though. Errloyn's brother and her sister's boyfriend were home. They punched the hole. For two nights the four of them waited to be rescued from the roof. They weren't. Rescue choppers flew by so close they could see the crew smile. But the pilots did not stop. Finally, out of food, out of water, sweltering in the rotten stench, men from the neighborhood took matters into their own hands. They jumped from the roofs, waded through the sewer below, commandeered a canoe from Tim's Marine, and went house to house picking up stranded survivors. Errolyn was one of them. She was six months pregnant.

The men ferried hundreds of stranded survivors to the dry levy behind St. Bernard Highway, across town from the opened floodgate. There they waited for days, again, no food, no water. A "big house" owned by a white family sat just beyond the levee. The owners invited the evacuees to take refuge from the scorching sun under their patio roof. But they weren't allowed inside the house. Coast Guard choppers eventually arrived to drop off bottled water. None of the pilots spoke with the stranded black survivors. They spoke only to the white woman who owned the "big house." They gave her the bottled water and told her to pass it on to the rest.

All five women eventually came to Houston as part of the massive evacuation from New Orleans. For weeks, they set up camp at the Houston Astrodome and Reliant Arena with tens of thousands of other evacuees. Some they knew. Some they did not. They now owned only what they wore on their back. Nothing else. Even that was torn, dirty, and reeked of stench. They slept one next to the other alongside unknown evacuees in rows of cots that stretched the length of a cement amphitheater. It was hot. It was cold. It was loud. It was dirty. It stunk. There were strangers on all sides of them, front and back, right and left. Snoring. Coughing. Pissing. Vomiting. Still, Rechelle was grateful to be alive in the Astrodome and for the angelic charity of individual volunteers who worked to make an intolerable situation—well, survivable. There were many: those who believed her when she said that no one came to rescue them, how they had to rescue themselves; those who listened when she said that Katrina did not destroy Violet, that someone opened the floodgates; those who didn't question. She worried, though, how long she and her daughter would have to stay in the Astrodome, when they could get back to New Orleans. She worried whether her daughter was safe. Men had repeatedly entered the women's restroom and showers, and she knew that it was only a matter of time before they came when her daughter was inside.

Linda Watson and her best friend Barbara Richards stayed at Reliant Arena after moving from the Astrodome when rumors started to surface that women there had been raped and sexually assaulted. They too were indebted to volunteers for humanizing their very inhumane ordeal, those who laughed with them and cried with them and brought toys for their kids. But they had no "next step." Linda and Barbara had two daughters and two grandchildren between them and still had not received a housing voucher from the Federal Emergency Management Agency (FEMA) by the time officials in Houston issued their own orders to evacuate the city in advance of Hurricane Rita. Linda was earning $12,000 a year as a school bus driver in New Orleans, hardly enough to pay for a new apartment, never mind rebuilding in an unknown city. In three weeks' time, Linda, Barbara, and four children in tow had moved from a rooftop, to the Convention Center, to the Astrodome, to Reliant Stadium, and they were now were being told over the public address system to find alternative shelter by the morning or board a bus to Fort Chaffee, Arkansas.

* * *

Multiple vulnerabilities uniquely endanger the lives of women in times of displacement and disaster, particularly poor women of color. Katrina was no exception. Black, white, rich, poor, young, old—the storm showed no one mercy. But Katrina was both a highly racialized and gendered event.[2] New Orleans had a higher proportion of blacks and greater number of women than any other metropolitan area within the hurricane's path.[3] According to the Institute for Women's Policy Research, more than half of the women in the city were single mothers, like Rechelle Carter, Linda Watson, and Barbara Richards, independently responsible for ensuring they and their children survived the storm. These women lived in greater poverty than men, owned fewer assets than men, had less formal education than men, and worked in less lucrative jobs than men. Women in New Orleans also comprised a majority of the elderly population, most of whom were destitute.

Physical limitations, economic hardship, and dependent care responsibilities made it impossible for many women to escape rising flood waters and compounded the difficulty of securing housing and employment once the storm took these things from them.[4] For this reason, black women, including Errolyn Warden, Brianna Cater, and thousands of others like them, made up the majority of those trapped in New Orleans during Katrina and confined to evacuation centers after the storm passed.[5] The resulting forced isolation—whether in an abandoned city or the anonymous vacuum of a mass evacuation shelter—amplified the violence of Katrina by subjecting these women to an unprecedented risk of rape and sexual assault. Though no woman is immune from sexualized violence following a natural disaster, dimensions of race and socio-economic status converged in Katrina, as they typically do, to increase the risk for black women.[6]

The sexualized violence black women disproportionately experienced at the hands of both intimate partners and strangers, both during and after the storm, is being submerged in the narratives arising out of New Orleans. This is in part reflective of most postdisaster discourse that marginalizes the experiences of women.[7] It is in part the product of patriarchy that promotes sexual exploitation with the aim of subordinating women.[8] It is in part proof that racism intersects with patriarchy to propagate stereotypes of black women as deceitful and promiscuous to an extent that denies their vulnerability to sexual assault.[9]

This essay testifies to structural inadequacies in the official Katrina relief effort that contributed to the sexual exploitation of black women during and after the storm. Part I discusses false reports of violence coming out of New Orleans that made it difficult to bring attention to the real risk of rape and sexual assault. Part II explains why rape, an already under-reported crime, was even less likely to be reported during and after Hurricane Katrina. Part III discusses the proven risk of sexual assault during natural disaster and exposes the tragic oversights that brought these risks to bear during Katrina and its aftermath. This essay concludes that race, sex, and class played a role in the government's failure to adequately predict, prevent, and respond to this risk of sexualized violence during Katrina.

No doubt this small effort to speak truth to power will meet resistance. Black women who speak out on their own behalf will be denied, disparaged, and disbelieved. White women and other allies who stand in solidarity with black women will be doubted, too, though we will be silenced in a different way. Charges of racism will be levied against those of us who seek to expose how sexualized racial violence was permitted to flourish in the aftermath of Hurricane Katrina. We will be blamed for mythologizing stereotypes of black men when we protest the rape of black women. We will be accused of conspiring a divide-and-conquer strategy that pits black men against black women in the service of a white racist patriarchy. Most ferocious in assailing us will likely be the very same reporters, politicians, media analysts, and law enforcement officials who themselves shamelessly disparaged black men through sensationalized violence after the storm. Still, the charges will sting. They are meant to. These charges divert attention away from the aim of this project, which is not to perpetuate black male stereotypes but to expose the indifference government officials displayed towards black women in failing to anticipate, prevent, and respond to the predictable epidemic of forced sex Katrina would bring to them. In short, we will be called racist for exposing racism in America. And still we bear witness.

PART I: CRYING WOLF

Raising awareness about sexual assault is never an easy task, but false reports of violence that erupted in the wake of Hurricane Katrina make it even more difficult to convince observers that allegations of rape and sexual assault taking place during and after the storm are legitimate.

Hurricane Katrina bulldozed the Gulf Coast of the United States on August 29, 2005. Leading up to the storm, the nation's attention was focused on New Orleans, a city of 1.2 million people situated below sea level directly in Katrina's path. Thousands of evacuees were ordered into the New Orleans Superdome and Convention Center without being told that the shelters did not have enough food or water. Within days supplies ran out, and desperate evacuees began to leave the shelters looking for a way to escape the city. They were not alone. Thousands of other survivors stranded on rooftops descended into the flood waters, taking it upon themselves to find a means to survive.

Reports of lawlessness spread like wildfire: thieves bulldozed a pharmacy, looters raided Wal-Mart, rioters sacked Baton Rouge, and gangsters waged open warfare in streets that had not flooded in the storm. The *Time-Picayune* reported that up to forty decomposing bodies were stuffed in freezers in the Convention Center, one of them being a 7-year-old girl "with her throat cut."[10] A later report told of rescue helicopters taking on sniper fire from stranded survivors perched on a rooftop. The *New York Times*, the *Los Angeles Times*, the *Chicago Tribune*, and every other major news outlet in the country repeated these accounts without verifying their accuracy. So, too, did media outlets across the globe. BBC News, the *Ottawa Citizen*, *The Australian* and *Al Jazeera* all reported how black evacuees stranded by Hurricane Katrina had descended into a state of madness. The unprecedented destruction and desperation of the storm dovetailed with, and seemed to explain, these accounts as survivors struggled to find food, water, and bodies of loved ones in a city Katrina had turned into a toxic wetland.

Confirmation by public officials made unverified reports of carnage and depravity all the more believable. Mayor C. Ray Nagin publicly pleaded for federal intervention to help evacuate people out of New Orleans: "They have people standing out there, have been in that frickin' Superdome for five days, watching dead bodies, watching hooligans killing people, raping people."[11] Appearing on the *Oprah Winfrey Show* on September 6, New Orleans Police Chief Eddie Compass reported on the conditions in the shelters: "We have individuals who are getting raped; we have individuals who are getting beaten. We had little babies in there, some of the little babies getting raped."[12]

The picture of New Orleans that ultimately emerged after the storm is now very different than what was first reported. Dead bodies were never found in the Convention Center freezers, and though it remains unclear whether helicopters in New Orleans took fire during the initial rescue effort, officials from the U.S. Air Force, Coast Guard, and Department of Homeland Security could not immediately confirm such incidents. The Louisiana State Department of Health and Hospitals reported in late September that ten people died at the Superdome, four at the Convention Center. Only two of those deaths appeared homicide related. Thankfully, no child was found with a slit throat.

Racism undoubtedly contributed to reports of widespread violence after Katrina; this in turn perpetuated invidious stereotypes about black storm victims, particularly black men. As one commentator noted, unverified accounts of black criminal behavior were taken as true because observers were "predisposed to accept the worst about poor, black flood victims." The aggregate of these initial reports coalesced into a single master narrative of mostly poor, mostly black flood victims "portrayed as beasts, raping and killing one another and even shooting at rescue workers trying to save them."[13]

This epidemic of false reporting reified criminal stereotypes about black men. Now, against this backdrop, those who shamelessly sensationalized a post-Katrina crime wave are eager to deflect charges of racism by demanding hard evidence, eyewitness testimony, or "official reports" before acknowledging an episode of Katrina-related violence, particularly sexual assault. Yet to presumptively discredit what is an objectively predictable form of violence against women—rape in times of natural disaster—reflects its own brand of racism. It relies on racism against black men to excuse the structural inadequacies of the relief effort that contributed to the sexual victimization of black women. It perpetuates stereotypes about black women—that they are deviant, deceptive, and promiscuous, and therefore to blame for their own victimization. It discredits the heroic efforts of the many black men who no doubt fought against individual predators and simultaneously shifts attention from those government officials whose malicious indifference ultimately provided those predators with an opportunity to assault black women. To deny the violently sexualized reality of Katrina on account of previous false reporting only compounds the horror of the storm for both black men and black women.

PART II: RAPE AND REPORTING

To deny that black women faced a disproportionate risk of sexual violence during Hurricane Katrina naively overlooks what is, at bottom, an unexceptional female life experience. According to the U.S. Department of Justice, nearly 18 million women are forcibly raped at some point in their lives.[14] A woman in this country is raped or sexually assaulted every 2.5 minutes,[15] most often at the hands of an acquaintance or intimate partner.[16] Whether characterized by violence, sex, or both, the sadistic subordination of rape and sexual assault perpetually endangers the everyday lives of women, especially women of color.[17]

Against this reality it is incomprehensible to suggest that sexual assault was anything but commonplace among black women already ravaged by Katrina's deadly strike on Louisiana and the other Gulf Coast states. Unlike acts of murder or other forms of alleged Katrina violence that can be discredited in the absence of hard evidence, the same cannot be said of rape and sexual assault. Rape scars are not always visible on the surface, and the emotional signs of sexual assault—anxiety, depression, sleeplessness—are indistinguishable from the posttraumatic stress induced by the death, destruction, and displacement of a natural disaster.[18] Sexual violence would compound the trauma of Katrina but would not be immediately apparent to an untrained eye.

Nor does the lack of an "official report," eyewitness testimony, or confirmation by police discount the violently sexualized aftermath of Katrina, as some news accounts and public officials suggest.[19] Experts agree that even in times of social stability less than one third of all rapes and sexual assaults are reported to police.[20] The majority of rape survivors never tell anyone about the incident. The intimate nature of rape and sexual assault, coupled with feelings of guilt, shame and stigmatization, make it extraordinarily difficult to track these crimes even under normal circumstances. Indeed, the worldwide attention paid to high-profile accounts of sexual violence, such as that of singer-songwriter Charmaine Neville, might have legitimized the experience for women subject to similar violence, but it exposed the risks of disclosure for those victims emotionally unprepared to relive the nightmare of assault.[21]

Witnesses to sexual violence also struggle to process the stigma, guilt, and helplessness that keeps victims themselves from reporting, and they are also not likely to come forward with an accusation of abuse in the immediate aftermath of a crisis when focused on their own survival and that of their loved ones.

Indeed, few women should be expected to take on the psychological stress of coming forward to reveal a rape when displaced by a natural disaster. This is particularly true for those women whose motivation for reporting might be retribution against an assailant. The likelihood of a sexual predator spending any amount of time behind bars is negligible. When a rape is reported, there is only a 16 percent chance the rapist will go to jail. Factor in unreported rapes and fifteen out of sixteen rapists will walk free. Only one will spend a day in jail.[22] Moreover, research suggests that perpetrators who sexually assault black women are least likely to be prosecuted or convicted.[23] The prospect of prosecution and punishment, therefore, is neither a significant deterrent against an act of sexual aggression nor an incentive for black women to report one.

The government's botched relief effort is itself in part responsible for the low number of official rape reports immediately following Katrina. Inadequate preparation and inexcusable delay showed government officials to be wholly indifferent to the survival needs of the mostly poor, mostly black residents of New Orleans hardest hit by the storm. Though scientists for years had predicted the destruction of New Orleans in the wake of even a category 3 hurricane,[24] Louisiana officials had no preestablished plan in place to evacuate the city's most vulnerable residents. Once the storm hit, state officials failed to marshal all available resources to accelerate evacuations and made an unwise and dangerous choice in directing stranded residents to either the Superdome or Convention Center as a refuge of last resort without an adequate supply of food or water.[25] Governor Kathleen Blanco's failure to make good on her promise to expedite evacuations using hundreds of available school buses showed her to be either incompetent or apathetic to the 150,000 residents who could not afford their own transportation out of New Orleans.[26] Convinced that government officials cared not whether those trapped inside New Orleans lived or died, why would black women trust law enforcement on the ground to respond to the victimization they experienced during or after the storm?

The federal government's deadly nonresponse likely killed whatever trust in government officials anyone should have had following the storm. Katrina made landfall on August 29, but as early as August 26, the likelihood of "unprecedented cataclysm" had been predicted.[27] President Bush nonetheless refused to convene a task force to coordinate a federal response until August 31, four days after Louisiana was placed under a state of emergency. That same day Bush told the nation, "I don't think anyone anticipated the breach of the levees."[28] Yet even before the storm made landfall, in addition to the public reports warning of catastrophic devastation, the Director of the National Hurricane Center personally briefed Bush and other high-ranking security officials that Katrina would overwhelm the deteriorating levee system meant to keep flood waters out of New Orleans.[29]

The Department of Homeland Security was also aware. Still, FEMA director Michael Brown waited three days before formally requesting that the Department dispatch rescue personnel to the Gulf,[30] and though the Louisiana National Guard requested 700 buses to speed evacuations, FEMA sent only 100.[31] All the while, on a Navy vessel stocked with hospital beds, food, and supplies, doctors and rescue workers waited offshore for dispatch instructions that never came.[32] The President vacationed in Crawford, Texas and posed for a birthday photo-op; the Secretary of State enjoyed a Broadway show in New York City; and the Secretary of Defense took in a San Diego Padres baseball game.[33] For high-ranking national officials, first-class entertainment appeared more important than saving the lives of destitute black hurricane victims abandoned in New Orleans. Surely such indifference at the federal level did little to persuade black women to trust the government officials and law enforcement agents they encountered in New Orleans.

Indeed, at the local level, allegations of racism and rampant corruption in the New Orleans Police Department led black residents of the city to distrust law enforcement long before Katrina hit. The devastation created by the storm only exacerbated existing tensions. Police officers themselves lived and worked in New Orleans and witnessed their own loved ones and livelihoods wash away in the storm. Looting among some members of the New Orleans police force was caught on video. Others abandoned their posts in the midst of the relief effort. At least two committed suicide.[34] Widespread reports of New Orleans descending into chaos, though now largely discredited, at the time prompted Governor Blanco to announce on television that local law enforcement agents deputized by the state "have M-16s and are locked and loaded. These troops know how to shoot and kill and I expect they will."[35] This hysteria influenced how law enforcement officials perceived and behaved towards black residents stranded by the hurricane, and it no doubt contributed to several shoot-outs, a videotaped beating, and at least one post-Katrina fatality at the hands of the police.[36]

Police from New Orleans were not the only law enforcement agents black citizens had reason to fear. To escape the intolerable conditions in the Convention Center, an estimated 800 black residents of New Orleans attempted to cross the Crescent City Bridge into neighboring Gretna where they were told buses would be waiting to take them to a safe shelter.[37] But before they reached the other side, members of the local sheriff's office halted them at gunpoint and turned them back to New Orleans. These experiences surely discouraged black women from engaging the help of law enforcement officials they already considered hostile to black men and distrustful of black women even before Katrina confirmed that impression.

Despite all this, some women did bravely attempt to elicit the help of law enforcement in response to rapes that did occur. Overwhelmed by the storm, however, some law enforcement officials were either unable or unwilling to take reports. When a forty-six-year-old home health care worker was raped at gunpoint in an abandoned New Orleans apartment building on the first night of the storm, a rape subsequently verified by a forensic nurse, members of the local National Guard refused to take her report a week later when they finally came with guns drawn to evacuate the building. The police were "stressed out themselves," the woman later recounted. "They didn't have no food. They didn't have water. They didn't have communication. They didn't have ammunition. The National Guards didn't want to hear it."[38]

Rescue operations taking place in cities outside of New Orleans also made it impossible for some Katrina evacuees to report a sexual assault that occurred in the

Superdome, the Convention Center, or anywhere else on the Gulf Coast for that matter. For the first several weeks following Katrina, police in Houston's major evacuation centers refused to take reports from shelter evacuees about sexual assaults that occurred outside their jurisdiction because, as a spokesperson for the police department explained, "there was no way to code reports from Katrina evacuees in the Department's computer system."[39] Though such "courtesy reports" are routinely collected to coordinate law enforcement activity, officers in Houston did not start doing so until September 13, after this appalling administrative defect was revealed in the press. In the meantime, the spokesperson stated, "Anything that happened in New Orleans [would] be reported to New Orleans" for possible investigation in the future.

Under these conditions, it is inconceivable to require a woman to produce an eyewitness, police officer, or "official" report to confirm that she was raped. Women raped in New Orleans were told in Houston to report the assault to law enforcement officers in a city whose infrastructure was submerged in water. In New Orleans itself, local police had either abandoned their post or remained on duty with orders to shoot on sight. Six regional domestic violence shelters were closed because of hurricane damage. Five were destroyed.[40] In the wake of Katrina to whom, exactly, were rape victims supposed to report?

PART III: RAPE AND NATURAL DISASTER

Law enforcement officials shrewdly denounced an epidemic of violence following the storm,[41] and as of December 21, 2005, the official number of post-Katrina rapes and attempted rapes stood at four.[42] What should have been apparent at the time is that additional rapes would eventually be reported. Predictably, they have.

The National Sexual Violence Resource Center has partnered with the Louisiana Foundation Against Sexual Assault and four other Gulf Coast rape crisis centers to track the number of unreported post-Katrina rapes through an internet database accessible to medical practitioners, social service providers, crisis counselors, and sexual assault victims. In its first six weeks of operation, and with almost no publicity, the group received forty-two reports of Katrina-related sexual assaults that occurred both inside and outside of New Orleans, including a disproportionate number of gang rapes and stranger rapes.[43] Witness Justice, a nonprofit victim services organization, received 156 reports of Katrina-related violent crimes in the first few days after the storm. About one third of those involved sexual assault. That number no doubt continues to grow.

The experience of the women who reported these crimes is entirely consistent with the violently sexualized aftermath of mass displacement worldwide. Reported sexual assaults rose 300 percent following the 1990 earthquake in Lome Prieta, California.[44] Similar trends were reported in the wake of the 2005 tsunami that pummeled South East Asia and the horn of Africa. The eruption of Mount Pinatubo in the Philippines, and Mt. St. Helens in Washington state triggered a spike in violence against women, as did the massive 1998 Canadian ice storm, the 1995 Exxon Valdez environmental oil disaster, and every one of the four major hurricanes that pummeled Florida in 2004.[45]

The reality of gender specific violence, particularly sexual assault, is so predictable during times of catastrophic upheaval that major human rights instruments that address the needs of refugees and displaced persons—including the Beijing Declaration and Platform of Action, and the United Nation's Guiding Principles on Internal Displacement—presume a heightened risk to women. The World Health Organization advises relief agencies, health workers, and other field staff working with displaced populations to "assume that sexual assault may be a problem unless they have conclusive proof to the contrary." The United Nations Population Fund advises that emergency responders treating refugees and others subjected to refugee like conditions "should act on the assumption that sexual violence is a problem, unless they have conclusive proof that it is not the case." The United Nation's Inter-Agency Standing Committee instructs relief workers that preventive measures should be implemented in the earliest stages of a disaster "regardless of whether the 'known' prevalence of sexual violence is high or low." Because of the difficulty in accurately measuring the magnitude of sexual violence in an emergency, "all humanitarian personnel should therefore assume and believe that [gender-based violence], and in particular sexual violence, is taking place and is a serious and life-threatening protection issue, regardless of the presence or absence of concrete and reliable evidence."

These organizations recognize that hardships created by displacement following natural disaster precipitate conditions ripe for sexual assault and violence. In his opening remarks to the Guiding Principles on Internal Displacement, Francis Deng, the Special Representative of the United Nations Secretary General emphasized, "Displacement. . .breaks up families, cuts social and cultural ties, terminates dependable employment relationships, disrupts educational opportunities, denies access to such vital necessities as food, shelter and medicine, and exposes innocent persons to such acts of violence as attacks on camps, disappearances and rape. . . .The internally displaced are among the most vulnerable populations, desperately in need of protection and assistance."[46]

Experts in the gendered dynamics of natural disaster agree that the destruction and disorder created by Katrina undermined social constructs of masculinity in a way that rendered women vulnerable to rape and violent assault. "Men without jobs and those unable to save family members and other victims may feel unmasked and unmanly. . . .Some men will cope through drugs, alcohol, physical aggression or all three, hurting themselves and putting the women and girls around them at risk."[47] In Katrina, sexual violence provided a mechanism for reasserting control and reestablishing patriarchy upset by social instability.

Natural disasters are particularly threatening to women when they obstruct law enforcement capabilities and destroy familial arrangements that naturally deter aggression. Katrina rendered large portions of New Orleans inaccessible to law enforcement officials and rescue workers. Communications breakdowns, logistical obstacles and personnel shortages delayed even volunteer efforts to rescue stranded survivors in those few areas that were marginally accessible. In the isolation of disaster, sexual abuse at the hands of a spouse or intimate partner becomes much more likely. At the same time, a disproportionate number of stranded survivors were single women, including those widowed or abandoned in the storm. Left to fend for themselves and their children without the protective buffer of a male spouse or partner, these women too become targets for assault from strangers who conceived of them as common property.[48]

Moreover, gendered economic disparities, childcare responsibilities, and divisions of labor render women increasingly dependent on abusive partners for support in times of crisis. This situational vulnerability undermines the ability of poor women to deny consent to sex and reduces the chance that a woman will report a sex crime that does occur.[49] It also increases the likelihood that a woman trapped by disaster will be forced to barter sex with men in exchange for food, shelter and children's needs.[50] An undetermined number of women from New Orleans will be forced into prostitution as a result of Katrina.[51]

Katrina forced these conditions not only on black women and their children trapped by rising floodwaters in New Orleans, but also on those stranded in sweltering, overcrowded shelters where they were forced to sleep and breathe in a miasma of human waste and disease. This is why leaders of the National Sexual Violence Resource Center and other anti-violence advocates called on government officials to prioritize evacuations out of the shelters as early as September 2.[52] Judy Benitez, Executive Director of the Louisiana Foundation Against Sexual Assault, warned that steps should immediately be taken to avoid sexual violence in light of the powerlessness and alienation evacuees were experiencing. Speaking of the inhumane conditions at the Superdome, Benitez warned:

> It is no secret that in our society, some people are strong and some are weak. Some of the strong help those who are weaker—and some prey on them. The animal-like circumstances of the evacuees in the Superdome—conditions in which no human being should ever have to live—caused frustration on a level that most people will never know. That sense of helplessness, lack of control, and powerlessness would make most people angry; for predatory people, the availability of someone over whom they can have power and control, on whom they can take out their anger, is all the excuse they need to commit rape.[53]

Shelters in Houston were only marginally safer than those in New Orleans, if at all. A spokesperson for the National Alliance to End Sexual Violence confirmed that "the Astrodome, while convenient and a first-step towards the goal of evacuation, [is] less than ideal. These large-scale shelters can become dangerous settings for so many crimes. We have already heard of many rapes being committed among those still waiting to be rescued and among those in the refugee areas."[54]

The population mix in the mass evacuation shelters established to house Katrina evacuees made them particularly dangerous settings for women. In anticipation of the storm, more than 15,000 registered sex offenders from New Orleans and surrounding parishes were sent to open shelters along with tens of thousands of other evacuees.[55] Though some states require that sex offenders be housed in separate facilities, no apparent effort was made in New Orleans either at the Superdome or Convention Center to segregate sex offenders from the general population during the initial evacuation effort. Nor did officials in Houston monitor initial access to the four shelters there. Privacy concerns prompted federal officials to withhold the identity of registered sex offenders from state and local officials operating shelters outside of Louisiana.[56] The absence of an accessible national database made it impossible to identify known sex offenders even among the understandably few evacuees who would have been able to show any form of identification. High rates of recidivism particular to untreated sex offenders, coupled with the increased stress created by the disaster itself, rendered the environment within the mass shelters particularly volatile for the many black women housed there.

Conditions within the mass evacuation centers that I witnessed firsthand in Houston aggravated the risk of assault by a sexual predator.[57] The people who came to Houston arrived on hot, crowded buses after spending days upon days trapped on a rooftop or packed into a distressed shelter in New Orleans. Many were elderly, many were sick. By the time the first few buses arrived, basic medical supplies had already run out. Several evacuees approached me begging for insulin to stave off diabetic shock—an elderly man, a sixty-year-old woman, a ten-year-old boy. The medical team had none, and together we tried to compensate with orange juice and chocolate bars that my colleague and I had brought anticipating a shortage of supplies.[58]

Inside the Astrodome, evacuees were promised a cot to sleep on. There were not enough. Blankets, sheets and pillows had either run out or were not being distributed. Clogged toilets spewed feces and other waste onto the bathroom floors that bare-footed evacuees tracked into the hallways, spreading disease where babies wrapped in sopping, soiled diapers slept on hard, cold, dirty concrete. The inhumanity of these conditions undoubtedly created frustrations that a sexual predator would unleash on a more vulnerable evacuee.

Though shelter conditions improved over time, their physical layout was inherently dangerous. International relief agencies recommend against "shared communal living space with unrelated families" in order to reduce the likelihood of sexual exploitation among displaced populations.[59] Despite this advice, women had no option but to share sleeping space with strangers while sheltering in the Astrodome and Reliant Park, as they did in the Superdome and Convention Center in New Orleans. This oversight was particularly troublesome in Houston given that protective arrangements were provided to families sheltered in the George R. Brown Convention Center, a downtown facility connected by skywalk to the Hilton of the Americas that operated its services under the watchful eye of international visitors.

Security lapses created additional problems. "The sheer size of the shelters and their many hiding places, coupled with a lack of lighting due to power outages, makes them less than ideal for emergency housing," said a representative for the Red Cross, the organization in charge of coordinating Houston's relief effort.[60] Though shelters were inexcusably overcrowded, they nonetheless provided ample opportunity for violence to take place without notice, especially in the most crucial hours of the evacuation when volunteers focused on rescue and relief over sex and security. In the Houston Astrodome, dark hallways, open storage rooms, and a labyrinth of secluded bleachers provided plenty of opportunity for a would-be rapist to perpetrate an assault beyond sight and out of earshot. A colleague and I proved as much by photographing these hotspots without ever being detected by law enforcement officers, evacuees or other volunteers. Broken furniture, open liquor bottles, the pungent stench of urine and feces and, most disturbingly, a wet, discarded baby's bunting, made it clear that someone had located these areas before we had. Surely one intent on perpetrating an act of violence could have surreptitiously accessed these areas, if they had not already.

Intermittent monitoring of restrooms and shower areas created additional dangers for women and girls housed in evacuation centers. At Reliant Arena, access to both male and female shower stalls was through a single entrance that opened to opposite facing hallways-showers for men to the left, showers for women to the right. Law enforcement officials monitored this area only sporadically. When they did, they stood outside the open entryway facing the shelter's main living area with their back towards the hallways that led to the shower stalls. Men and boys entering the showers took advantage of this arrangement by making a quick turn behind the officers, down the right-facing hallway, into the women's bath. I observed this on more than one occasion and chased down several boys trying to sneak into the women's showers. This was more than child's play. Many female evacuees I spoke with at Reliant Arena, as well as other shelters in Houston, refused to enter the showers because men and boys had made their way inside. The same was true for female restrooms. A girl about twelve years old recounted to me that a man opened the door to the ladies room and attempted to pull her into the men's room where she could see a number of other men waiting behind the open door. She was able to escape before the man grabbed her.

PART IV: CONCLUSION

Gendered aspects of inequality endanger all women in a natural disaster. They cruelly intersected with race and class to particularize the danger for black women trapped in New Orleans during Hurricane Katrina and housed in Houston's shelters following the storm. But more than situational vulnerability is to blame for the circumstances black women faced following Katrina.

Government officials inexcusably failed to anticipate and prevent hurricane-related sexual violence throughout the evacuation and sheltering process. They also failed to respond when concerns about sexual assault were brought directly to their attention. More than once I personally informed law enforcement officials in Houston that women had relayed to me their fear of vulnerability, that they or other women they knew had been harmed. Yet these concerns were dismissed. One officer acknowledged that he too "heard rumors about sexual assault," but could do nothing until an attack was "officially reported." Another blithely conceded that "anywhere you have 25,000 people, sexual assaults are bound to happen," so there was little he could do to prevent them from occurring. Only after photographs of security lapses inside the Astrodome were posted on the internet did law enforcement officials in Houston block access to at least some secluded rooms and passageways.

To me, there is no doubt that the bulk of this indifference was influenced by race—by the fact that a majority of the evacuees sent to Houston from New Orleans were black—and poor. Just as quickly as law enforcement officers and other shelter officials dismissed my concerns for the security of evacuees, they queried whether I had been "attacked," whether I felt "unsafe." These exchanges became routine, and I suspect that similar exchanges with white volunteers took place in New Orleans as well. I wonder how they would have responded had I answered yes to their questions. I wonder if they understood, as I did, how racism, conscious or otherwise, influenced their interactions with me, how it triggered a spontaneous emotional concern for the well-being of a white woman while at the same time dampened their willingness to acknowledge, much less address, the very real risk of rape and sexual assault facing thousands of stranded black women. Perhaps they cannot see, and would never admit, the pernicious influence of sex, class and race on the decisions they made, and action they neglected to take, during Katrina. Yet some of us know. Because this we witnessed.

NOTES

1. Several evacuees recounted to me how Violet was destroyed by waters released by the floodgates. See also Cameron McLaughlin, "New Orleans: Deliberate Act of Sabotage Was the Opening of Floodgate," *Global Research*, September 11, 2005, http://www .globalresearch.ca/PrintArticle.php?articleId=930; "To ABC's Surprise, Katrina Victims Praise Bush and Blame Nagin," *Cyberalert*, September 16, 2005, http://www .mrc.org/cyberalerts/2005/cyb20050916.asp.

2. Elaine Enarson, "Women and Girls Last? Averting the Second Post-Katrina Disaster," *Social Science Research Council* (November 15, 2005), http://understandingkatrina.ssrc .org/Enarson; Joni Seager, "Natural Disaster Exposes Gender Divides," *Chicago Tribune*, October 3, 2005, http://www.thedailytimes.com/sited/story/html/21906.

3. "The Women of New Orleans and the Gulf Coast: Multiple Disadvantages and Key Assets for Recovery," *Institute for Women's Policy Research*, October 2005, http://www.iwpr.org/pdf/D464.pdf; John Logan, "The Impact of Katrina, Race and Class in Storm-Damaged Neighborhoods," *Brown University*, 2006, http://www.s4 .brown.edu/katrina/report.pdf; Member scholars of the Center For Progressive Reform, "An Unnatural Disaster: The Aftermath of Hurricane Katrina," *Center For Progressive Reform*, 2005, http://www.progressivereform. org/Unnatural_Disaster_512 .pdf.

4. Joni Seager, "Natural Disaster Exposes Gender Divides," *Chicago Tribune*, October 3, 2005, http://www.thedailytimes.com/sited/story/html/219062; Loretta J. Ross, "A Feminist Perspective on Katrina," *Sister Song*, October 10, 2005, http://zmag.org/ content/showarticle.cfm? SectionID=72&ItemID=8912. See also Elaine Enarson, "Surviving Domestic Violence and Disasters," *The FREDA Center for Research on Violence Against Women and Children*, http://www.harbour.sfu.ca/freda/reports/ dviol.htm.

5. "Survey of Hurricane Katrina Evacuees," *The Henry J. Kaiser Family Foundation*, September 2005, http://www.kff.org/newsmedia/upload/7401.pdf.

6. Shelby A. D. Moore, "Understanding the Connection Between Domestic Violence, Crime and Poverty: How Welfare Reform May Keep Battered Women from Leaving Abusive Relationships," *Texas Journal of Women and the Law* 12 (2003): 455; Adriene Katherine Wing, "A Critical Race Feminist Conceptualization of Violence: South African and Palestinian Women," *Albany Law Review* 60 (1997): 948.

7. Alice Fothergill, "The Neglect of Gender in Disaster Work: An Overview of the Literature," in *The Gendered Terrain of Disaster: Through Women's Eyes*, ed. Elaine Enarson and Betty Hearn Marrow (Westport, CT: Praeger, 1988), 11–25.

8. Mary Ann Becker, "The Prostitution of Sexuality," *De Paul Law Review* 52 (2003): 1043–44.

9. Marilyn Yarbrough and Crystal Bennett, "Cassandra and the 'Sistahs': The Peculiar Treatment of African American Women in the Myth of Women as Liars," *Journal of Gender, Race and Justice* 3 (2000): 625, 634–55; Lisa A. Crooms, "Speaking Partial Truths and Preserving Power: Deconstructing White Supremacy, Patriarchy, and the Rape Corroboration Rule in the Interest of Black Liberation," *Howard Law Journal* 40 (1997): 459, 469–90; Sarah Gill, "Dismantling Gender and Race Stereotypes: Using Education to Prevent Date Rape," *U.C.L.A. Women's Law Journal* 7 (1996): 27, 40–46; Morrison Torrey, "When Will We Be Believed? Rape Myths and the Idea of a Fair Trial in Rape Prosecutions," *U.C. Davis Law Review* 24 (1991): 1013, 1053–55.

10. Brian Thevenot, "Bodies Found Piled in Freezer at Convention Center," New Orleans *Times-Picayune*, September 6, 2005, http://www.nola.com/newslogs/tporleans/index .ssf?/mtlogs/nola_tporleans/archives/2005_09_06.html; "New Orleans Mayor Orders Looting Crackdown," *MSNBC News*, September 1, 2005, http://www.msnbc.msn .com/id/9063708; Robert E. Pierre and Ann Gerhart, "News of Pandemonium May Have Slowed Aid," *Washington Post*, October 5, 2005.

11. Martin Savidge, "Separating Fact From Fiction in Katrina's Wake," *MSNBC News*, September 27, 2005, http://www.msnbc.msn.com/id/9505408.

12. Michelle Roberts, "Reports of New Orleans Mayhem Probably Exaggerated, Police Say," *Houston Chronicle*, September 28, 2005, http://www.freerepublic.com/focus/f-news/1492418/posts; Martin Savidge, "Separating Fact From Fiction in Katrina's Wake," *MSNBC News*, September 27, 2005, http://www.msnbc.msn.com/id/ 9505408/.

13. Brian Thevenot, "Myth-Making in New Orleans," *American Journalism Review*, December 2005/January 2006, http://www.ajr.org/Article.asp?id=3998; see also Danny Duncan Collum, "Reporting Through the Grapevines," *Sojourners* (January 2006): 40.

14. Patricia Tjaden and Nancy Thoennes, U.S. Department of Justice, Office of Justice Programs, *Extent, Nature, and Consequences of Rape Victimization: Findings from the National Violence against Women Survey*, Pub. No. NCJ 210346 (2006), 1.

15. Rape, Abuse & Incest National Network, http://www.rainn.org/statistics/minutes.html.

16. Shannon M. Catalano, U.S. Department of Justice, Bureau of Justice Statistics, *Criminal Victimization 2004*, Pub. No. NCJ 210674 (2004); Patricia Tjaden and Nancy Thoennes, National Institute of Justice, *Full Report of the Prevalence, Incidence and Consequences of Violence against Women*, Rept. NCJ 183781 (2000); Ida M. Johnson and Robert T. Sigler, *Forced Sexual Intercourse in Intimate Relationships* (New York: New York University Press, 1997), 35–49.

17. Despite statistically insignificant differences in the absolute number of rapes and sexual assaults against black and white women in the general population, statistics suggest that black women face an increased danger of assault from an intimate partner. C. M. Rennison, U.S. Department of Justice, Bureau of Justice Statistics, *Prevalence and Consequences of Violence against Women Survey, Violent Victimization and Race, 1993–98*, Pub. No. NCJ 176354 (2001).

18. Diane Myers, "Disaster Response and Recovery: A Handbook for Mental Health Professionals," *Empowerment Zone*, 1997, http://www.empowermentzone.com/ disaster.txt.

19. Michelle Roberts, "Reports of New Orleans Mayhem Probably Exaggerated, Police Say," *Houston Chronicle*, September 28, 2005, http://www.freerepublic.com/focus/f-news/1492418/posts; Gary Younge, "Murder and Rape–Fact or Fiction?" *The Guardian*, September 8, 2005, http://www.freerepublic.com/focus/f-news/1478276/ posts; Nancy Cook Lauer, "Rape-Reporting Procedure Missing After Hurricane",

Women's eNews, September 13, 2005, http://www.womensenews.org/article.cfm/dyn/aid/2448.

20. Catalano, *Criminal Victimization*; Mary P. Koss, "Hidden Rape: Sexual Aggression and Victimization in a National Sample of Students in Higher Education," in *Rape and Society: Readings on the Problem of Sexual Assault*, ed. Patricia Searles and Ronald J. Berger (Boulder, CO: Westview, 1995), 35, 42–46.

21. On the impact of victim disclosure, see Patricia H. Davis, "The Politics of Prosecuting Rape as a War Crime," *International Lawyer* 34 (2000): 1223, 1245; Terry Nicole Steinberg, "Rape on College Campuses: Reform Through Title IX," *Journal College & University Law* 18 (1991): 39, 46–47.

22. "Every Two and a Half Minutes," *Rape, Abuse & Incest National Network*, http://www.rainn.org/statistics/minutes.html.

23. Gail Elizabeth Wyatt, "The Sociocultural Context of African American and White American Women's Rape," *Journal of Social Issues* 48 (1992): 77–91; Kimberle Crenshaw, "Mapping the Margins: Intersectionality, Identity Politics, and Violence Against Women of Color," *Stanford Law Review* 43 (1991): 1250–51 and n. 35.

24. Joel K. Boerne, Jr., "Gone With the Water," *National Geographic*, October 2004, http://magma.nationalgeographic.com/ngm/0410/feature5/?fs=www3.nationalgeographic.com; John McQuaid & Mark Schleifstein, "In Harm's Way," New Orleans *Times-Picayune*, June 23–27, 2002, http://www.nola.com/hurricane/index.ssf?/washingaway/harmsway_1.html; Mark Fischetti, "Drowning New Orleans," *Scientific American*, October 1, 2001, http://www.sciam.com/article.cfm?articleID=00060286-CB58-1315-8B5883414B7F0000&sc=I100322.

25. Lara Jakes Jordan, "Docs Show City Opened Shelter Without Food," *Houston Chronicle*, January 31, 2006, http://www.sfgate.com/cgi-bin/article.cgi?f=/n/a/2006/01/31/national/w162827S06.DTL.

26. Brian DeBose, "Blacks Fault Lack of Local Leadership," *Washington Times*, September 10, 2005, http://washingtontimes.com/national/20050909-113107-3180r.htm.

27. Jose Antonia Vargas and Brendan Loy, "The Guy of the Storm," *Washington Post*, October 23, 2005.

28. Joby Warrick, "White House Got Early Warning On Katrina," *Washington Post*, January 24, 2006.

29. Tamara Lush, "For Forecasting Chief, No Joy in Being Right," *St. Petersburg Times*, August 30, 2005, http://www.sptimes.com/2005/08/30/State/For_forecasting_chief.shtml.

30. Ted Bridis, "FEMA Chief Waited Until After Storm Hit," *SFGate.com*, September 6, 2005, http://www.sfgate.com/cgi-bin/article.cgi?f=/n/a/2005/09/06/national/w16331 1D97.DTL. Brown's memo to Department of Homeland Security Chief Michael Chertoff is available at http://wid.ap.org/documents/dhskatrina.pdf.

31. Keith O'Brien, "Chronology of Errors: How a Disaster Spread," *Boston Globe*, September 11, 2005, http://www.boston.com/news/nation/articles/2005/09/11/chronology_of_errors_how_a_disaster_spread/?rss_id=Boston+Globe+-+National +News; Matthew Cella and Robert Redding Jr., "New Orleans Evacuation Lapses Hold Planning Lessons: Officials Likely to Focus Efforts on Those Who Are Poor, Without Vehicles," *Washington Times*, September 9, 2005, http://nl.newsbank.com/nl-search/we/Archives?p_action=list&p_topdoc=111.

32. Stephen J. Hedges, "Navy Ship Nearby Underused," *Chicago Tribune*, September 4, 2005, http://www.chicagotribune.com/news/nationworld/chi-0509040369sep04, 0,3659600.story?track=mostemailedlink.

33. Susan Page, "Hurricane's Aftermath Whips Up New Deal For President," *USA Today*, September 27, 2005, http://www.usatoday.com/news/washington/2005-09-27-bush-roosevelt-cover_x.htm; James Ridgeway, "Crude Manipulation," *Village Voice*, September 13, 2005, http://www.villagevoice.com/news/0537,mondol,67761,6.html; Larry Eichel, "Hesitancy Deadly, Seen at All Levels," *San Diego Union Tribune*, September 11, 2005, http://www.signonsandiego.com/uniontrib/20050911/news_1n 11response.html; *The White House*, http://www.whitehouse.gov/news/releases/2005/08/images/20050829-5_p082905pm-0125-515h.html.

34. Manuel Roig Franzia, "New Orleans Fires 51 Police Officers," *Washington Post*, October 29, 2005; Adam Nossiter, "New Orleans Probing Alleged Police Looting," *Washington Post*, September 30, 2005; Keith O'Brien, "Amid Horror Two Officers Commit Suicide," *Boston Globe*, September 5, 2005, http://www.boston.com/news/nation/articles/2005/09/05/amid_horror_2_officers_commit_suicide/.

35. Felicity Barringer, "Troops Bring Food, Water and Promise of Order to New Orleans," *The New York Times*, September 2, 2005, http://www.nytimes.com/2005/09/02/national/nationalspecial/02cnd-storm.html?ex=1283313600&en=a144858313 da66be&ei=5088&partner=rssnyt&emc=rss.

36. James Rainey, "Doubts Now Surround Account of Snipers Amid New Orleans Chaos," *Los Angeles Times*, November 24, 2005, http://www.informationclearinghouse.info/article11135.htm.

37. Andres Buncombe, "'Racist' Police Blocked Bridge and Forced Evacuees Back at Gunpoint," *The Independent*, September 11, 2005, http://news.independent.co.uk/world/americas/article311784.ece.

38. John Burnett, "More Stories Emerge of Rapes in Post-Katrina Chaos," *National Public Radio*, December 21, 2005, http://www.npr.org/templates/story/story.php?storyId=5063796.

39. Nancy Cook Lauer, "Efforts to Track Rape Emerge Between Hurricanes," *Women's eNews*, September 23, 2005, http://www.womensenews.org/article.cfm?aid=2460.

40. Jordan Flaherty, "Notes From Inside New Orleans," *Left Turn Magazine*, September 2, 2005, http://www.leftturn.org/Articles/Viewer.aspx?id=670&type=W; *Sexual Assault and Trauma Resource Center of Rhode Island*, http://www.satrc.org (accessed February 5, 2006).

41. "Shelter Security Balances Order with Accommodation," *Houston Chronicle*, September 4, 2005, http://www.chron.com/disp/story.mpl/metropolitan/3339312.html.

42. John Burnett, "More Stories Emerge of Rapes in Post-Katrina Chaos," *National Public Radio*, December 21, 2005, http://www.npr.org/templates/story/story.php?storyId=5063796.

43. *Relief Fund For Sexual Assault Victims*, http://relieffundforsexualassaultvictims.org/ (accessed April 4, 2006); "Katrina Rapes Reported; U.S. Cuts Social Programs," *Women's e-News*, December 24, 2005, http://www.womensenews.org/article.cfm/dyn/aid/2583/context/outrage.

44. "Fact Sheet: Women, Natural Disaster and Reconstruction," *Women's Edge Coalition*, http://www.womensedge.org/index.php?option=com_kb&Itemid=91&page=articles &articleid=5.

45. Elaine Enarson, "Violence Against Women in Disasters," *Gender Disaster Network*, http://www.gdnonline.org/resources/VAW%20in%20Disasters%20Fact%20Sheet%202006.doc.

46. Report of the Representative of the Secretary-General, "Guiding Principles on Internal Displacement," Introductory Note, delivered to the UN Commission on Human Rights, UN Doc. E/CN.4/1998/53/Add.2, February 11, 1998.

47. Elaine Enarson, "Women and Girls Last? Averting the Second Post-Katrina Disaster," *Social Science Research Council*, November 15, 2005, http://understandingkatrina.ssrc .org/Enarson.

48. UN Population Fund, *Reproductive Health In Refugee Situations, An Inter-Agency Field Manual* (1999): 37–38.

49. World Health Organization, *World Report on Violence and Health* (2002): 96, 156.

50. *Reproductive Health*, 38.

51. Loretta J. Ross, "A Feminist Perspective on Katrina," *Sister Song*, October 10, 2005, http://zmag.org/content/showarticle.cfm?SectionID=72&ItemID=8912.

52. "Urgent Call to Evacuate Victims of Katrina by Nation's Anti-Sexual Violence Leaders," National Sexual Violence Resource Center, September 2, 2005, http://www .ncdsv.org/images/UrgentCallEvacuateVictimsKatrina.pdf.

53. Lucinda Marshall, "Were Women Raped in New Orleans? Addressing the Human Rights of Women in Times of Crisis," *Dissident Voice*, September 14, 2005, http:// www.dissidentvoice.org/Sept05/Marshall0914.htm.

54. "Urgent Call To Evacuate Victims of Katrina By Nation's Anti-Sexual Violence Leaders," National Sexual Violence Resource Center, September 2, 2005, http://www .ncdsv.org/images/UrgentCallEvacuateVictimsKatrina.pdf.

55. Monica Novotny, "Sex Offenders Scattered after Katrina," *MSNBC.Com*, September 21, 2005, http://www.msnbc.msn.com/id/9426538/.

56. Hugh Aynesworth, "FEMA Slow to ID Sex-Offender Evacuees," *Washington Times*, November 5, 2005, http://www.washtimes.com/national/20051104-111846-7487r.htm; Kevin Johnson, "Sex Offenders Gone Missing After Katrina," *USA Today*, October 12, 2005, http://www.usatoday.com/news/nation/2005-10-12-katrina-sex-offenders_x.htm.

57. Tracy McGaugh and Kathleen Bergin, "It's Been a Long Time Coming," *Whitewashing the Black Storm*, December 8, 2005, http://whitewashingblack.blogspot.com.

58. Leigh Hopper, "Astrodome Medical Volunteers Overwhelmed," *Houston Chronicle*, September 2, 2005, http://www.chron.com/disp/story.mpl/special/05/katrina/33369 27.html.

59. *Reproductive Health*, 39.

60. Nancy Cook Lauer, "Rapes in New Orleans Chaos Were Avoidable," *Women's eNews*, September 4, 2005, http://www.womensenews.org/article.cfm/dyn/aid/2439/context/ cover. Ironically, the representative praised Houston for its response, even though the characteristics of its major relief centers created the very conditions for abuse she warned against.

CHAPTER 13

THE IMPACT OF HURRICANE KATRINA ON THE RACE AND CLASS DIVIDE IN AMERICA

THOMAS J. DURANT JR. AND DAWOOD SULTAN

INTRODUCTION

In the days immediately following the devastation caused by Hurricane Katrina, media attention primarily focused on the rescue and relief operations, the extent of the physical damage, and the slow response by local, state, and federal agencies to one of the worst disasters in recent American history. However, it would not take long before journalists, researchers, and political analysts turned their attention to the social problems and issues created by the destruction that resulted from Hurricane Katrina. Two social issues that recently have been vigorously debated by the media, journalists, academicians, and the general public are whether or not black residents received more exposure to damage from Hurricane Katrina than white residents and whether or not lower-class residents received more exposure to damage than middle- or upper-class residents. These issues are related to a larger issue concerning the extent to which Hurricane Katrina exacerbated the race and class divide in New Orleans, as well as other large American cities in the United States, and what can be done to remedy the problem of inner city poverty exposed by the hurricane.

This chapter focuses on the impact of Hurricane Katrina on concentrated black poverty in New Orleans and on the race and class divide in New Orleans and other large American cities. The implications for policy solutions to the problem of concentrated black poverty and the race and class divide in America are discussed. We argue that there is a continuing race and class divide in America and that this divide was exposed by the devastation caused by Hurricane Katrina in New Orleans. The city of New Orleans was selected as the focus of this chapter because

it affords the opportunity to explore the social effects of a major disaster as a naturally occurring phenomenon and to compare pre-Katrina New Orleans with post-
Katrina New Orleans in order to understand the effects of a major disaster on
poverty, social inequality, and the race and class divide.

A major premise of this chapter is that in order to fully understand the effects of
Hurricane Katrina on the racial and class divide in New Orleans, one must understand the conditions that prevailed in the city prior to the hurricane. In this chapter,
we illustrate that pre-Katrina New Orleans was already on the brink of a "social disaster" and that this powerful storm exposed and exacerbated the problem of chronic,
concentrated black poverty in the city of New Orleans. We also show that impoverished black neighborhoods in New Orleans were highly vulnerable to the disaster
caused by Hurricane Katrina, and, in a sense, these communities were a tragedy just
waiting to happen with the right precipitant. Hurricane Katrina was that precipitant.

DISASTERS IN AFRICAN AMERICAN HISTORY

African Americans have historically been victimized by a variety of disasters (both
natural and manmade). Thus, it is important to place the Katrina disaster in historical perspective before conducting an assessment of race and class in pre-Katrina and
post-Katrina New Orleans. Disasters historically experienced by African Americans
helped to create and perpetuate conditions such as urbanization, poverty, racial segregation and inequality, and social isolation, all of which directly or indirectly contribute to the racial and class divide in America. Moreover, the history of disasters
among African Americans contribute to their perceptions of the importance, relevance, causes, and effects of disasters. In this regard, it is important to understand
the historical effects of disasters, like Hurricane Katrina, on the social and psychological status of African Americans.

The history of African Americans is replete with catastrophic events, ranging
from manmade disasters, such as the enslavement "holocaust" or "maafa" (a word
from Kiswahili, spoken in east Africa, that means "great disaster") to vigilante terrorism, lynchings, race riots, and massacres. To illustrate, between 1889 and 1932,
almost 4,000 people were lynched in the United States, the majority of whom were
African Americans. During the early 1900s, thousands of African Americans were
killed by white mobs and vigilantes, as well as militias, in so-called "race riots."

During the great Mississippi Flood of 1927, many residents of black communities
were forced to build or repair levees and were often the victims of racial discrimination and brutal attacks by white militants who sought to maintain Jim Crow laws of
racial segregation. During this period, a levee was dynamited to protect upper-class
and commercial areas, resulting in flooding and damage to poor black neighborhoods
in New Orleans. Many blacks in New Orleans believed that in 2005, during
Hurricane Katrina, a levee was intentionally breached by city officials in order to save
upper-class mostly white neighborhoods resulting in greater flooding and damage to
black neighborhoods in the Ninth Ward. During the civil rights movement of the
1960s, violence against African Americans was rampant, as blacks fought to eliminate
poverty, racial discrimination, segregation, inequality, injustices, and violence by

whites, especially in large cities across America. These illustrations reveal that African Americans have frequently been plagued with "disasters" over an extended period of their history. The widespread, persistent, and cumulative effects of these disasters over more than a century contributed to the current racial and class divide in America, which was exposed and exacerbated when Hurricane Katrina struck New Orleans.

THE CONTEMPORARY BLACK-WHITE DIVIDE IN AMERICA

The racial and class divide in America is not a new phenomenon. In 1967, almost forty years ago, in the wake of the Newark, Detroit, and Cleveland riots, President Lyndon Johnson appointed a special commission of distinguished Americans, under the leadership of Governor Otto Kerner of Illinois, to search for the root causes of urban discontent and the widening gap between white and black Americans. After seven months of "painstaking investigation" the National Advisory Commission on Civil Disorders found that the underlying cause of the riots was "white racism." The Commission went on to explain that pervasive racial discrimination and segregation resulted in urban black poverty and its effects, including black migration and white exodus, black ghettos, frustrated hopes, violence, powerlessness, and confrontations with the police. The Commission also warned of a growing inner-city black population and noted that "Our nation is moving towards two societies, one black, one white—separate and unequal" (National Advisory Commission on Civil Disorders 1968, 1).

More than a decade after the National Advisory Commission on Civil Disorders issued its report, Andrew Hacker similarly concluded that America had become two nations, black and white, separate, hostile, and unequal (Hacker 1992, ix). These findings suggest that American society is characterized by a close interplay between race and class. Accordingly, on one side of the divide are blacks living in poverty, within inner-city areas perpetuated by social isolation, institutional racism, and socioeconomic inequality; and in stark contrast, on the other side of the divide, are whites living in affluence, primarily in buffeted and fortified suburbs, perpetuated by the intergenerational transmission of wealth, power, and white privilege. These two contrasting groups at opposite ends of the continuum define the extreme ends of the race and class divide in America.

RACE AND CLASS IN PRE-KATRINA NEW ORLEANS

The interplay of race and contemporary poverty in the city of New Orleans has been rather unfavorable to black people, who comprised more than two-thirds of the pre-Katrina population. Official census data reveal that over the five years that preceded the arrival of Hurricane Katrina, about one-third of the total black population consistently earned incomes below the poverty benchmark. What is rather paradoxical about this unfortunate connection is that in 2004, the percentage of poor African Americans in New Orleans was the same as the percentage of poor American blacks in 1974. U.S. Census Bureau statistics reveal that from 2000 to 2004, the percentages of African Americans in New Orleans with incomes below the poverty threshold gradually declined, but averaged about 30 percent of the total city black population and about 19 percent of the total city population. Yet Census Bureau statistics for the same period show that from 2000 to 2004, the percentage of African Americans nationwide with incomes below the poverty level averaged about 24.6 percent.

Despite a very gradual decrease in the indicators of individual and household poverty between 2000 and 2003, the economic fortunes of African Americans in New Orleans took a turn for the worse in the year preceding Hurricane Katrina. Census data for 2004 revealed a 4.8 percent increase in the number of poor blacks in the total black population and a 3.5 percent increase in the percentage of African Americans with incomes below the poverty threshold in the total city population. These conditions occurred with practically no change in the percentages of black people in the total black and city labor force populations. That is, relatively more black people were ushered into the ranks of the poor while the relative share of black people in the total labor market of the city remained unchanged.

The economic histories of the State of Louisiana, Orleans Parish, and the city of New Orleans have much to do with the stagnation in employment rates among African Americans. On a variety of recent income and poverty measures, the State of Louisiana appears to have a rather poor performance. In 2000, the median household income in Louisiana was among the lowest in the United States (at $32,566, only third from the bottom above Mississippi and Arkansas). Louisiana ranked third from the bottom (with 19.6 percent and 15.8 percent of individuals and families, respectively), above Mississippi and the District of Columbia, in the percentage of both individuals of all ages and of families with incomes below the poverty level. Official statistics reveal that in 2004, the economic fortunes of Louisiana were not much better, and its rankings remained more or less the same. Overall, New Orleans did not fare well, either. The median family income levels in the city were below national averages, and unemployment and poverty rates were well above the national averages. Furthermore, in 2004, only "a very small portion of jobs in the metropolitan area and an even smaller portion in the city were in the high-paying manufacturing sector. . . .By contrast, a large portion of metropolitan employment and an even larger share of city employment were located in the hospitality and entertainment sectors. The average wage in these sectors was the lowest among sectors."

Still, higher collective financial gains from a service-oriented and rather weak economic structure are possible for any social group if a significant portion of its population has a disproportionately higher educational attainment profile. However, the figures on the educational attainment of African Americans twenty-five years and older in New Orleans show that despite gradual improvements, the numbers of black people who possessed the educational credentials necessary for entitlement to higher occupational earnings remained very modest. Between 2000 and 2004, the percentage of black adults twenty-five years and older with college degrees ranged from 11 to 18 percent and averaged about 15.3 percent, well below the national average of 24.4 percent. The bulk of the adult black population (averaging more than 80 percent for each of the five years under consideration) did not possess the educational means to competitively access whatever high-paying service sector jobs were available. The overall performance and dynamics of the economies of both the city and metropolitan area might be responsible for generating formidable forces that consistently pressured down black incomes and inadvertently kept a considerable number of black people in the ranks of the poor.

Between 2000 and 2004, an average of slightly less than 40 percent of the New Orleans black workers were employed in full-time occupations; an average of 33 percent of full-time black workers earned less than $17,499 in annual real incomes; and the median annual real earnings of black workers were about 11 percent less than the median annual real earnings in New Orleans. When coupled with the fact that the structure of the New Orleans economy was primarily constituted of low-paying jobs, these figures suggest that the black work force in New Orleans not only lacked the protection provided by tenure of full-time occupations against economic recessions and downturns in business cycles, but that the majority of full time workers were concentrated in low-paying jobs. Moreover, the fact that annual rates of participation in full-time, year-round employment varied narrowly from 2000 to 2004, further suggests that black people in New Orleans consistently faced structural impediments to tenure-protective employment.

Median annual individual black earnings stood anywhere from 11 percent to 19 percent below the estimate of the median real earnings for the city as a whole. The same trend seems to appear in the household income statistics, with the estimated median black household annual real income being 12.6 percent to 27.3 percent below the corresponding estimates for the city. These figures indicate considerable income disparities and reveal the relatively low economic standing of both individual African Americans and black families in New Orleans. What is rather troubling about this finding is the fact that the distribution of income in New Orleans became increasingly polarized over time. In fact, in 1999 the top 10 percent of New Orleans city households accounted for an estimated 41 percent of all income earned in the city while "those households in the middle of the income distribution experienced a significant decline in their income share." There is no evidence to suggest to that this pattern improved in the 2000 to 2004 period.

We can get a closer view of the socioeconomic disparities in pre-Katrina New Orleans by observing the neighborhoods with the highest concentration of urban black poverty. The neighborhoods with the highest concentration of urban black poor were the housing projects, including Fisher, Iberville, Florida, Calliope, St. Thomas, St Bernard, and Desire. Although these neighborhoods vary in size of population, they were almost completely black, with the percentage of poor ranging from 62.5 percent to 88.2 percent. Thirty-three of the sixty-nine neighborhoods had a percentage of poor that exceeded that for the city as a whole (27.9 percent), and 54 neighborhoods in New Orleans exceeded the national average of 15.5 percent poor.

Census data compiled by the Brookings Institution reveal that in 2000, New Orleans ranked second among U.S. cities on concentrated poverty, with a rate of 37.7 percent. Concentrated poverty is not unique to New Orleans and can be found in many large cities across the nation. To illustrate, Fresno, California ranked first of ten major U.S. cities, with a rate of 43.5; Louisville, Kentucky, third (36.7); Miami, Florida, fourth (36.4); Atlanta, Georgia, fifth (35.8); Long Beach, California, sixth (30.7); Cleveland, Ohio, seventh (29.8); Philadelphia, Pennsylvania, eighth (27.9); Milwaukee, Wisconsin, ninth (27.0); and New York, New York, tenth (25.9); compared to the U.S. total of 10.3. New Orleans ranked third in the rate of concentrated black poverty (42.6) behind Miami (67.7) and Louisville (53.2).

These statistics show that the rate of concentrated black poverty is a much more serious problem in large American cities than the rate of concentrated poverty in the total population of these cities. What makes the problem more serious in New Orleans is that poor black households were highly concentrated in high-poverty zones: of the 131,000 poor people residing in New Orleans in 2000, almost 50,000 (38 percent) lived in high-poverty zones. Research has well documented that concentrated poverty exacts multiple costs on individuals, families, and society, including reduced private-sector investments and job opportunities, inflated food prices for the poor; higher levels of crime, negative impacts on mental and physical health, low-quality neighborhood schools, heavy burdens on local governments; and high out-migration of middle-class households, all of which adversely influence the life chances and quality of life of residents of high-poverty neighborhoods.

Although much attention was given to the poor or lower class in the aftermath of Hurricane Katrina, pre-Katrina New Orleans had a substantial number of middle-class blacks residing in middle-class neighborhoods, notably Village de L'Est, Pontchartrain Park, Read Boulevard East, Read Boulevard West, Gentilly Woods, Edgelake/Littlewood, Fairgrounds/Broad, Pines Village, Dillard, and St. Anthony. Other black middle-class residents were scattered over different neighborhoods that contained a relatively low percentage of poor residents. An interesting pattern that existed in pre-Katrina New Orleans, due in part to high population density, was the close proximity of lower-class and middle-class neighborhoods, a condition that helped to reduce the isolation of the poor. At one popular restaurant in the inner city, one could dine with a full view of a housing project located just across the street. However, most middle-class neighborhoods were buffered by a major thoroughfare, drainage canal, lake, park, business area, wall, or natural barrier. In spite of the close proximity of middle- and lower-class neighborhoods, most residents who had lived in the city for some time were readily aware of neighborhood boundaries, as well as the type of people who lived in different neighborhoods. However, Hurricane Katrina did not spare middle-class black neighborhoods, especially in the New Orleans East and Gentilly communities, which suffered substantial damage.

THE EFFECTS OF KATRINA ON RACE AND CLASS IN POST-KATRINA NEW ORLEANS:
SOME EMPIRICAL INVESTIGATIONS

New Orleans lost more than three-fourths of its total population due to Hurricane Katrina, mostly from the massive evacuation that occurred before and after the storm. Although the storm adversely affected a wide range of people across different races, classes, and neighborhoods, a number of studies have shown that the real or perceived effects of Hurricane Katrina were not uniform across racial, class, and residential groups in different neighborhoods in New Orleans. In a study of the impact of race on the social construction of reality by those who responded to Hurricanes Katrina and Rita, Austin and Miles (2006, 2) found that two separate realities existed in the Gulf Coast following Hurricane Katrina—one for whites and one for blacks. Based on data gathered from ethnographic observations and personal interviews of black and white residents included in a purposive sample, two contrasting realities were identified: the reality for whites, which included the perceptions that "there were dangerous black men looting, marauding and raping," and "race did not play a role in rescue, relief and recovery efforts"; and the reality for blacks, which included the perceptions that "race was and is a factor in the white authority's rescue, relief and recovery efforts" and "white authority criminalized blacks due to racism." Austin and Miles reported that lower class residents of the Ninth Ward of New Orleans were more likely to feel that "race did matter" while middle class whites of the Garden District were more likely to feel that "race did not matter" in the rescue, relief, and recovery efforts following Katrina. In general, they concluded that whites were less likely and blacks were more likely to perceive that race was a factor in the rescue, relief, and recovery efforts following Hurricane Katrina.

In support of the conclusions reached by Duke and Austin, Elliot and Pais explained that in times of crisis, class differences are likely to shrink and racial differences expand as individuals define, interpret and respond to the situation before them. Accordingly, the perception that race matters in times of disasters may be shaped by a variety of factors, including past experiences, historical memory/knowledge, et cetera. The rationale is that because blacks are more likely than whites not only to experience social inequality, racism, injustices, and poverty, but also to possess a collective memory of black victimization by white supremacy, white power, authority, racism, and discrimination, their definitions and perceptions of the rescue, relief, and recovery efforts following Katrina are likely to be different from that of whites, thus creating two separate realities of Hurricane Katrina. This notion is similar to what Du Bois observed over one hundred years ago in his seminal research, *The Philadelphia Negro*, in which he concluded that among Negroes there was a "widespread feeling of dislike for his blood, something that keeps him and his children out of decent employment, from certain public conveniences and amusements, from hiring houses in many sections, and in general, from being recognized as a man." These findings suggest that the racial and class divide may be characterized by differential perceptions or definitions of the reality between blacks and whites, in addition to socioeconomic disparities.

In testing general hypotheses regarding racial and class differences in responses to Hurricane Katrina, Elliot and Pais found that blacks were less inclined than whites to evacuate before the storm, mostly because they did not believe that the hurricane would be as devastating as it eventually was; blacks were more likely to report "leaning on the Lord" while whites were more likely to report relying on friends and family; black workers were seven times more likely to have lost their jobs than the average white worker and thus would be less likely to return to the city, leaving New Orleans with a different racial and class composition, with fewer blacks; poor blacks had fewer cars to get out of the city; and low-income home owners needed the most assistance with housing and new jobs in order to return to the city and rebuild. Elliott and Pais concluded that both race and class appeared to have mattered in response to hurricane Katrina.

In a study of the social and demographic effects of Hurricane Katrina, using official statistics from secondary sources, sociologist John Logan found that although some neighborhoods received greater damage than others, the damage caused by the hurricane was disproportionately borne by African Americans compared to whites, renters compared to home owners, and the poor and unemployed compared to the wealthy and employed. Accordingly, damaged areas were 45.8 percent black, compared to only 26.4 percent in undamaged areas; 45.7 percent of homes in damaged areas were occupied by renters, compared to 30.9 percent in undamaged areas; in damaged areas, 20.9 percent of households had incomes below the poverty line, compared with 15.3 percent in undamaged areas; and 7.6 percent of persons in damaged areas were unemployed, compared to 6.0 percent in undamaged areas. These results reveal that the typical neighborhood or group that received the most damage from the storm was comprised of black, poor, unemployed renters, a combination that indicates multiple jeopardy (that is, multiple disadvantages) in poor neighborhoods in New Orleans. This means that in the event of a disaster, the more vulnerable neighborhoods are likely to be those with a relatively high proportion of residents with multiple disadvantages, and these neighborhoods are also likely to be more severely affected and to have greater difficulty in recovering from disasters like Hurricane Katrina. Consequently, these factors may also contribute to the widening of the race and class divide.

Policy Implications and Solutions

Mostly everyone would agree with the observation made by the United Nations human rights panel that the United States must better protect poor people and African Americans in natural disasters to avoid problems such as those that occurred after Hurricane Katrina. Although official responses to the Hurricane Katrina disaster revealed a critical need for improvement of rescue, relief, and recovery operations, the storm exposed epidemic proportions of concentrated urban black poverty that existed in New Orleans, and "laid bare many of the disparities that continue to separate Americans by race and class." Thus it seems rather obvious that in order to lessen the detrimental social effects of natural disasters in urban areas, policy solutions must be fashioned and implemented to reduce, if not eliminate, the roots causes of concentrated urban black poverty.

Although many public officials, including the president of the United States, acknowledged that inner-city poverty exposed by Hurricane Katrina runs much deeper due to the destruction caused by the storm, little serious attention has been given to the development of policies and action plans aimed at resolving the problem of concentrated urban black poverty that was exposed by Hurricane Katrina. However, policy recommendations are gradually beginning to emerge from "think tank" policy institutes and research centers around the country. For example, the Brookings Institution recommends that "Congress should consider policy options to put the nation back on track toward alleviating concentrated poverty, by supporting choice and opportunity for lower-income residents in distressed neighborhoods." The following policy recommendations were offered: restore funding to the HOPE VI program, increase support for housing vouchers, pilot a "housing-to-school voucher initiative," adopt President Bush's proposed tax credits for homeownership, target affordable housing to low-poverty areas with the assistance of regional housing corporations, and expand the EITC (Earned Income Tax Credit) to help working families afford housing in better neighborhoods.

Total Community Action Inc. (TCA) found that there has been little progress in reducing poverty in New Orleans in the past three decades. Thus, there is a need to resolve the chronic socioeconomic problems in New Orleans, including economic stagnation, sub-employment, excessive unemployment, low labor force participation rates, a disproportionate level of poverty, extreme income inequality, low educational attainment, and a high geographic concentration of the poor are long-term problems in New Orleans. In addition, New Orleans lags behind peer cities; blacks lag behind whites; males lag behind females; poverty is heavily concentrated in poor black neighborhoods that have been isolated by the flight of middle-class black and white families; and the higher educated, working population has moved to the suburbs. On top of these problems are failing schools, high crime rates, and substandard housing.

A critical part of the problem of concentrated urban poverty that needs to be addressed is the worsening condition of undereducated, disconnected, unemployed, young black males in inner cities across America. For example, in New Orleans in 1999, only 49 percent of black males age sixteen to twenty-four were in the labor force, compared to 62 percent of the white male population. The jobless rate for male high school dropouts in their twenties soared to 72 percent by 2004, compared with only 34 percent for white dropouts and 19 percent for Hispanic dropouts. Due to these findings, TCA calls for an official policy aimed at reducing poverty and improving the economic conditions among urban black males.

TCA also argues that there is a need for a poverty-reduction policy at both the state and city levels that includes formally trained professionals who would systematically formulate and administer a plan to help reduce poverty, instituting a wide range of considerations of interventions tried in many cities, regions, and countries. TCA offers an eight-point poverty reduction plan, specifically directed at New Orleans and the State of Louisiana. It recommends that the city and state (1) adopt an official policy on poverty reduction as a part of their economic development strategy; (2) use a percentage of the city's Community Development Block Grant Funds to induce private/public partnerships to tailor their fringe benefits to help income-eligible workers of six years to build assets; (3) formally link and coordinate with public and private entities to intensively and consistently promote proven interventions that both grow and sustain growth out of poverty; (4) weigh the pros and cons of instituting an Earned Income Tax Credit (EITC) at local and state levels; (5) enact procurement policies that drastically reduce the underutilization of minority and female entrepreneurs in public procurement; (6) undertake an aggressive, intensive, and ongoing campaign to encourage the private sector to purchase goods and services from minority entrepreneurs as a normal course of business; (7) encourage universities in Louisiana to offer curricula that prepare students and current public employees on policies and methods for averting and reducing poverty, as well as sustaining growth from poverty, and also to offer undergraduate and graduate degrees with concentrations in poverty reduction; and (8) urge the media to use their considerable resources to assist in educating citizens, themselves, and policymakers as to the causes and consequences of chronic poverty.

In consideration of the findings reported in this chapter, it is reasonable to conclude that the reduction of urban black poverty must be is a key factor in closing the race and class divide. It is also reasonable to conclude that the reduction in racism in all of its forms and its attendant consequences is critical to closing the race and class divide. However, in order to accomplish this monumental task, a monumental effort is required, one that includes a comprehensive and effective combination of policies, strategies, actions, and resources aimed directly at the heart and roots of the problem. In essence, there is need to establish a permanent structure of government, in partnership with the private sector, with the mandate to develop and implement official policies aimed at the reduction of urban black poverty. This mandate and commitment must be pursued with the same vigor, determination, and tenacity as those made by the national program to conquer outer space. This mandate must include the most basic elements that are essential to improving the quality of life of the masses, including education, health, employment, and housing.

A major challenge is to find meaningful ways to involve grassroots individuals and agencies in the development and implementation of policies aimed at the reduction of poverty in their communities. Whereas it seems plausible to adopt strategies or programs with "proven" success, there is also a need for fresh, innovative ideas from a variety of sources that could be considered or utilized, based on merit. Although it may not be possible to eliminate all poverty or completely close the race and class divide, it does seem plausible that the quality of life of people who live in substandard conditions in America's largest cities can be vastly improved. Hurricane Katrina exposed a small part of the larger problem of urban poverty and the malaise that accompanies it. However, Hurricane Katrina created a great challenge and opportunity to reduce poverty in America. Hurricane Katrina also tested the will and soul of America. The question that remains to be answered is, will America rise to the challenge?

SUMMARY

Poverty remains one of the most critical social problems in both the State of Louisiana and the nation. The problem becomes even more serious when combined with race, in the sense that poverty is more heavily concentrated among blacks who reside in America's largest cities. When race and poverty meet, they contribute to the race and class divide in America, a divide that has widened in recent years. The race and class divide was graphically exposed by Hurricane Katrina in the city of New Orleans. However, it could well have taken place in any of the large cities in America where the problem of concentrated urban black poverty exists. If Hurricane Katrina taught us any lesson, beyond the need to take swift and certain action before and after a major natural disaster, it is that we must also take swift and certain actions to resolve the problem of urban black poverty in America.

REFERENCE LIST

Austin, Duke, and Michelle Miles. 2006. *The construction of reality along the color line following Hurricanes Katrina and Rita.* Unpublished Paper, University of Colorado, Boulder, CO.

Bankole, Katherine. 1995. *The Afrocentric guide to selected black studies terms and concepts, an annotated index for students.* Lido Beach, New York: Whittier Publications.

Barry, John. 1998. *Rising tide: The great Mississippi flood of 1927 and how it changed America.* New York: Simon and Schuster.

Berube, and Katz. 2005. *Katrina's window: Confronting concentrated poverty across America.* Washington, DC: The Brookings Institutition.

Bobo, James. 1975. *The New Orleans economy: Pro bono publico? Research study no. 19.* New Orleans, LA: Division of Business and Economic Research, College of Business Administration, University of New Orleans, HC107.L58, no. 19 SPEC.

Brinkley, Douglas. 2006. *The great deluge, Katrina, New Orleans and the Mississippi Gulf Coast.* New York: Morrow, HarperCollins.

Brookings Metropolitan Policy Program. 2005. *New Orleans after the storm: Lessons from the past, a plan for the future.* Washington, DC: Brookings Institution.

Dangerfield, Peter W., William B. Oakland, and KL&M Company of New Orleans. 2006. *Poverty reduction, a plan for Louisiana.* New Orleans: Total Community Action.

Du Bois, W. E. B. 1899. *The Philadelphia negro.* New York: Schocken Books.

Elliot and Pais. 2006. *Race, class and Hurricane Katrina: Social differences in human responses to disaster.* Social Science Research 35:295–321.

Fothergill, A., E. Maestra, and J. Darlington. 1999. Race, ethnicity and disasters in the United States: A review of the literature. *Disasters* 23 (2): 156–173.

Hacker, Andrew. 1992. *Two nations, black and white, separate, hostile, unequal.* New York: Charles Scribner's Sons.

Haynes, Gerald David, and Robin M. Williams Jr. 1989. *A common destiny: Blacks and American society.* Washington, DC: National Academy Press.

Hine, Darlene Clark, William Hine, and Stanley Harrold. 2006. *An African-American odyssey.* 3rd ed. Upper Saddle River, NJ: Pearson Education.

Klapper, Bradley. 2006. U.N. panel addresses shortcomings: Effects of Katrina reveal need for U.S. to ensure rights of poor, it says. *The Advocate,* Baton Rouge, July 29, p. 7A.

Logan, John. 2006. *The impact of Katrina: Race and class in storm-damaged neighborhoods.* Unpublished Manuscript, Spatial Structures in the Social Sciences Initiative, Brown University.

National Advisory Commission on Civil Disorders. 1968. *Report of the national advisory commission on civil disorders.* New York: New York Times Company.

Page, Clarence. 2006. President's address to NAACP left unsaid issues with black poverty. *The Advocate,* Baton Rouge, July 27, p. 13B.

U.S. Census Bureau. 2000. *Income and poverty in 1999–2000.* Census 2000 Summary File 3 (SF3)—Sample Data. Available at http://www.census.gov.

U.S. Census Bureau. 2000–2004, *American community survey.* Available at http://www.census.gov.

CHAPTER 14

KATRINA'S SOUTHERN "EXPOSURE": THE KANYE RACE DEBATE AND THE REPERCUSSIONS OF DISCUSSION

ERICA M. CZAJA[*]

In the weeks that followed Hurricane Katrina, many scholars and media profession-als expressed the view that devastating racial inequality had been "exposed," "revealed," and "laid bare" by the hurricane and its aftermath (DeParle 2005; Frymer, Strolovitch, and Warren 2005; Gilman 2005; Jackson 2005; Stevenson 2005; Weisman 2005). A discernable air of hope was present in their writings, hope that this natural disaster would increase understanding of and concern for the social disaster that existed in the Ninth Ward of New Orleans and many other predomi-nantly black neighborhoods around the country long before Hurricane Katrina hit the Gulf Coast (Massey and Denton 1993). However, nationwide public opinion polls conducted a week after the hurricane on September 6 and 7, 2005, found that 71 percent of African American respondents, but only 32 percent of white respon-dents, said that they "felt that the disaster revealed the persistence of racial inequal-ity" (Pew Research Center 2005). Later polling, conducted October 28, 2005, through November 17, 2005, as part of the "2005 Racial Attitudes and the Katrina Disaster Study" at the University of Chicago, found an even greater divide between

*For their helpful advice at various stages of this work, I thank Kim Babon and Martha Van Haitsma. I am also grateful for the intellectual stimulation of Elisabeth S. Clemens, Melissa Harris-Lacewell, and Michael C. Dawson. Finally, for his limitless generosity, encouragement, and direction, I am especially indebted to my advisor, Dorian T. Warren.

blacks and whites on this same issue: 90 percent of African Americans agreed that "Katrina shows there's a lesson to be learned about continued racial inequality," compared to only 38 percent of white respondents (Dawson, Cohen, and Harris-Lacewell 2006).

These surveys highlight the drastically different interpretations that most blacks and whites took away from the disaster. Though a hurricane by itself is not a racialized event, the lower rates at which whites perceived Katrina's aftermath as evidence of racial inequality compared to blacks mirror the racial gaps in public opinion about issues of race consistently found by numerous studies over the last several decades (Schuman et al. 1997; Sears, Sidanius, and Bobo 2000). Hurricane Katrina did not teach black or white Americans anything new. Scholars and journalists were evidently too hopeful about the impact Hurricane Katrina could have on mass public opinion. But why? Why did what seemed so obvious to some resonate so little with the white majority? To determine why what I call the "exposure hypothesis" (that the hurricane exposed racial inequality in the minds of ordinary Americans) fell flat, I examine public discussions about race that occurred in Hurricane Katrina's aftermath. In doing so, I aim to build a bridge between studies of public opinion and studies of public discourse. Specifically, this project attempts to answer the following: How did Americans publicly communicate about race in Katrina's aftermath? More broadly, how does contemporary race discourse influence continuing racial inequality in the United States? These are important questions because widespread public discussion of race in the aftermath of Hurricane Katrina has the potential to foster greater understanding of the historical and societal forces that influence African Americans' life chances, thereby altering prevailing public opinion. Gaining a more structural understanding of the legacies and persistence of racism, particularly among white Americans, could then be the first step toward increasing public support for government policies that attempt to ameliorate the problem of racial inequality.

On the whole, my findings are troubling. Most problematic for those who had hoped that Katrina would reveal something about racial inequality is that, while nine of the fifty-four letter writers in my sample reference race along with class, only two letter writers recognize the relationship between colored skin and poverty as problematic; the remaining seven seem to take this correlation for granted (*USA Today*, September 14, 2005, p. A10; September 15, 2005, p. A12). In addition, only one letter writer demonstrates any understanding of the structural and historical forces that determine life chances (*USA Today*, September 14, 2005, p. A10). My results confirm those of survey researchers: while Hurricane Katrina may have exposed something to the American public, it was not racial inequality as traditionally conceived by progressives. In the absence of acknowledgement of racial inequality and without discussion of social structural forces, changes in racial attitudes and policy preferences as a result of this disaster are unlikely. Under these conditions, societal change seems nearly impossible.

The rest of this paper will proceed as follows. First, I review previous research on the racial opinion gap, its potential sources, and the influence of the media on whites' racial attitudes and policy preferences. I also explore recent literature on how and in what contexts people use conversations to interpret media information. In an attempt to bridge the divide between public opinion scholarship and recent work on public political discourse, I then suggest how both lines of research might apply to citizens' discussions of the role of race in Katrina's aftermath. Using letters to the editor of *USA Today* as my data, I explore why assumptions among scholars and journalists that Katrina's aftermath exposed continuing racial inequality in the United States did not hold true. Finally, I propose strategies that the government, the academy, the media, and the American public might adopt to overcome current obstacles to informed, productive discussions about racial issues.

LITERATURE REVIEW

Public opinion often drives government policy (Gilens 1999; Hetherington 2005; Kinder and Winter 2001; Schuman et al. 1997; Sears, Sidanius, and Bobo 2000), and the American public often bases its opinions on information it receives from the media (Delli Carpini and Keeter 1996). If media coverage of Katrina exposed racial inequality in the public mind as many have claimed, then longstanding differences in the opinions of blacks and whites should not have been reflected in Americans' interpretations of the Katrina disaster as the above surveys demonstrate. How ordinary Americans discussed what, if any, lesson Hurricane Katrina's aftermath taught is an important topic of study because it may help us learn something about the mechanisms of and necessary conditions for public opinion shifts and therefore policy change.

RACIAL INEQUALITY, PUBLIC OPINION, AND POLICY PREFERENCES

While surveys over the last four decades have found steadily increasing support among the American public for the principle of racial equality, white support for a government role in making such equality a reality has consistently registered at much lower levels than that of African Americans (Kinder and Winter 2001; Schuman et al. 1997; Sears, Sidanius, and Bobo 2000). However, there is much disagreement among scholars regarding the source of this racial opinion gap. In *Racialized Politics: The Debate about Racism in America*, editors David Sears, Jim Sidanius, and Lawrence Bobo outline three theoretical approaches to explaining the racial opinion gap; the divergence of whites' policy preferences from those of African Americans has three potential sources: "prejudice, social structure, [and] ordinary politics" (Sears, Jim Sidanius, and Lawrence Bobo 2000, 2).

Prejudice or "new racism" models assume that whites are socialized into the prejudiced belief that African Americans are less committed to American values, such as individualism and hard work (Sears, Sidanius, and Bobo 2000). According to these models, perceptions of African Americans as values-deficient and undeserving are largely responsible for whites' opinions about racial inequality and what to do about it, particularly when it comes to welfare (Gilens 1999). Alternatively, scholars who argue that social structure is the source of whites' public opinion assert that inter-group competition for limited resources and self-interest explain whites' lower levels of support for government policies that address racial inequality. Finally, some scholars claim that the source of the racial opinion gap is merely a result of regular political processes. They maintain that, because "public opinion on racial policy is now primarily motivated by values and ideologies that are race-neutral. . .whites' opinions are strongly influenced by the exact nature of the policy proposals under consideration" (Sears, Sidanius, and Bobo 2000, 16).

In his study of public opinion and support for welfare, Martin Gilens (1999) finds evidence for "new racism" models, but he also complicates the story by highlighting the nuanced opinions of whites regarding anti-poverty policies. He finds that whites overwhelmingly think that government welfare spending should be reduced but also that the government should be doing more to help the poor. These seemingly contradictory beliefs stem from the false, but prevailing, perception that most welfare recipients are African American and the stereotype that blacks lack work ethic and are not committed to the American value of individualism. Accordingly, whites dislike welfare because they believe that predominantly black welfare recipients would rather milk the system for benefits than work to provide for themselves. Because of blacks' perceived laziness, welfare recipients are seen as undeserving, which then leads to unfavorable opinions about welfare. However, Gilens argues that poor individuals who are perceived to be working toward self-sufficiency, including African Americans, can join the ranks of the deserving poor in the public mind. Employing a type of "ordinary politics" argument, he advocates for anti-poverty policies that help the poor to help themselves, which he claims should receive the widest public support (Gilens 1999). However, despite the sweeping welfare reforms of 1996, including time limits and work requirements, other researchers have found that public opinion regarding poverty and government assistance has not changed (Shaw and Shapiro 2002). Thus, negative stereotypes about African Americans continue to undermine support for welfare even under the conditions that Gilens suggests.

Recent studies suggest that whites use a color-blind ideology to maintain their privileged position in society, especially when arguing against race-based policies designed to reduce racial inequality such as affirmative action. Eduardo Bonilla-Silva (2003), for example, finds through interviews with white college students that many passionately deny the existence of racism, claiming that society is now color-blind; the disadvantaged position of blacks is therefore considered their own fault (Bonilla-Silva 2003). Joe Feagin and Eileen O'Brien (2003) also find evidence of a color-blind ideology during their interviews with wealthy, white male professionals (Feagin and O'Brien 2003). These findings, along with evidence that negative stereotypes and racist attitudes still exist (Gilens 1999; Hetherington 2005; Myers 2005), suggest that whites may have found denying racism with a color-blind ideology to be an effective strategy for maintaining their dominant group position. Denying racism (even as it persists) and thereby any societal responsibility for racial inequality while still supporting modest assistance for the poor, who are disproportionately African American, could simply reflect greater sophistication on the part of whites in their competition for scarce resources.

THE INFLUENCE OF MEDIA BIAS AND FRAMING EFFECTS ON RACIAL ATTITUDES AND POLICY SUPPORT

Based on the above literature, negative stereotypes and prejudiced beliefs that African Americans lack commitment to traditional American values appear to account for whites' opinions and policy preferences. But how do these negative stereotypes persist in a self-proclaimed color-blind culture? Gilens (1999) explains that stereotypes originally used by whites to justify slavery lived on in the minds of media professionals even after formal equality was established during the civil rights era. Journalists unconsciously employed their prejudices, particularly during Lyndon B. Johnson's War on Poverty, which ushered in a new era of negative media coverage of the poor, depicted as predominantly black and undeserving (Gilens 1999). As Shanto Iyengar (1996) notes, the consequences of such media coverage have been and continue to be counterproductive to the goal of racial equality: "The frequent association of African Americans with economic failure. . .encourages and justifies the expression of racist attitudes" (Iyengar 1996, 70). In terms of policy, he also finds that racial cues, such as images of the black poor, elicit individualized rather than societal solutions among media consumers more frequently than do images of impoverished whites (Iyengar 1996).

While Gilens (1999) and Iyengar (1996) find that media attention to social problems affecting blacks makes things worse, Eric Klinenberg (2002) demonstrates that completely race-neutral media coverage can actually be biased and inaccurate, too. In his social autopsy of the deadly Chicago heat wave of 1995, he shows how journalistic practices and the framing of the heat wave as a natural phenomenon that affected all Chicagoans equally directed public attention away from the issues of race, poverty, social isolation, and neighborhood breakdown, which were highly relevant to an accurate representation of the disaster. Though the heat wave disproportionately affected poor, elderly African Americans, local news coverage actually served to naturalize the disaster and mask the underlying societal conditions that contributed to the most devastating outcomes for the most vulnerable (Klinenberg 2002). Klinenberg's findings along with those of Iyengar (1996) point to a kind of paradox in which media coverage produces a negative result, in the form of public unawareness of social problems or reduced policy support, regardless of whether it glosses over or showcases issues of black poverty.

The conclusions of Paul Kellstedt's (2000) study, based upon his content analyses of 4,040 *Newsweek* articles about race printed between 1950 and 1993, offer a solution to this apparent paradox. His study provides indirect support for Iyengar's (1996) findings but also sheds light on ways that the media could overcome its current contribution to negative stereotypes and racial inequality. Through his content analysis, Kellstedt measures the potential media framing effects of two core American values, egalitarianism and individualism, on the enactment of liberal race-conscious policies (Kellstedt 2000). Because media framing operates not by changing people's core beliefs but by calling to mind pre-existing beliefs and thereby making them more salient (Nelson, Oxley, and Clawson 1997), Kellstedt's study offers important insight into when U.S. citizens might prioritize one core American value over another, such as equality over individualism, or vice versa. Kellstedt counted articles as employing egalitarian frames when they "portray[ed] American society as one in which blacks are (or have been) treated as less-than-full citizens, as unequals." He coded articles as individualistic when they contained some reference to the following merit-based criteria for deservingness: "Those who live the Protestant work ethic are exalted as virtuous, whereas those who depend (especially on the government) lack character" (Kellstedt 2000, 253).

Kellstedt (2000) finds that frames highlighting individualism have neither a positive or negative effect on policy preferences, which indirectly confirms Iyengar's (1996) findings that racial cues alone bring to mind individualistic attitudes. Together, their studies suggest that media frames highlighting individualism are redundant among Americans, many of whom already adhere strongly to individualistic values. However, media frames that highlight egalitarianism increase aggregate support for race-based government policies (Kellstedt 2000), demonstrating that media emphasis on equality is one effective method for overriding Americans' initial, individualistic reactions to racial cues. In addition, Kellstedt (2000) finds that the amount of coverage about race matters little for the level of public support for race-based policies. This finding must be interpreted carefully, though, since race-neutral coverage of events and social conditions that disproportionately affect African Americans, as Klinenberg (2002) has shown, can obscure racial inequalities and prevent public realization of the presence of social problems.

In sum, while Gilens (1999) and Iyengar (1996) show the potentially detrimental effects of portraying black poverty in the media, the findings of Klinenberg (2002) and Kellstedt (2000) illustrate that a certain amount of media attention given to race is a necessary, but not sufficient, condition for positively influencing citizens' opinions regarding racial inequality and a need for race-conscious government programs. In addition, Kellstedt's (2000) study provides hope that media coverage of racial issues does not inherently perpetuate racial stereotypes. Rather, certain types of media coverage, such as that which employs egalitarian frames, can counteract the powerful combination of negative stereotypes and American individualism that other scholars have termed "new racism," which often influences public opinion and policy preferences.

THE ROLE OF PUBLIC DISCUSSION

As stated earlier, this project aims to build a bridge between the public opinion literature and studies of public discourse. Other scholars have already begun to lay the foundation. Though it seems like common sense when stated so plainly, Nina Eliasoph (1999) illuminates what many interested in eliminating racial inequality seem to overlook: "Changing the structure of *communication* has to be part of the process of social change. But to do that, whites would have to acknowledge, together, that racial inequality is indeed a problem" (Eliasoph 2000, 483; emphasis in original). In relation to Hurricane Katrina, the survey data indicate that the hurricane did not expose racial inequality as a problem to most white Americans (Pew Research Center 2005; Dawson, Cohen, and Harris-Lacewell 2006). Why?

While "new racism," a color-blind ideology, and the influence of the media all likely play a role in perpetuating racial inequality and the racial opinion gap, Eliasoph (1999) explains that there is a real need for social scientists to examine the interactive discussion processes through which people reproduce racial inequality: "Social research often focuses either on internal subjective states or objective patterns of segregation and oppression in competitive markets. But this focus on civil society could overcome this seeming dichotomy, by examining the implicit rules in play in institutions that members themselves assume are 'free'. . .conversations are. . .the muscles and tendons that make the bones of social structural racism move. Understanding them requires understanding how people create civic institutions that so often silence the most well-meaning whites and amplify voices of indifference or bigotry" (Eliasoph 1999, 484).

As "the muscles and tendons that make the bones of social structural racism move," discussions among citizens in Hurricane Katrina's aftermath are crucial pieces needed to solve the puzzle of why racial inequality was not exposed in the way that so many assumed (Eliasoph 1999, 484). Eliasoph (1999) finds that whites with good intentions often do not speak up about racial issues, which are widely viewed as political, because talk is not seen as an effective method for enacting change in racial, as well as numerous other, matters of politics. In addition, most of Eliasoph's (1999) participants constantly distinguished between "speech contexts" by evaluating "*what is sayable where*" (Eliasoph 1999, 485); accordingly, they were much more likely to express anti-racist sentiments in private settings but appeared indifferent or racist in public when they remained silent in response to others' racism. In essence, a few bad apples, in combination with the silence of the rest of the bunch, spoiled the overall speech environment by making it "more racist than the sum of its parts" (Eliasoph 1999, 488). Based upon this finding, the letters to the editor section of a newspaper seems promising as a forum for discussing racial inequality in that it allows people who might remain silent in a group setting to formulate and express their thoughts in private before making them public.

A national newspaper may also be useful as a public discussion forum because it is the public's main source of information (Delli Carpini and Keeter 1996) and often assists people in gauging the relative importance of issues and events (Nelson, Oxley, and Clawson 1997). However, based upon her ethnographic study of two groups of white senior citizens living in a Midwestern town, Katherine Cramer Walsh cautions that the influence of media on public opinion should not be overemphasized. She participates in a women's craft guild and a (mostly) men's daily discussion group and finds that, while mass media stories did direct the issues and information that the groups discussed, the interactive processes through which people make meaning of the news significantly impact their media consumption (Walsh 2004).

Using the perspective of their collective social identity, the men's group "transformed" media frames by adding social categories, such as race, that were not suggested by news stories in order to make sense of the news (Walsh 2004). When the men discussed a newspaper story about a recent high school graduate who drowned in a nearby pond after a graduation party, for example, they organized their discussion around the fact that the young man was black and therefore an "other." This racial detail was evident from a photograph accompanying the story, but the text of the newspaper article did not include race as a salient category. Similar to the men's group in Walsh's (2004) study, Kanye West interpreted early media coverage of Katrina's aftermath based on his racial group identity, even though the media frames did not initially highlight race:

I hate the way *they* portray *us* in the media. *You see a black family, it says, "They're looting." You see a white family, it says, "They're looking for food.";* And, you know, it's been

five days [waiting for federal help] because most of the people are black. And even for me
to complain about it, I would be a hypocrite because I've tried to turn away from the
TV because it's too hard to watch. I've even been shopping before even giving a dona-
tion, so now I'm calling my business manager right now to see what is the biggest
amount I can give, and just to imagine if I was down there, and *those are my people
down there.* So anybody out there that wants to do anything that we can help—with
*the way America is set up to help the poor, the black people, the less well-off, as slow as pos-
sible.* I mean, the Red Cross is doing everything they can. We already realize a lot of
people that could help are at war right now, fighting another way—and they've given
them permission to go down and shoot us!. . .*George Bush doesn't care about black people!* (de Moraes 2005; emphasis added)

Though only West, his family, and his friends know for sure, it seems likely that he
would have discussed the post-Katrina media coverage with other African Americans
within his social circle. These conversations may then have contributed to his understandings of the events in New Orleans. Perhaps also because of his African American
social identity, West interpreted the racial cues present in the media images not indi-
vidualistically, as many Americans may have done, but in egalitarian terms (Iyengar
1996; Kellstedt 2000). West's critiques of the media and the government for not
treating blacks equally to whites became a part of the media coverage itself, both during the live NBC telethon when it originally aired and in subsequent news controversy, producing an egalitarian frame for American media consumers to interpret.
Looking only at the reaction of West, the idea that media coverage of Hurricane
Katrina exposed racial inequality seems like a valid assumption. However, given what
we already know about black public opinion and African Americans' greater sense of
racial group interest (Dawson 1994), it may also be that such exposure did not reveal
anything new to West. Previous public opinion polls have demonstrated that white
Americans are those that most needed to learn the lesson of racial inequality, but they
likely approached Katrina with diverse conceptions of their own social identities and
varied pre-existing racial attitudes. Therefore, it seems safe to assume, then, that ordinary Americans, particularly whites, interpreted the hurricane's aftermath in multiple
different ways in comparison to West and each other.

DATA/METHODS

Though survey data show that Hurricane Katrina did not impact Americans' beliefs
about the persistence of racial inequality in the manner that many progressives
assumed would occur, how ordinary Americans' discussed race in Katrina's aftermath
has yet to be examined. That is the purpose of the present project. To accomplish
this, I analyze a rather unique, nonrandom sample of "discussions": letters to the editor written to *USA Today* about Hurricane Katrina in the immediate aftermath of its
landfall in New Orleans (published Aug. 29 through Sept. 2, 2005, Sept. 5 through
Sept. 9, 2005, and Sept. 12, through Sept. 16, 2005). I contend that a content
analysis of these letters will provide insight into the ways in which ordinary
Americans from across the nation discussed Hurricane Katrina within their own
social circles.

I chose to examine letters written to *USA Today*, as opposed to other newspapers, for multiple reasons. First, *USA Today* is a national newspaper, which is consistent with my goal of assessing the national landscape of public discourse in Katrina's aftermath. Second, it is the most frequently purchased and read of American national newspapers (see Table 14.1). Finally, readers of *USA Today* are more representative of the American public than the readers of other large national newspapers, namely *The Wall Street Journal* and the *New York Times*.

A comparison of readers' household incomes, a standard measure of socioeconomic status, given the national median household income of $44,389 (U.S. Census Bureau 2004) strongly suggests that Americans who read *USA Today* are much more ordinary than the elite readers of *The Washington Post* and the *New York Times* (see Table 14.1).

I collected these letters to the editor electronically using an online database of newspaper text, *Proquest Newsstand*. During my initial review, I found sixty letters written in response to Hurricane Katrina. From these sixty, I decided to exclude one letter because it focused entirely on gas prices and energy efficiency rather than the events in New Orleans. Since my study focuses on the ways in which Americans talk about race, I also excluded five other letters based on the foreign citizenship of their writers. In total, I analyzed fifty-four letters. Through both deductive and inductive reasoning, I coded all fifty-four letters about Hurricane Katrina along three variables: identity, responsibility, and action (see Table 14.2). I then performed a more in-depth, textual analysis of the eleven letters that directly mentioned race or racial categories.

ANALYSIS

The near absence of discussions about values-deficiency in the hurricane's aftermath demonstrates that "new racism" models cannot explain the opinion gap between blacks and whites regarding whether Hurricane Katrina showed the persistence of racial inequality. New Orleans residents were identified overwhelmingly as fellow Americans in writers' usage of "we," "our," and "us." While 11 percent blamed the victims at least partially for their fate, only two letter writers (4 percent) justified themselves by questioning victims' adherence to the traditional American values of self-reliance and personal responsibility. In light of previous research, the strong sense of collective identity that results in times of crisis on American soil may be enough to drum up the core American value of egalitarianism and override the individualism that is often elicited when white Americans view African Americans in need (Iyengar 1996; Kellstedt 2000). Alternatively, the idea of a *natural* disaster perpetrated by Mother Nature might make it more difficult for whites to propose individualized rather than societal solutions to problems that resulted from Katrina. However, it is unclear from the letters what, if anything, Americans believe should be done about social problems that pre-dated the hurricane.

In addition, conceptions of American culture and notions of what it means to be an American were overtly stated by some writers and used to generate individual acts of assistance. Ginny from Pennsylvania reminded readers of their identities as follows:

Table 14.1 Audience size and reader income distribution of the three largest national newspapers

	Daily Audience		Percent of Readership by Household Income						
	Circulation	Readership	$25,000–$34,999	$35,000–$49,999	$50,000–$74,999	$75,000–$99,999	$100,000–$149,000	$150,000 and up
USA Today	2,590,695	6,980,074	7.7	17.8	21.9	18.3	15.6	11.4
Wall Street Journal	2,100,760	5,242,129	4.3	11.5	15.5	18.3	20.0	26.8
New York Times	1,682,644	5,065,303	5.7	13.6	15.4	16.8	17.5	23.8

Source: Circulation figures represent the largest number of copies sold in a day as reported by the Audit Bureau of Circulations during the six-month period between April 1, 2005, and September 30, 2005. Readership figures represent the number of readers on an average weekday. Readership and income figures are from the Newspaper Association of America's 2006 Newspaper Audience Database Report, which reported Scarborough Research data from telephone interviews conducted between February 2004 and March 2005.

Table 14.2 Katrina discussion content

Identity: Who are New Orleans residents?	
Fellow Americans	69%
African Americans or blacks	20%
Persons lacking American values	4%
Responsibility: Who or what is to blame?	
Government, any level or leader	39%
Social factors (i.e., poverty, sickness, old age)	17%
New Orleans residents	11%
Action: What will help?	
Talking about politics	50%
Individual acts (i.e., volunteer, donate, etc.)	31%
Not talking politics or playing "blame game"	17%
Both talking politics and individual acts	5.5%

Note: Percentages in each category ≠ 100%.

"Americans are a great people who always help those in need" (*USA Today*, September 1, 2005, p. A12). In line with this spirit of compassion and cooperation, Joshua from New York scolded other letter writers who blamed those who did not evacuate New Orleans on the grounds that these writers were "doing a disservice to what it should mean to be an American" (*USA Today*, September 9, 2005, p. A10). Accordingly, many used the concept of collective American identity to garner individual acts of support for Katrina victims, such as volunteering at shelters, donating money, or giving blood. These appeals drew on a rosy view of American history in which Americans were, by virtue of their identity, cooperative and generous. Ronald from Pennsylvania, for example, issued the following rallying cry for help: "Now is the time for all Americans to help their fellow countrymen, as we have done consistently since the inception of this great nation" (*USA Today*, September 2, 2005, p. A20). While put to an arguably good purpose, these kinds of historical fictions may also serve to blur our country's ugly racial past in the minds of Americans, thereby reducing understanding of the historical and structural forces that contribute to racial inequality (*One America in the 21st Century* 1998). Furthermore, the sentiments of Ginny, Joshua, and Ronald echo those of Eliasoph's (1998, 1999) volunteers, whose idea of good citizenship hinged on the ability to cheerfully roll up one's sleeves or get out one's checkbook to do something or make a difference as an individual.

While a substantial proportion of letter writers (31 percent) advocated such individual acts of assistance, a full half of writers understood the disaster in political terms, contradicting my expectations given Eliasoph's findings (Eliasoph 1998, 1999). Those who favored talking politics as a response in and of itself to the Katrina disaster used the national newspaper opinion forum to criticize the government (39 percent), to discuss the social structural factors present in New Orleans that may have inhibited evacuation (17 percent), or to make suggestions for government action. Again, in opposition to Eliasoph's research (Eliasoph 1998, 1999), I found that only 17 percent of letter writers expressed distaste for "playing" politics or the "blame game," which for these folks meant "name-calling," "pointing fingers," and just generally being critical and negative (*USA Today*, September 6, 2006, p. A20; September 16, 2005, p. A10). While these results might initially appear to disagree with Eliasoph's findings, her argument is largely based on context (Eliasoph 1998, 1999). It is probable that the context of a face-to-face group meeting within one's own community is sufficiently different from an anonymous discussion forum in a national newspaper to produce these divergent findings. Coupled with the self-selection of more politically oriented individuals into the group of letter writers, the context of a public newspaper forum likely makes it more conducive to generating greater discussion around the issues of race and politics. It is noteworthy that nine writers (17 percent) implicated social structural factors, such as poverty, old age, and illness, in preventing many New Orleans residents from evacuating before Katrina sideswiped the city and broke the levees. With no basis for comparison, however, it is difficult to say whether the hurricane revealed previously unknown, social problems to the American public.

THE KANYE RACE DEBATE

Overall, eleven out of the fifty-four letter writers, or about 20 percent, referenced race directly, using words such as "race," "black," "white," and "African-American." Given the demographic makeup of the city of New Orleans and those that did not evacuate, most of the media images that circulated in Katrina's wake depicted African Americans (Frymer, Strolovitch, and Warren 2005). Combined with the media controversy over Kanye West's September 2 comments, particularly his statement that "George Bush doesn't care about black people," the information environment during the three-week period immediately after Hurricane Katrina was replete with racial cues and frames (de Moreas 2005). Even given the influence of *USA Today*'s opinion editors' selection processes, it is significant that only 20 percent of the published letters during this time made mention of race. Previous research has shown that people forego public discussions of race for a number of reasons. They might ascribe to a color-blind ideology and consider race irrelevant (Bonilla-Silva 2003; Feagin and O'Brien 2003); they might be afraid of revealing racist beliefs (Myers 2005) or being called a racist (Wachtel 1999); or they might place race in the same category as other sociopolitical problems, which they consider too discouraging, and therefore unbecoming of good citizens, to discuss (Eliasoph 1998, 1999).

Of these eleven letter writers, eight entered the debate about whether race played a role in the federal response or commented upon blacks' and whites' divergent perspectives on the disaster. Four writers supported the idea that race factored into the federal response or at least expressed understanding of why African Americans might hold this opinion. The other four writers expressed strong feelings of displeasure related to the introduction of race as a relevant issue. Given the survey data showing a divide along racial lines in interpretations of the role of race in Katrina's aftermath, this equal division between those who generally support the position of Kanye West and those who disagree with him likely reflects journalistic practices rather than an actual opinion distribution. Despite these letter writers' divergent opinions regarding the role of race in the disaster, a common thread ran throughout the majority of their writings: class. In fact, nine out of the eleven letter writers who mentioned race also made reference to class, suggesting that these issues, particularly blackness and poverty, are strongly linked in the public mind.

KANYE'S RIGHT!

Drawing upon the models of other scholars (Eliasoph 1998, 1999; Gilens 1999; Walsh 2004), I carefully analyze the letters that directly mention race or racial categories in order to illuminate the patterns of discussion that Americans use to talk about race. I begin with Steve, "a middle-age, middle-class, white guy" from Massachusetts, who finds the racial divide in opinion regarding the federal response predictable. To Steve, it is obvious that the government would have acted differently had the hurricane occurred in a wealthy area, which he believes "by definition would be white." His association of whiteness with wealth and his usage of the term "poor people" to reference New Orleans residents are evidence that perhaps Steve also believes that the poor are by definition African American:

> I am a middle-age, *middle-class, white guy.* I find it amazing and disheartening, but expected, that there is such a *divide along racial lines* regarding the response to the Gulf Coast disaster. *Of course the federal response would have been different had the hurricane struck a predominately rich area—which, I believe, by definition would be white.* After all, the current political credo is "of the rich, by the rich and for the rich." Having lived in Florida, I know that the primary disaster avoidance technique is "get in your car and drive away." This could not work in areas with a concentration of *poor people who do not own cars* and where public transportation is not designed for evacuation. As New Orleans Mayor Ray Nagin has made clear, evacuations to the convention center and Superdome were desperate measures. . . .And Nagin based his decisions on the expectation that the cavalry would be there within a day, not days. No federal plans seem to have been developed to take care of the *most vulnerable in our society.* (*USA Today,* September 15, 2005, p. A12; emphasis added)

Connie of Washington, who identifies herself as "a sixty-year-old white person," makes a very similar argument by hypothetically changing the race and class standing of the hurricane victims: "On the third day after Katrina struck, with no relief in sight, I said to my husband, '"I'll bet if New Orleans were made up of *mostly rich white people,* they'd be getting help in there sooner.' I also told him I thought the response would *worsen race relations.* . . .Would our government have allowed thousands of *upper-crust white folks* to sit in that holding tank for days on end without food or water, in those deplorable conditions? I don't think so" (*USA Today,* September 15, 2005, p. A12; emphasis added).

Although they are among the most progressive letter writers in my sample, both Steve and Connie seem to take the correlation between blackness and poverty for granted; like their more conservative counterparts (see the letters of Becky and Ed below), they unquestioningly accept it as a social fact.

Lending support to those who had hoped that the hurricane would expose the persistence of racial inequality, Greg of Utah goes beyond just associating race with class by pointing out that there is something wrong with this close relationship: the poor are "disproportionately" African American. Greg also offers evidence that some Americans did gain larger, contextual understandings of racial issues when thinking about Hurricane Katrina.

> The disjunction of views that blacks and whites have of the federal response to Katrina may relate to the different views they have of *black American poverty*. This Republican administration has undeniably favored the *rich and disproportionately white*, while attempting to cut the social safety systems relied upon by the *poor and disproportionately black*. When this is factored in as part and parcel of the federal response, it is a logical conclusion that there is some racist principle at work. Perhaps, as many are doing now, the debate begun by Katrina should be expanded to how we, as a society, deal with the invisible poor. (*USA Today*, September 15, 2005, p. A12; emphasis added)

Greg's critical thinking around the issues of race, policy, and politics is encouraging. Unlike Ed of Texas, who criticizes African American leaders for failing to recognize Bush for positive actions (see below), Greg shows an understanding of why this might be the case given the Republican Party's policy platform. He argues that it makes sense for African Americans to view the federal response as racist in light of the current administration's attempts to shrink safety net policies. In addition, unlike Eliasoph's volunteers (Eliasoph 1998, 1999), Greg seems to value public political talk among citizens in his proposal to continue the national conversation on black poverty that he perceives Katrina to have started.

KANYE'S WRONG!

An examination of letters that disagreed with the introduction of race into the post-Katrina discussion paints a much different picture than the letters above. Analyzing these letters, however, helps to explain why more letter writers did not join the Katrina race debate. Letter writers who insisted that race was not a factor in Katrina's aftermath utilized arguments based on two types of evidence: the positive proof of individual acts of help on the part of many Americans and the lack of evidence for intentional discrimination along racial lines. Eric from Kansas, for example, takes offense in response to the comments of Kanye West and others:

> I am *outraged* at rapper Kanye West and other critics who have turned the aftermath of Hurricane Katrina into an issue of *ethnicity and race*. . . . *We, as Americans*, have come together and supported the survivors of Hurricane Katrina with food and water. . . . *Let us not confuse a natural disaster with an ethnic issue*. This comparison is a slap in the face for those of us who have poured out our hearts and given donations. (*USA Today*, September 12, 2005, p. A22; emphasis added)

Becky from Texas finds the race debate similarly upsetting:

> How *distressing* it is to hear the continued focus on the division Hurricane Katrina cre-ated in this country vs. the unity 9/11 brought about. . . . *For those of us who are actu-ally working with Katrina survivors*, these are times not of division, but of healing.

These are times of crossing *racial and economic lines* to help one another and perhaps to heal some of the *old wounds. USA TODAY's poll focusing on the racial issue creates a problem. It also may dissuade some from doing volunteer work.* (*USA Today*, September 15, 2005, p. A12; emphasis added)

Both Becky and Eric use evidence of individual acts of charity to deny the salience of race in the hurricane's aftermath. Many Americans certainly made donations and volunteered to provide assistance to Katrina victims; however, it is unclear why people believe that after-the-fact help by individual citizens counters government criticism or charges of structural racism. Eric's proclamation of American identity and Becky's volunteer-style attitude offer a window into why this might be the case: their ideas about what it means to be a good citizen include individual acts of help but exclude racial/political talk. These letter writers confirm Eliasoph's (1998, 1999) conclusions in that they perceive racial/political talk as "distressing" in Becky's case or as "a slap in the face" in Eric's case. Indeed, Becky plainly states her volunteer orientation when she highlights one of the worst aspects of "focusing on the race issue": it "may dissuade some from doing volunteer work." While Eliasoph's (1998, 1999) volunteers never said that talking about race or politics was anything but pointless and ineffective, Becky asserts that talking about race actually "creates a problem." Interestingly, she contradicts herself by acknowledging the persistence of racial problems, such as "racial and economic lines" that are not usually crossed and "old wounds" which may not be healed. Still, she sees no value in talking about racial divisions or past harms since discussing such matters only makes things worse. To Becky, volunteering is the best way to fix old problems without creating new ones.

Another Texan, Ed, directs his "frustration" directly at African Americans. He does not understand why African Americans would perceive President Bush's efforts as less than positive, because, for him, the "outpouring of selfless love and hospitality" that citizens of the president's home state have bestowed upon Katrina survivors somehow negates the lacking federal response:

> I read with frustration USA TODAY's report on the perceptions of African Americans toward the president for his role or lack thereof in the Katrina relief effort. . . . And I find it hard to recall when any African-American leader has stepped up and recognized President Bush for anything positive. Never before has the president's home state of Texas—considered a bastion of conservatism—shown such an outpouring of selfless love and hospitality. As a Texan, I don't care in the least whether victims are white or black, poor or rich, Democrat or Republican. Katrina didn't discriminate, and neither will most of those helping our neighbors to the east. . . . To those who see this catastrophe as an opportunity to gain political ground, stop your bashing. And to USA TODAY and other media, why not direct polls at Katrina's victims now living in Texas to determine what they think? My guess is that you'd get quite different results. (*USA Today*, September 15, 2005, p. A12; emphasis added)

Ed demonstrates his understanding of racism in terms of acts of individual prejudice when he claims that "Katrina didn't discriminate." He does, however, acknowledge at least residual racism by saying that "most" Texans will not discriminate in their efforts to help victims of Hurricane Katrina. Even still, Ed seems convinced that a poll of survivors who evacuated to Texas would show that experiences with hospitable, helpful

Texans had changed African Americans' views of the President and government's response. In stark contrast to conservative Ed's homegrown sympathy for the president, Michael of Ohio demonstrates that Americans do not have to be Bush supporters in order to come to his defense against allegations of racism:

> *These absurd assumptions that race played a role in the federal relief effort following Hurricane Katrina are preposterous.* I agree the federal response to Katrina was initially lethargic, but the ineptitude of the federal government did not discriminate. Both whites and blacks were dreadfully impacted by the government's initial ineffectiveness. I concur that *President Bush should be criticized—not for any racism. . .President Bush is not a racist. He's merely incompetent. (USA Today,* September 15, 2005, p. A12; emphasis added)

Like Ed, Michael understands racism as an act of intentional malice. Even though he is much less sympathetic to President Bush than Ed, he is appalled that people might call the president a racist. However, the originator of the Katrina race debate, Kanye West, did not call President Bush a racist, either; he called him indifferent. While alleging indifference to human suffering based on skin color is certainly not an attempt at flattery, it is definitely distinct from calling someone a racist (Wachtel 1999). It may be that some people equate talk of race with talk of racism and therefore immediately transformed the terms of debate in the aftermath of Hurricane Katrina from passive indifference to active racism. Paul Wachtel (1999), a psychologist who studies interracial dialogue, metaphorically makes the negative consequences of such a transformation abundantly clear:

> Accusing a guilty man of the wrong crime is one of the greatest gifts one can bestow upon him. It fosters an orgy of self-righteous conviction of innocence, and conveniently diverts his attention from the offense of which he is truly guilty. In similar fashion, the ubiquitous claim that racism is the cause of the grievous circumstances of life in our inner cities is, ironically, enabling white America to slough off its responsibility for the shameful neglect of the least privileged members of our society. The real crime of which white America is now most guilty is not racism. It is indifference. Understanding the difference between the two is a crucial step in liberating ourselves from the sterile and unproductive impasse that has characterized the dialogue on race relations in recent years. (Wachtel 1999, 37–38)

As was the case in Katrina's aftermath, such unproductive dialogue limits opportunities for increasing understanding of structural inequalities.

In summary, the public discourse that immediately followed Hurricane Katrina provided little evidence for "new racism" models. While Katrina's predominantly African American victims were frequently referred to as fellow Americans, the rosy view of history that accompanied writers' attempts to generate material support for the victims may have served to further erase our racial past from collective memory. The vast majority of letter writers failed to address race at all, but of the eleven who did, nine also made reference to class standing. Still, just two of these nine letter writers acknowledged the association between race and class as a problem (*USA Today,* September 14, 2005, p. A10; September 15, 2005, p. A12). Finally, just one letter writer showed an understanding of the longstanding societal forces, which some scholars have termed structural racism, that drive racial inequality (*USA Today*, September 14, 2005, p. A10). In general, letter writers seemed to understand racism purely in terms of individual acts. Hence, these writers disagreed with the introduction of race into the post-Katrina discussion because neither Mother Nature, the federal government, nor Americans who volunteered as part of relief efforts were seen as being discriminatory on the basis of race. Indeed, focusing on race was discouraging and insulting to those who were trying to help through individual acts, such as volunteering and making donations. Furthermore, my data provide evidence that a debate that originally began with accusations of indifference was "transformed" into one about racism (Walsh 2004).

IMPLICATIONS

If a tragedy like Hurricane Katrina cannot alert Americans to the realities of racial inequality in the United States and prompt widespread public dialogue aimed at greater interracial understanding, is there really any hope for a racially equitable society? How can policy makers gain support for programs directed at reducing racial inequality if the public is unwilling or unable to collectively acknowledge racial inequality as a societal problem? What mechanisms might be employed in the future to structure public discussions in ways that would facilitate heightened understanding of racial inequality? Some scholars have suggested the utility of newspapers for informing the public about contemporary political issues (Delli Carpini and Keeter 1996). Despite extensive news coverage of Hurricane Katrina, my research demonstrates that media efforts alone are not enough. The complex processes through which letter writers interpreted and discussed the Katrina disaster did not produce the result for which optimistic journalists and scholars, including this researcher, had hoped.

Walsh (2004) demonstrates that the use to which people put media information often depends upon the group in which they discuss and interpret it. Many members of the men's group that Walsh observed were former high-school classmates who had lived together in the same town for their entire lives and shared similar worldviews. These men frequently discussed politics in reference to their similar social identities, which they partially constructed based upon membership in their cohesive, homogeneous group. By interacting with each other, the men reinforced their political views and social identities as well as their notions of "we" and "us." Along with the sense of community on which they built their social identities, however, the men also distinguished themselves from those they identified as "others," "they," or "them." The men "otherized" liberals, university students and faculty, and particularly, African Americans. At the same time that it produced social and democratic desirables, like solidarity, social capital, and increased political engagement, membership in this homogenous group also had a dark side: the exclusion of other social categories of Americans from the group's sphere of concern (Putnam 2005; Walsh 2004).

To counteract the in-group, out-group identity perspectives that undermine the social and democratic utility of traditionally homogenous, voluntary associations, Walsh offers the solution of local, inter-group dialogue programs. In these programs, individuals of diverse backgrounds volunteer to participate in a series of facilitator-led discussions regarding pressing social issues, such as racial tensions or educational achievement gaps (Walsh 2004, 2005). These sessions "often end in 'action forums' in which the participants propose and plan ways to address the issues they have discussed" (Walsh 2004, 191). "Through the dialogues, participants lay open their different experiences, testify to their perceptions of other groups, and thereby gain awareness of the perspectives through which other people view public problems. . . .Self-reports from participants in community dialogues suggest that these programs succeed in. . .achieving a more pluralistic conception of community" (Walsh 2004, 192–93). At the national level, Bill Clinton's Initiative on Race, entitled *One America in the 21st Century*, facilitated various types of community dialogues about racial issues around the country with the goal of establishing "common ground" (*One America in the 21st Century* 1998, 2).

Other scholars emphasize the potential utility of combining such dialogue groups with media as a strategy for increasing interracial understanding. In a recent, quasi-experimental study, Hernando Rojas and his colleagues (2005) examine the effects of media consumption on political dialogue using a television documentary of the racially motivated murder of James Byrd, Jr. by three white men in Jasper, Texas. These researchers find that watching the documentary and participating in an interracial dialogue group about it increased awareness of and willingness to discuss racism (Rojas et al. 2005). Overall, the above findings suggest that combining media and discussion in a heterogeneous group setting may be the most successful strategy for increasing interracial understanding.

CONCLUSIONS

In their efforts to understand and respond to the Katrina disaster, ordinary Americans who wrote to *USA Today* overwhelmingly identified New Orleans residents as fellow Americans in need. Of writers, 20 percent identified Katrina victims as blacks or African Americans and 17 percent acknowledged pre-existing social factors that contributed to the scope of the disaster. Furthermore, nine of the eleven letter writers who mentioned race also mentioned class issues, illustrating a robust link between these two concepts in the public mind. Unfortunately for those who had hoped that Hurricane Katrina would reveal continued racial inequality in the United States, only two of these nine letter writers viewed the overrepresentation of African Americans within the lower tiers of society as a problem that ought to be solved. In addition, just one letter writer demonstrated any understanding of the structural and historical forces that determine life outcomes. These findings show that the public discourse in Katrina's aftermath was not one of racial inequality and suggest that the ways in which Americans discussed race prevented the "exposure" that many believed would occur.

The American public failed to see racial inequality in Katrina's aftermath for two main reasons. First, the strong notions of collective American identity that Katrina produced drew on a rosy, fictional view of history; the idea that Americans have always helped one another likely served to further obscure public awareness of negative historical and structural forces that influenced the lives of African Americans even before Hurricane Katrina made landfall in New Orleans. Second, Kanye West's introduction of race into the post-Katrina debate, though mainly a structural argument about government indifference and maltreatment, was transformed into a discussion of individual acts of racism. When any discussion of race ultimately turns into one of racism, my analysis of letters shows that Americans, particularly whites, become distressed, insulted, and outraged. Such defensive reactions halt productive dialogue and make greater understanding of structural racism and recognition of racial inequality impossible. This was particularly true in Katrina's aftermath when perceived charges of individual racism proved largely incompatible in the public mind with Americans' outpouring of individual acts of help.

If the United States is ever to live up to its egalitarian ideals, the government, the academy, the media, and the American public all need to work together to confront our nation's ugly racial past *and* present. Combining media that uses racial frames of equality with well-facilitated, interracial dialogue groups is one method that we already know works. In the tradition of President Bill Clinton's Initiative on Race, this method should be used to begin an ongoing, national dialogue about racial inequality in order to educate the American public about the historical and structural forces that continue to shape the lives of African Americans today. To do their part, social scientists should develop interracial dialogue experiments that will strengthen this method and allow for the discovery of other variables that influence racial understandings. From these experiments, "pre-packaged" dialogue programs could be established and distributed to government organizations and civic groups.

While the present study has shed light upon how Americans from across the nation publicly talked about race in Hurricane Katrina's immediate aftermath, the ways in which citizens will continue to discuss race as rebuilding efforts in New Orleans progress is an equally important topic of study. Specifically, ethnographic research of race talk among those directly involved with planning and rebuilding activities in New Orleans, such as members of community organizations, local officials, and residents, would greatly increase scholarly understandings of the repercussions of discussion—how public discourse perpetuates or redresses racial inequality.

REFERENCE LIST

Appiah, Anthony, and Amy Gutmann. 1996. *Color conscious: The political morality of race.* Princeton, NJ: Princeton University Press.

Bonilla-Silva, Eduardo. 2003. *Racism without racists: Color-blind racism and the persistence of racial inequality in the United States.* Lanham, MD: Rowman and Littlefield.

Dawson, Michael. 1994. *Behind the mule: Race and class in African-American politics.* Princeton, NJ: Princeton University Press.

Dawson, Michael, Cathy Cohen, and Melissa Harris-Lacewell. 2006. *2005 Racial Attitudes and the Katrina Disaster Study.* Initial report available at http://www.melissaharrislacewell.com.

Delli Carpini, Michael X., and Scott Keeter. 1996. *What Americans know about politics and why it matters.* New Haven, CT: Yale University Press.

de Moraes, Lisa. 2005. Kanye West's Torrent of Criticism, Live on NBC. *The Washington Post,* September 3.

DeParle, Jason. 2005. Broken Levees, Unbroken Barriers. *New York Times*, September 4.

Druckman, James N., and Kjersten R. Nelson. 2003. Framing and deliberation: How citizens' conversations limit elite influence. *American Journal of Political Science* 47 (4): 729–45.

Dyson, Michael Eric. 2006. *Come Hell or High Water*. New York: Basic Civitas.

Eliasoph, Nina. 1998. *Avoiding politics: How Americans produce apathy in everyday life*. New York: Cambridge University Press.

———. 1999. "Everyday racism" in a culture of political avoidance: Civil society, speech, and taboo. *Social Problems* 46 (4): 479–502.

Feagin, Joe and Eileen O'Brien. 2003. *White men on race: Power, privilege, and the shaping of cultural consciousness*. Boston: Beacon Press.

Frymer, Paul, Dara Z. Strolovitch, and Dorian T. Warren. 2005. Katrina's political roots and divisions: Race, class, and federalism in American politics. *SSRC Forum on Understanding Katrina: Perspectives from the Social Sciences*. http://understandingkatrina.ssrc.org/frymer strolovitchwarren.

Gilens, Martin. 1999. *Why Americans hate welfare: Race, media and the politics of antipoverty policy*. Chicago: University of Chicago Press.

Gilman, Nils. 2005. What Katrina teaches about the meaning of racism. *SSRC Forum on Understanding Katrina: Perspectives from the Social Sciences*. http://understandingkatrina .ssrc.org/gilman.

Hetherington, Marc. 2005. *Why trust matters: Declining political trust and the demise of American liberalism*. Princeton, NJ: Princeton University Press.

Iyengar, Shanto. 1996. Framing responsibility for political issues. *Annals of the American Academy of Political and Social Science* 546:59–70.

Jackson, Stephen. 2005. Un/natural disasters, here and there. *SSRC Forum on Understanding Katrina: Perspectives from the Social Sciences*. http://understandingkatrina.ssrc.org/jackson/.

Kellstedt, Paul M. 2000. Media framing and the dynamics of racial policy preferences. *American Journal of Political Science* 44 (2): 245–60.

Kinder, Donald R., and Nicholas Winter. 2001. Exploring the racial divide: Blacks, whites, and opinion on national policy. *American Journal of Political Science* 45 (2): 439–53.

Klinenberg, Eric. 2002. *Heat wave: A social autopsy of disaster in Chicago*. Chicago: University of Chicago Press.

Kluegel, James R. 1987. Macro-economic problems, beliefs about the poor and attitudes toward welfare spending. *Social Problems* 34 (1): 82–99.

Massey, Douglas S., and Nancy A. Denton. 1993. *American apartheid: Segregation and the making of the underclass*. Cambridge, MA: Harvard University Press.

Myers, Kristen. 1998. Allegiances, coups, and color wars: A strategy for breaking the silence on race issues in the classroom. *Transformations* 9:183–95.

———. 2005. *Racetalk: Racism hiding in plain sight*. Lanham, MD: Rowman and Littlefield.

Nelson, Thomas E., Zoe M. Oxley, and Rosalee A. Clawson. 1997. Towards a psychology of framing effects. *Political Behavior* 19 (3): 221–46.

One America in the 21st century: Forging a new future. 1998. President's initiative on race, advisory board's report to the president. Washington, DC: Government Printing Office.

Orbe, Mark, and Tina M. Harris. 2001. *Interracial communication, theory into practice*. Belmont, CA: Wadsworth/Thomson Learning.

Pew Research Center. 2005. Commentary: The black and white of public opinion: Did the racial divide in attitudes about Katrina mislead us? October 31.

Putnam, Robert. 2000. *Bowling alone: The collapse and revival of American community*. New York: Simon and Schuster.

Rojas, Hernando, Dhavan V. Shah, Jaeho Cho, Michael Schmierbach, Heejo Keum, and Homero Gil-de-Zuniga. 2005. Media dialogue: Perceiving and addressing community problems. *Mass Communication & Society* 8 (2): 93–110.

Schuman, Howard, Charlotee Steeh, Lawrence Bobo, and Maria Krysan. 1997. *Racial attitudes in America: Trends and interpretations.* Cambridge, MA: Harvard University Press.

Sears, David O., Jim Sidanius, and Lawrence Bobo. 2000. *Racialized politics: The debate about racism in America.* Chicago: University of Chicago Press.

Shaw, Greg M., and Robert Y. Shapiro. 2002. The polls—Trends: Poverty and public assistance. *Public Opinion Quarterly* 66 (1): 105–28.

Stevenson, Richard W. 2005. Amid the ruins, a president tries to reconstruct his image, too. *The New York Times,* September 16.

U.S. Census Bureau. 2004. *Income, poverty, and health insurance coverage in the United States: 2004.* Washington, DC: Government Printing Office.

Wachtel, Paul L. 1999. *Race in the mind of America: Breaking the vicious circle between blacks and whites.* New York: Routledge.

Walsh, Katherine Cramer. 2004. *Talking about politics: Informal groups and social identity in American life.* Chicago: University of Chicago Press.

Walsh, Katherine Cramer. 2005. Communities, race, and talk: An analysis of occurrence of civic intergroup dialogue programs. Extension of a paper presented at the 2003 annual conference of the Political Communication section of the American Political Science Association, Annenberg School for Communication, University of Pennsylvania, Philadelphia, August 27.

Weisman, Jonathan. 2005. Economists, housing experts question proposals. *The Washington Post,* September 16.

CHAPTER 15

ORAL HISTORY, FOLKLORE, AND KATRINA

ALAN H. STEIN AND GENE B. PREUSS

The hurricane season of 2005 reached historic proportions in the sheer number of tropical storms and hurricanes, people displaced, homes and businesses destroyed, and lives lost to Hurricanes Katrina, Rita, and Wilma. The Associated Press overwhelmingly voted the Gulf Coast hurricanes as the number one story of 2005. The growing number of town meetings, conferences, and congressional subcommittees that have convened to hear the testimony and stories of both the disaster itself and the condemnations of the slow response of local, state, and national relief efforts provide an overwhelming amount of oral history material.

Dyan French Cole told the members of the House Select Committee investigating the response to Hurricane Katrina, "I have witnesses that they bombed the walls of the levee, boom, boom!" Cole, a community activist known to many of New Orleans's Seventh Ward residents as "Mamma D," was among several witnesses to address the Committee in early December 2005. Other witnesses told U.S. Representatives of the harsh treatment they received from U.S. military personnel, who kept refugees penned behind barbed wire, refusing requests for health care and first aid. "They set us up so that we would rebel, so that they could shoot us," another woman complained. "At one point they brought in two truckloads of dogs and let the dogs out." Another woman was even more emphatic: "We was treated worse than an animal. People do leave a dog in a house, but they do leave him food and water. They didn't do that. And that's sad." She explained her frustration:

> I'm from New Orleans, Louisiana and I was caught into the storm. I never thought New Orleans would have done us the way they done us. I didn't realize what was going on until maybe the third day after I was trying to get out of that place—they would not let us out. I was on top of the Interstate, the Interstate in front to the Superdome and some guys came along in an Ozone Water truck and picked up a lot of people and we got near as far as getting out. They turned us around with guns. The army turned

us around with guns. Policemen. And I realized then that they really was keeping us in there. And you want me to tell you the truth, my version of it? They tried to kill us. When you keep somebody on top of the Interstate for five days, with no food and water, that's killing people. And there ain't no ands, ifs, or buts about it, that was NOPD [New Orleans Police Department] killing people. Four people died around me. Four. Diabetes. I am a diabetic and I survived it, by the grace of God, but I survived it. But they had people who were worse off than me, and they didn't make it. Old people. One young woman couldn't survive it because of the dehydration. So I mean, this is what you call NOPD murder. Murder. That's what I call it. What else would you call it?

Certainly, historians writing about the effect of the hurricanes and the historical significance of the 2005 hurricane season, the damage visited on the Gulf Coast, and the changed lives of the people of New Orleans and the American South will assess and analyze the oral testimony, but how should we respond presently? Mamma D's story of levee bombings are among the stories circulating that lend credence to the growing belief that racism and a lack of concern for victims in poorer sections of the Crescent City influenced the government's reaction to the hurricane's victims. For historians, oral history methodology helps with conducting, collecting, and evaluating personal narratives that can shed the light of experience on a path increasingly darkened by the overwhelming number of accounts that emerge on a daily basis.

WHAT IS ORAL HISTORY AND HOW DOES IT PRESERVE COMMUNITY?

This chapter explores the uses of oral history in documenting and interpreting Katrina. As a general matter, oral history is the application of modern technology to history's most ancient technique of gathering historical material: using a recording machine to preserve both the questions to, and answers of, eyewitnesses or participants concerning selected aspects of history. For centuries, stories and traditions were passed orally from one generation to the next. One of the stories that emerged from the massive tsunami of 2004 that killed hundreds of thousands is the account of how relatively few died on the Indonesian island of Simeulue. Residents remembered the stories their grandparents told them of the "semong"—big waves—that followed earthquakes and so headed for high ground when they first felt the earthquake.

Oral history became a tool of significant historical inquiry and documentation in both the literate and nonliterate worlds, but it was with the development of recording technology and the ability to capture the actual voices of narrators that the practice became more popular and legitimate. During the Great Depression, the Federal Writers' Project employed writers and folklorists to document the stories of ex-slaves, immigrant families, famines, and floods.

While the field was legitimized in 1948 by the creation of the Columbia University Oral History Program, it was not until the 1960s and the development of the new social history that the oral history movement came into its own, as it focused on the lives and experiences of "ordinary people." History "from the bottom up" became favored by social historians and activists who viewed oral history in the community as an empowering process for interviewer and narrator alike: it brought to light the voices of those who had been silenced by the dominant historiography due to race, gender, and class biases.

Aided by the world of television and digital media, oral history is a way of documenting urgent events and insights that otherwise might not be recorded. The numerous Katrina projects cited in this chapter illustrate how evacuees are active participants in their own historical drama. Through rich audio-video interviews, their life stories present enormous opportunities for uses from pedagogy to research. But they also present profound challenges: How reliable are the interviewers and the oral sources? How will the interviews be accessed in their audio and video form? When should we even begin collecting "urgent" oral history of a catastrophe? Finally, how do we interpret what we are collecting?

WHEN SHOULD WE BEGIN COLLECTING ORAL HISTORY?

Oral historians, a multidisciplinary group that includes folklorists, anthropologists, sociologists, and historians, have worked for many years with testimony, family stories, and oral accounts like those arising from the ruined areas of the Gulf Coast region. The techniques and methodologies oral historians have developed may help unravel the tangle of confused, horrific, and contradictory testimony from evacuees and survivors displaced by the hurricanes.

An important question oral historians (unlike their journalism colleagues) wrestle with is when to begin collecting the narratives. How our media responded is at the same time reflective of our interest in catastrophes and the media's need for ratings, but there is human compassion as well. *USA Today*'s Peter Johnson wrote "The human side of Katrina—tales of agony and misery that thousands of Katrina's victims still endure a month after the storm—also has gripped many reporters, who want to stay on the story indefinitely."[1] Reporters covering the 9/11 terrorist attacks and the Shuttle Columbia accident expressed similar sentiments. How should the oral historian respond? The gut-level instinct is to grab a recorder and get "in the field" as soon as possible. But historians, like journalists, are always concerned about maintaining an objectivity about their research and writing. Most historians achieve this "historical distance" by writing about the past, not contemporary events.

The discussion about Katrina oral history projects became one of the year's leading discussions on H-ORALHIST, an H-NET discussion list [http://www.h-net .org/~oralhist], with oral historians divided in their opinion. The subsequent discussion begged the question: When do current events become history? Historians usually prefer the temporal distance presented by the past. "I have never come nearer to contemporary history than a perspective of 25 years," Barbara Tuchman wrote. On September 8, 2005, Holly Werner Thomas wrote: "It may be too soon to ask, but it seems important that the stories of the refugees from Hurricane Katrina be told. Does anyone have any information about a possible oral history project regarding the hurricane and its aftermath?" One respondent admitted that he felt vexed about beginning to collect narratives too rapidly. After rushing to the refugee center located in Houston's Astrodome to collect oral history of this disaster, he admitted, "Yet when I arrived, I found myself wondering if we should allow some time for reflection." Would rushing in to collect oral histories while people were still displaced, sheltered in auditoriums, churches, and public buildings, be adding more traumas to their lives? "People will need time to settle down, time to reflect, and time to put their lives back together before they will be able to discuss how the disaster affected them. . . .They are still 'in the midst' of their story." He worried that rushing in and conducting interviews too close to the traumatic incident would be "working at the intersection of grief and history," as public historians have described in their experience working with 9/11 survivors.

Columbia University Oral History Research Office Director Mary Marshall Clark argues, "If projects can be well-organized, funded well enough, and there is an archive willing to accept tapes and transcripts and to disseminate them when the time comes—it is possible to begin early." Getting an early start could also be important not only to historians, but also to survivors as many interviewed after 9/11 "experienced relief in knowing that someone was willing to listen to personal accounts of horror."

Another dilemma facing oral historians and others working with people whose lives were shattered by disasters and tragedy is maintaining a professional distance. How involved should one become? Public historians working after the September 11 tragedy in New York explained that they confronted this dilemma. How should historians collect historical information for the future, and still provide for the needs of the present? Mary Marshall Clark described the importance of interviewing soon after an event in order to avoid the problems of contaminating individual memories with those of others, or a larger "public narrative." While conducting interviews for Columbia's September 11, 2001, Oral History Narrative and Memory Project, Clark reported that many interviewees expressed confusion about the meaning and significance of the attacks. There were multiple interpretations based upon the interviewee's ethnicity, sex, and culture. Clark and her colleagues found themselves asking, "Is this history yet? Is it memory? And. . . .Is it therapy?" Clark goes on to describe the challenges of doing Katrina-related oral history projects:

> I have a perspective on the questions raised about documenting Katrina. It's in part philosophical perspective, which is based on understanding how important oral history can be to those who tell. My perspective is this: I think it is both "never too soon" to tell, and "never too late" to tell. The story of Katrina and the failure of the relief effort [as well as] the impact on African Americans and others and the poorest of the poor is still unfolding. Oral histories of Katrina will involve not only the event and its immediate aftermath but its still unfolding legacy: the story of the largest displacement of people since the Civil War and maybe the largest displacement of children ever.
>
> These stories of displacement and resettlement, the decisions about how and whether to return, the redefinition of home, etc. are all stories that can be documented over time, and should be carefully planned so as to respect the human struggles that will be involved. We found with 9/11, though, it can't compare to Katrina in most ways, as the complete destruction and devastation of every kind of support is so much more extreme, that people who were displaced from their homes/lost jobs/were disrupted in major ways couldn't really focus on the oral history process. So we never pushed them to describe the difficulties that were ongoing unless they chose to describe these specific struggles themselves.

How Involved Should One Become when Collecting
"Urgent" Oral History of a Catastrophe?

In the spirit of "never too soon to tell," volunteers began conducting oral history research within weeks of the Katrina crisis, even as the crisis event unfolded. Most notably is the "Alive in Truth: The New Orleans Disaster Oral History and Memory Project" [http://www.aliveintruth.org], run by an all-volunteer group of interviewers/recorders, transcribers, translators, therapists, donors, and community members. The project began in early September 2005 outside the Austin Convention Center, which sheltered some 6,000 New Orleans evacuees, mostly African American residents who were trapped in the city after the storm. It is one of the first projects to utilize a "life history" approach, focusing on the entire life of the interviewee, not only their hurricane-experience stories: "I'm from the 7th Ward, and I was raised in New Orleans. I was raised and born in New Orleans. I worked as a babysitter. Went to school there, finished the 10th grade. Went to Warren Easton High School. At the age of 17, I decided to go to Job Corps, but so far I didn't make it there, so I wound up getting pregnant with my first daughter, which is named Dionnka T."

This kind of bottom-up approach also helps evacuees find their voice. By interviewing residents from New Orleans's Lower Ninth Ward, the project seeks to document individual lives, restore community bonds, and uphold the voices, culture, rights, and history of New Orleanians.

Some members of the Oral History Association expressed concern over Alive In Truth's volunteers because they become directly involved as advocates for the interviewee. Interviewers have driven narrators to sign up for Medicare, to access warehouses of clothing and furniture, and to file claims with the Federal Emergency Management Agency (FEMA). In the meantime, volunteers continue to encounter people who do not have furniture, who are missing family members and are unaware of resources to help locate them. In many cases, the volunteers encounter survivors with untreated medical conditions and who are not in contact with preliminary case management services. Volunteers are prepared to help evacuees find social services for help or to accompany them to FEMA information centers. In the minds of some oral historians, the Alive in Truth interviewers are too close to the problems, and so their personal "bias" will intrude on the interview since the interviewer takes an activist role by empathizing with the narrator who is contributing their story to the project.

Alive in Truth Project Director and former New Orleanian Abe Louise Young has been a Research Fellow for the Jewish Women's Archive, for The Project in Interpreting the Texas Past, and for the Danish-American Dialogue on Human Rights. She describes the importance of active oral history in shaping public policy by networking with other organizations—grassroots, nonprofits, oral history, human rights, state and national, people of color–led groups—in order to connect with a broader social change movement. She also believes that the legacy of Alive in Truth will be in preserving "the archive of accounts [that have] achieved rapid dissemination, educating and informing various constituencies: this is evidence of the broad scope possible with multiple media liaisons, a vision of justice, and belief in the speakers." In February 2006, Young also introduced an interpretative photography exhibit documenting Katrina at the Carver Library in East Austin, Texas. Entitled

"Surviving Katrina: Sharing Our Stories," it was one of the first oral history exhibits documenting the experience of Katrina evacuees. Through text and photographs, it tells the story of six New Orleanians and their experiences coming to Austin.

The project also collected over sixty interviews that are, on average, one to two hours long. Young has contributed the interviews to the U.S. Human Rights Network reports, as well as the Katrina Task Force, established at the Ben L. Hooks Center for Social Change at the University of Memphis.

How Do We Interpret What We Are Collecting?

In response to the critical issues laid bare by Hurricane Katrina and its aftermath, the Open Society Institute (OSI) offered a fellowship competition for projects exposing the persistent problems of poverty, racism, and government neglect. OSI, part of the George Soros Foundations Network, recognized the importance of documenting the hurricane aftermath by creating the Katrina Media Fellowships in 2006, a one-time award (averaging $15,000–$35,000) for journalists, photographers, and documentary film makers who are using sound recordings collected from oral history interviews for media production. Along with OSI, other national and state arts and humanities councils have supported oral history research projects in the past, but there exists a greater sense of urgency for funding these special projects, a fact recognized by the Oral History Association (OHA), the national organization of oral history professionals headquartered at Pennsylvania's Dickenson College.

The increases in recent disasters and tragedies have forced oral historians to reevaluate the timely collection and interpretation of "urgent" interviews. The OHA is primarily concerned with the slow response time from foundations for funding meaningful interpretative projects. It recognized the importance of emerging crises in oral history research by creating an Emerging Crises Oral History Research Fund in 2006. The fund is primarily designed to provide a more expedient source of funding and a quick turnaround time for oral historians to undertake "crisis research," as well as field work in the United States and internationally.

While Katrina has provided an opportunity for gathering documentary photography and oral history recordings, the real challenge has been the dissemination of the digital data. The digital revolution means that crucial aspects of information can be organized, searched, extracted, and integrated with relative speed and accuracy, leading historians like Roy Rosenzweig into the age of digital history. Rosenzweig is the founder and director of The Center for History and Media at George Mason University and Executive Producer of The Hurricane Digital Memory Bank (HDMB), which uses electronic media to collect, preserve, and presents the stories of Hurricanes Katrina, Rita, and Wilma. The George Mason Center for History and New Media (http://chnm.gmu.edu) and the University of New Orleans, in partnership with the Smithsonian Institution's National Museum of American History and other partners, organized this project. HMDB is an ongoing effort by historians and archivists to preserve the record of these storms by documented firsthand accounts, on-scene images, blog postings, and podcasts.

A project partner with the HDMB is the University of Southern Mississippi Center for Oral History and Cultural Heritage. They are engaged in a large-scale

"Hurricane Katrina Oral History Project" that will seek to capture the larger human experience of this landmark event. To date this project includes a partnership with scholars in six states (South Carolina, Louisiana, Tennessee, Utah, Virginia, and Arizona) trained in oral history. The Center is conducting over 1,000 interviews to capture the breadth of the experiences of those impacted, including emergency management officers, local officials, residents, volunteer relief workers, and those displaced by the storm. Other partner projects include:

The Photojournalists of Hurricane Katrina. This oral history and book project features a dozen of the photojournalists who covered New Orleans and the Mississippi coast post-Katrina. The project will feature unpublished photographs and personal narratives of the aftermath of the storm, allowing these journalists to relate a powerful story through a blend of words and images.

Archivists and Hurricane Katrina Project. Working with archivists and libraries along the Mississippi Gulf Coast, this project will use focus groups and individual interviews to assess the disaster preparedness of historical and cultural institutions. This initiative will examine existing protocols, how these were implemented, and identify issues that need to be addressed in planning. The outcome of the project will be to develop and disseminate documented, workable procedures to fulfill community information needs of archival processes when disaster strikes. This project is a partnership with Solinet, Mississippi Library Commission, and the School of Library and Information Science at The University of Southern Mississippi.

Hurricane Katrina and the Coastal Vietnamese Community. One of the most at-risk communities following the storm are the coastal Vietnamese. This project, working with humanitarian relief agencies along with local churches and temples, strives to capture the stories of these members of the Gulf Coast community, many of whom lost everything in the storm. Through gathering the personal narratives of the Vietnamese, this project offers a unique opportunity not only to affirm the strength of their community but also share with them as they rebuild.

Mississippi Nurses and Hurricane Katrina Project. This is an effort to interview nurses who worked during the immediate pre- and post-Katrina period on the Mississippi Gulf Coast and in the Hattiesburg area. Nurses, as frontline caregivers, were the ones at the "point of care," attempting to improvise and prioritize care as they could. These stories are critical to capture, for, as time goes by, many of these nurses who are already in a state of flux are becoming harder to identify and locate.

Hurricane Katrina Exhibit. A partnership with the Mississippi Sound Historical Museum to develop a museum exhibit at their facility in Gulfport, MS. The exhibit will feature oral narratives, photographs, and artifacts in one of the first efforts by a cultural institution to place the storm in historical perspective. One documentary project, entitled "People Power: Citizen Responses to Hurricane Katrina," is a photo-driven, electronic media project that tells the dramatic stories of how Louisianans responded with bravery, improvisation, and humanity to save lives, evacuate survivors, and continue to care for those affected by the hurricane and subsequent flooding. Each person interviewed will provide a case study of the recovery process, access to resources and information, and the interviewee's personal well-being. The narrators will represent the diversity of people living in southern Louisiana pre-Katrina. Their stories will shed light on the complexities of the social inequalities and shared histories of this region. The project combines photography, recorded interviews, and written stories to stimulate debate about the effectiveness of civil society's response to the disaster.

Dr. Lance Hill is the Executive Director of the Southern Institute for Education and Research at Tulane University. Hill worked as a community activist and labor organizer for twenty years before embarking on an academic career. From 1989 to 1992, he served as the Executive Director of the Louisiana Coalition Against Racism and Nazism (LCARN), the grassroots organization that led the opposition to former Klansman David Duke's Senate and gubernatorial campaigns. Hill is a consultant and appears in the New Orleans Documentary Project, produced by Organic Process Productions. The film uses oral histories to interview "characters" like Lewis Taylor, a retired farmer and fisherman, and to document the impact of uprooting communities, as well as the effect this will have on future land development and gentrification plans for the area. The film examines controversial issues of governmental power, such as the use of eminent domain and privatization.

Oral history can be used as a tool in collecting social history and folklore. Folklorists must confront "tall tales" or urban myths. Can memories always be trusted? How do we respond to stories so fantastic that they tax credulity, even when they come from dispassionate sources? For example, Tulane historian Douglas Brinkley witnessed Federal Emergency Management Agency (FEMA) officials prohibiting private citizens and people in unauthorized trucks from trying to aid others. Rescuers pulled people from the water, he reported, but medical aid was absent; instead, FEMA officials stood idly by. Brinkley's new book, *The Great Deluge: Hurricane Katrina, New Orleans, and the Mississippi Gulf Coast*, draws upon hundreds of oral history interviews and arrives at the conclusion that it was not a natural disaster at all but a failure of government—"one that, through breached levees and massive government incompetence, the country brought upon itself."

A coalition of Louisiana- and Mississippi-based scholars, coordinated by Maida Owens, director of the Louisiana Folklife Program, has developed two projects designed to empower evacuees and help contribute to public policy decisions. One program is entitled "In the Wake of the Hurricanes: A Coalition Effort to Collect Our Stories and Rebuild Our Culture."[2] Folklorist Susan Roach, an English professor and folklorist at Louisiana Tech University in Ruston, began talking to other folklorists around the state. Roach teamed up with Shana Walton, former associate director of Tulane's Deep South Humanities Center to design a project and sought a grant to help pay interviewers and hire unemployed hurricane survivors to conduct interviews. They model their program on the New Deal Writers' Project, which paid unemployed writers and artists to do interviews and life histories with ex-slaves and survivors of the Dust Bowl. Together, they have recruited about 100 people, from Mississippi to California. Sponsored by the American Folklife Center at the Library of Congress, these projects involve seven universities and faculty throughout Louisiana.

Walton had conducted several interviews at a New Orleans–area shelter shortly after the hurricane struck but realized that it was too soon after the event. Because of the chaos and stress, it was not the best place to do formal interviews or field work. She then began collecting oral histories for her project on Little Black Creek, a FEMA Camp in rural Mississippi, one of dozens of temporary housing camps established by FEMA to house people displaced from New Orleans and the Gulf Coast after the hurricanes. The project is based on relief work and oral histories with camp residents between November 2005 and January 2006, giving an overview of demographics of the approximately 200 camp residents; how their individual trajectories led them to be in a FEMA camp; their assessment of the local and federal government; the biggest challenges facing them; and the attitudes of most toward returning to where they lived before the hurricane. Walton also documents the emergence of community in the camp, and the long-term view from the residents' perspective. Interviews show that camp residents are not only displaced from their homes, but also isolated from reliable sources of information about the rebuilding or from having any input in the process. This overview will be generally compared with others whose research looks at residents who lost their homes but have returned to the city and those who have chosen to relocate.

A similar effort is underway in Texas. As Houston's post-hurricane population swelled by 250,000, Carl Lindahl, a folklorist and English professor at the University of Houston, started conducting several interviews with evacuees, eventually expanding it into a project entitled "Surviving Katrina and Rita in Houston," with his colleague Patricia Jasper, a folklorist with over twenty years of experience. Their goal is to create as many as 3,000 narratives by teaching volunteers how to interview evacuees, because, as he explained, "We've found that a person who has gone through this is a much better interviewer than those who have not. I found the survivors to be heroes rather than victims." He added that the project will not stress the "traumatizing effects" of the storms themselves, nor dwell on the "horror stories," but instead will focus on the "cultural richness" in Houston's community, pointing to the influx of Louisiana evacuees after the 1927 Mississippi River flood that created a section of the city called "French Town."

"Narrating Katrina Through Oral History" (a project sponsored by the Albert Gore, Sr., Research Center at Middle Tennessee State University) is the title of a new initiative under the direction of Lisa Pruitt. Teams of student interviewers from Middle Tennessee State University, working under Pruitt's direction, are conducting oral history interviews with individuals who were forced to evacuate to middle Tennessee from coastal Louisiana, Mississippi, and Alabama before, during, and after Hurricane Katrina. The students are also interested in interviewing volunteer responders from the Middle Tennessee region. Several thousand people relocated to Middle Tennessee, either temporarily or permanently. Furthermore, hundreds of people from Middle Tennessee have traveled to the Gulf Coast as volunteer responders. The overall goal of this project is to create a documentary record of the experiences of as many of these people as possible through the medium of oral history. Tapes and transcripts of the interviews will become a permanent part of the Middle Tennessee Oral History Collection at the Gore Center. Teams of student interviewers will ask participants to describe their experiences evacuating; staying in shelters or with family or friends; and re-establishing their lives in Middle Tennessee (whether temporarily or permanently). It will also cover their perceptions of media coverage of the events, their evaluation of various agencies and organizations that responded to the disaster and with whom they had direct experience, and their hopes for their own futures and the future of the affected region. Volunteer responders will be asked to describe their motives for volunteering, the logistics of volunteering, the details of their work, their perception of the scope and impact of the disaster, their perception of media coverage, their perception of the performances of various agencies and organizations involved in responding with whom they had direct contact, and their feelings about the experience of volunteering in response to a major disaster.

The Historic New Orleans Collection (HNOC), a museum and research center located on the dry ground of the French Quarter, found itself at the center of the deluge and established oral history projects with the first responders from the New Orleans Police Department and the New Orleans Fire Department. The interviews with NOPD and NOFD members, entitled "Through Hell and High Water," focuses on the period between August 29, when Katrina made landfall, and September 15, when President Bush gave his address to the nation in front of the St. Louis Cathedral. It was difficult to gain the trust and cooperation of the departments due to the intense media scrutiny of the event. The breakdown of the social structure of the city is a focus of the NOPD interviews, as are the tensions between personal and professional responsibilities, as well as interaction with the citizens who remained there. But memories of both departments' past hurricane preparedness plans were also discussed. How these plans were implemented during the hours before the storm and how they unraveled during it and its aftermath are major themes of the project. Within the NOFD interviews is the documentation of the rescue operations of about twenty-five firemen, who, based on their own memories of Hurricane Betsy in 1965, took it upon themselves to bring their own recreational boats to the NOFD prestorm staging areas on the upper floors of downtown hotel parking garages and began boat rescue operations immediately following the storm. It is estimated that in the days following the storm that these twenty-five boats rescued between 12,000 and 15,000 people who were stranded on rooftops or trapped in attics. In addition to the interviews, HNOC has collected approximately 1,000 images taken during this two-week period by members of both the NOFD and NOPD.

Phyllis E. Mann coordinated the volunteer criminal defense attorney efforts in Louisiana to assist evacuated prisoners. She described her experiences in an article for *The Advocate*, the Louisiana Association of Criminal Defense Lawyers newsletter. "There has been so much bad this month in Louisiana," she recalls, and then "there has been so much good." In her own "oral history," she relates the stories she heard from people in jail cells:

> The first stories we heard were from men evacuated out of the many Orleans Parish Prison buildings. They received their last meal on Sunday night—it was a cold sandwich. During the night, the power began to go on and off as the Hurricane began to make landfall. Inside the jail, it was dark and the air soon became hot and stale. Without electricity, the controls for the cell doors, dormitory doors, and main doors were inoperable—everyone was trapped. Guards had been required to stay in Orleans during the storm and had been encouraged to bring their families to the jails, so there were children in those buildings also. And the guards were stretched to the breaking point—worrying about their own safety, worrying about the safety of their families, and not doing quite so much worrying about the safety of the people locked inside the cells. . . . The people who were locked inside were so very much like you and I and especially like our children. There was one young man who had been arrested for reading

Tarot cards without a permit. . . .There were college age folks who had come to New Orleans for a good time, but had the misfortune of getting a little too drunk just a day or two before a hurricane. There were young women who were pregnant;. . .middle-aged soccer moms who just had not gotten around to paying that speeding ticket; an older grandmother who was visiting her grandchildren and overstayed her visa from Jamaica; and then there were the poor of New Orleans who were arrested for sleeping on the street. . .and the stories go on and on. What all of these people had in common. . .is that they were all trapped together, inside of a building with no lights, with no air, and they had no food to eat or water to drink. . . .These are the stories that have broken our hearts."

HOW CAN WE LEARN FROM THE MISTAKES?

The debate over oral histories has often focused on the issue of historical accuracy. Like the dramatic accounts of the bombing of the New Orleans levees Mamma D remembers in the aftermath of Hurricane Katrina, other witnesses and officials have denied deliberate acts of destruction and malevolence. Instead, they note, the levees were bombed during the Great Mississippi Flood of 1927, and suggest that perhaps the collective memory of these two events long separated chronologically were somehow merged in the memory of the witness. Validation is a common phenomenon oral historians face. While some reminiscences might not be factually accurate, they are accurate in the psyche of the interviewee.

Captain Francis J. Arnona Jr. was a ship-docking captain for Crescent Towing and Salvage in New Orleans. Two months before Katrina, he had been through tropical storm Cindy. He had a different impression about the collapse of the levees. He confirmed this in an oral history interview he gave at the Gore Center:

Basically, New Orleans did well. Yeah, there was a lot of wind damage and all from the storm. A lot of windows, the roof on the Superdome was blown off. But, as far as the storm, that didn't cause it. That problem came when the levees breached. . .after the fact. Now, seems to me. . .they have levee boards down there and, you know, that's run by politicians again. Well, if those levees breached, there was a reason they breached. And they're supposed to be monitoring those levees and shoring them up constantly, but they got a little sitting back on their haunches, "Oh, we're all right. We're all right." Well, guess what? It wasn't all right. . . .

All they are is dirt and they stick a little slab of concrete on them. That's all they are. I'm going to tell you—a lot of those levees, there were so many broke loose barges and boats drifting around. They hit those levees, they cracked the cement, you know. And they said, "Oh, St. Bernard [Parish] and all got flooded because the levee gave away back there." But, if they ever looked real close, there was a barge sitting on the other side of the levee in somebody's front yard.[3]

One woman who left New Orleans and was interviewed in Austin tells a story where her fear, exaggerated by exhaustion, played tricks with her imagination:

We wanted to stay. We thought maybe the water was going to go down. We didn't know it was that damaged. You know, all over the city because for days we couldn't get out, you know, go nowhere, 'cause of the flood. So they took us, we got on the bus, took us to a bus, took us to the Louis Armstrong airport, and flew us here to Austin. Mm-hm.

'Cause I seen so much, so much happening, hearing so much happening; I was dev-astated. I was walking in the water, trying to find a way, you know, to get out? And I could 'a sworn I felt a body, or, something; somebody's body wrapped around my leg! But I know it couldn't 'a been true! [laughing]

Careful analysis has provided several explanations for the problems in relying upon multiple versions of oral testimony. Psychologists term the tendency for hostages to identify with their captors "Stockholm Syndrome," and interviews with former pris-oners of war, kidnapping victims, and torture survivors reveal that some victims do not assess any personal blame upon the perpetrators. Historian Alessandro Portelli explains that for some people, certain "climatic moments" overwhelm their ability to place these events into perspective; instead, they are "wholly absorbed by the total-ity of the historical event of which they were a part." For them, these events become epic stories in their memories.

Another method historians have used to overcome the conflict between narrative and documentary evidence is to compare oral traditions with the commonly accepted scholarly version of history. James G. Blight, a cognitive psychologist who is professor of international relations at the Watson Institute at Brown University, has pioneered what he terms "Critical Oral History." Critical oral history seeks to reevaluate historical events by bringing together academic scholars, leading actors and policy makers, in the event, and historical documents, in an attempt to arrive at a fuller understanding of how the historical event in question developed. By com-bining documents, scholars, and participants, Blight explains, the goal is to "learn to collaborate in an effort to get a more comprehensive picture of the historical reality during whatever events we are studying."

In the aftermath of Hurricane Katrina, many have criticized the treatment African American Louisianans received. Among the dozen or so hurricane research projects listed on the Louisiana Folklife Web site is one offering a social and environmental interpretation of events, entitled "Katrina Narratives of African Americans in an Unprecedented Diaspora: A Social and Environmental Oral History Project," coordinated by Dr. Dianne Glave from Tulane University's Bioenvironmental Research Department (which relocated to Atlanta following Katrina). Glave's proposal re-enforces the need for oral historians to expand on the news media's impressionistic reporting. She believes oral history interviewers share responsibilities with news media:

In the aftermath of Hurricane Katrina, the fragmented and harrowing pieces of many narratives of African Americans who were trapped in the Superdome, Convention Center, and their flooded homes have emerged on television and the internet. Some evacuated immediately while others were forced to wait many days to be rescued; most migrated to points across the United States; and many are now attempting to return to the Gulf region. As a result, the news media has opened an insightful dialogue across the United States and throughout the world concerning race, racism, and class. Scholars now have an opportunity to add to this exchange of ideas—not merely replicating the

news—as a catalyst for analyzing the historical context for this natural disaster by looking at African influences, the Middle Passage, enslavement, freedom, migration, the
Civil Rights movement and more. Out of this tragedy, I propose an oral history project that would give the Katrina narratives by African Americans scope, adding to what
is in the news [by] emphasizing the social and environmental implications.[4]

Ninety-year-old folklorist Stetson Kennedy infiltrated the Ku Klux Klan during the
1940s and exposed it in a series of documentary books like *Southern Exposure* (still
doing business as an investigative magazine). As the director of the Florida Writers'
Project Folklore unit in the 1930s, Kennedy worked with Zora Neal Hurston to
gather Negro folklore, interview ex-slaves, and expose slave-labor conditions throughout the South. Kennedy wasted no time in making his voice heard in support of the
African Americans who were trapped inside the Superdome. Watching the catastrophe
unfold on CNN angered Kennedy, who would later write, "It is up to every American
to assess, according to his or her conscience, our individual and national shame for
what happened before and after Katrina." He wondered what the men, women, children, infants, aged, and infirm victims of Katrina did to deserve such "bestial" treatment in the Superdome and elsewhere. "Absolutely nothing," he concluded:

> Much like the victims of the Holocaust, theirs was the misfortune of being the wrong
> kind of people. They were poor and, for the most part, dark-skinned. When Katrina
> took aim at New Orleans, they were included out of preparations for evacuation. And
> when push came to shove, and everybody else had gotten out, they were left stranded
> and abandoned in the tsunami of debris and excrement. And all that the great city,
> state, and nation offered them was to go to the Superdome.
>
> With the backdrop of a devastated city, it was all too obvious that whoever took
> them in would be stuck with them for a very long time—perhaps even forever.
> Evacuating Japanese-Americans from our Pacific Coast during WW II was quite dif-
> ferent. We wanted them out of there, ostensibly for reasons of security, and none of us
> objected, because Uncle Sam would be feeding them behind barbed wire out in the
> desert for the duration, and meanwhile, some of us could take over their homes and
> businesses. No such incentives in New Orleans.
>
> Was racism/bigotry an added factor on top of the economic tab? This observer was
> reminded of the time, a half century earlier, when a shipload of one thousand Jewish
> refugees from Nazi Germany sailed up and down the coast of America, hoping in vain
> for us to take them in. No nation in the world would open its doors, so most of them
> ended up in the ovens. It was shame on us then and shame on us now.

Narratives and public statements reflect deep-seated resentment against years of
poverty and discrimination. There is even fear that as communities rebuild, the discrimination will persist. Sibal Holt is the first African American and first woman to
head the Louisiana AFL-CIO. She expressed her concerns to an interviewer about the
social transformation and gentrification of poor communities devastated by the storm:

> A socially engineered community is what's being done here! They want to engineer
> what the community should look like. That is wrong! I don't care what anybody else
> says, that is wrong and that is not paranoia speaking. From the Governor on down
> they're talking about who should be allowed to come back. They gave certain people

one way tickets out of this state, OK? Well some of this money may be able to be used
to give them a ticket back into this state. This is their home. But let's be real, what people are booking on is that if these people aren't able to come home within a year, [then]
they're going to get settled somewhere else. But what areas of New Orleans and
Louisiana are we talking about? The French Quarter, Uptown, and the Garden
District—we're talking about the casino areas—what does that sound like to you? Did
you hear any names of any predominately African-American poor communities that
they're talking about? Now I'm not going to say that they're never going to get to
them—but it won't be tomorrow, trust me.

CONCLUSIONS

What will the "new" New Orleans look like? Who rebuilds, and who does not? The
answers to questions such as these will pique the attention of social commentators,
politicians, community activists, and historians long into the future. Sibal Holt
reflects the concerns shared by many displaced workers from the Gulf Coast.

Her concerns remind us of the importance of perspective: How we deal with one
disaster will teach us lessons that will help us prepare for the future.

> And all these issues have been brought up, but we need to keep them in the face of
> America because people easily forget, we forget so easily, but what has been happening
> down here in New Orleans and Louisiana is nothing new—people chose not to see it
> before. Katrina has brought out some of the ugliest things in this state, and that's the
> embarrassment—that the outside world has seen it. Katrina has brought out how our
> government treats the poorest of its poor—and that's an embarrassment. Katrina has
> brought it out, you know. I hope that all of America, that their hearts are touched, their
> conscience is raised and we need to do right by all our citizens because it's Louisiana
> today but there could very well be an earthquake in California tomorrow, or it could
> be a tornado in Oklahoma, you know. It could be anything, mudslides or snowstorms,
> where they would feel the same type of tragedy, not necessarily a hurricane but the dev-
> astation of that tragedy. Would you want the people of your state to be treated the way
> they treated the people in New Orleans?

Using oral history will help prepare not only for future Katrinas, but future race and
class struggles. The past connects us to the present, and to the future; it is part of
that human need for immortality of some sort. How investigators, lawmakers, and
the public wade through the mountains of oral witnesses to evaluate what went
wrong during and after Katrina will inform their decisions as they make plans for
future responses.

How Americans weave this tragedy into their collective history and the lessons we
take from the disaster will preoccupy the minds of pundits, historians, and both
political and community leaders for generations. However, the impact of Hurricane
Katrina is not limited to Louisiana, Mississippi, and the Gulf Coast. "Katrinaland"
has become a national and international phenomenon that has struck both the
nation and the world.

REFERENCE LIST

Alive in Truth. 2005a. Lynette T, interview, September 20, 2005. The New Orleans Disaster Oral History & Memory Project. http://www.aliveintruth.org/stories/lynette_t.htm.

Alive in Truth. 2005b. Clarice B., interview, October 12, 2005. The New Orleans Disaster Oral History & Memory Project. http://www.aliveintruth.org/stories/clarice_b.htm.

American Folklife Center, Folklife Program. 2006. List of hurricane research projects. http://www.louisianafolklife.org/hurricane_projects.html.

Brinkley, Douglas. 2006. New Orleans: Diary of a disaster. *Vanity Fair*. June.

Chronicle of Higher Education. 2002. "Critical oral history" as a scholarly tool. http://chronicle.com/colloquylive/2002/10/blight/.

Clark, Mary Marshall. 2003. The September 11, 2001, Oral History Narrative and Memory Project. In *History and September 11*, ed. Joanne Meyerowitz, 125. Philadelphia: Temple University Press.

Clark, Mary Marshall. 2005. E-mail message to H-Oralhist. http://h-net.msu.edu/cgi-bin/logbrowse.pl?trx=vx&list=h-oralhist&month=0509&week=c&msg=fipOoibMC7xz4lboKrUTSg&user=&pw=.

Dallas Morning News. 2005. Hurricanes' havoc overwhelming choice as top '05 news story, December 22.

Gardner, James B., and Sarah M. Henry. 2002. September 11 and the mourning after: Reflections on collecting and interpreting the history of tragedy. *Public Historian* 24 (Summer): 41.

Hodges, Leah. 2005. Testimony before the Select Bipartisan Committee to Investigate the Preparation for and Response to Hurricane Katrina, December 6, 2005. http://katrina.house.gov/hearings/12_06_05/hodges_120605.rtf.

Johnson, Peter. 2005. Media mix: Katrina could forever change how TV news covers storms. *USA Today*, September 25, Life section.

Kennedy, Stetson. 2005. Shame on us! Unpublished manuscript.

K'Meyer, Tracy E. 2005. E-mail message to H-Oralhist. http://h-net.msu.edu/cgi-bin/logbrowse.pl?trx=vx&list=H-Oralhist&month=0509&week=b&msg=bgZiWbA%2bLBxeRfPRAfGLZw&user=&pw=.

Louisiana Division of the Arts. 2006. In the wake of the hurricanes: A coalition effort to collect our stories and rebuild our culture. Folklife in Louisiana. http://www.louisianafolklife.org/katrina.html.

Mann, Phyllis E. 2005a. E-mail message to Chester Hartman, November 29.

Mann, Phyllis E. 2005b. "Hurricane Relief Aid," *The Advocate: Louisiana Association of Criminal Defense Lawyers* 2, no. 4 (Fall): 3, 50–51.

Mead, Jerry. 2005. *Labor Express Radio*, "Hurricane Katrina Reveals the True Race & Class Divide in U.S. Society: An Interview with Sibal Holt, President of the Louisiana AFL-CIO." 21 December, 2005. http://www.laborexpress.org/page7.html (accessed).

Myers, Lisa. 2005. Were the levees bombed in New Orleans? MSNBC. http://www.msnbc.msn.com/id/10370145.

Oral History Association. 2006. Emerging Crises Oral History Research Fund. http://www.dickinson.edu/organizations/oha/org_aw_ecoh.html.

Portelli, Allesandro. 1991. *The death of Luigi Trastulli and other stories: Form and meaning in oral history*. Albany: State University of New York Press.

Preuss, Gene B. 2005. E-mail message to H-Oralhist. http://h-net.msu.edu/cgi-bin/log-browse.pl?trx=vx&list=H-Oralhist&month=0509&week=b&msg=A%2bY3nNi WmbdMejZ5qoPGTQ&user=&pw=.

Scudder, John R. Jr., and Barbara Gulick. 1972. History's purpose: Becker or Ortega? In *History Teacher* 5, no. 4 (May): 41–45.

Shilds, Gerard. 2005. Project to record storm stories. *The Baton Rouge Advocate*, November 6.

Thomas, Holly Werner. 2005. E-mail message to H-Oralhist http://h-net.msu.edu/cgi-bin/logbrowse.pl?trx=vx&list=H-Oralhist&month=0509&week=b&msg=kevhqGOTIW8fC dHxg0DZkg&user=&pw=.

Tuchman, Barbara. 1996. Distinguishing the significant from the insignificant. In *Oral history: An interdisciplinary anthology*, ed. David K. Dunaway and Willa K. Baum, 2nd ed. Walnut Creek, CA: AltaMira.

Young, Abe Louise. 2005. Alive in truth. http://www.aliveintruth.org/a_pages/our_process .htm.

NOTES

1. Author John R. Tisdale wrote in *The Oral History Review* that "the journalist and the oral historian, however, are concerned with leaving information in a physical format to be studied later. Both are concerned with recording information, both are concerned with accuracy, and both rely on the interview as the primary source of information and credibility." John R. Tisdale, "Observational Reporting as Oral History: How Journalists Interpreted the Death and Destruction of Hurricane Audrey," *The Oral History Review* 27, no. 2 (Summer–Fall 2000): 41.

2. According to Maida Owens in an e-mail to co-author Stein, her organization had the first of four anticipated budget cuts in 2005–06. State government will greatly be affected by the hurricanes, and learned society meetings will not likely be supported at all. Even travel expenses to the American Folklore Society will not supported, forcing staff to take vacation days to attend. No changes are anticipated in the 2006 budget.

3. For more information, please contact Gore Center Director Lisa Pruitt (lpruitt@ mtsu.edu) or Project Coordinator Sarah Elizabeth Hickman (seh2r@mtsu.edu).

4. In the wake of Hurricane Katrina, cable news networks experienced huge spikes in ratings. See "Eyes On the Storm," *Broadcasting & Cable* 135, no. 36 (September 5, 2005): 6.

PART IV

REIMAGINING THE PAST AND RECONSTRUCTING THE FUTURE

What Happens When the Footprints Shrink: New Orleans and the End of Eminence

Julianne Malveaux

When people talk about New Orleans these days, they talk about the footprint shrinking. What they seem to mean is that the city needs to be content to be smaller and more compact, more "safely" built and on higher ground. This concept is openly discussed, with no hesitation or embarrassment, as if a city that was once half a million people strong should now be half that size and as if it were OK that those who can't afford to come back don't.

If someone posited, openly, that the coming Chinese economic revolution would eclipse the United States, you would have an organized response and resistance aimed at dealing with the reality that China's economy is thriving, that a quarter of the world's consumers are Chinese, and that as China develops there are competitive issues that the United States must confront. There is no casual talk of a shrinking footprint when we look forward to shifting global realities. Instead, the United States behaves as if we will maintain our preeminence and role.

Considering the double standards and hypocrisies that define the United States of America, this chapter examines the concept of the shrinking footprint in New Orleans, as well as our nation's inevitable shrinking footprint in the world. In particular, this chapter considers whether our nation will be willing to allow New Orleans to shrink and fade away. Or will we attempt to fiercely defend our global position by invading countries we cannot possibly defeat, making global pronouncements we cannot possibly defend, and breaking international agreements because we don't play well with others? All the while learning the hard way that we must play on a field that we no longer dominate. It seems to me that much of our eroding global position hinges on the inadequate manner in which we have dealt with race matters—with those who lie on the margins.

My thesis here is that the United States responded so poorly in New Orleans that we have raised questions about our principles, efficiency, and integrity. In addition, our behavior raises questions about our fitness as a world leader. Questions of our fitness have been raised, naturally, because of the war in Iraq and the erosion of our international esteem. These issues are exacerbated by our response to our own citizens in the wake of a devastating hurricane and the preventable breach of levees.

To explore this thesis, this essay is divided into two parts. First, I explore the aftermath of Hurricane Katrina; then I look at the U.S. position in the world and ways that our kaleidoscope shifts because of the flaws illustrated from our response to the hurricane. I conclude that Hurricane Katrina and the levee breach is, perhaps, a metaphor for our nation's challenges around race, growth, engagement, and integrity.

KATRINA AND ITS AFTERMATH

When I link footprints in New Orleans to footprints in the rest of the world, my thesis largely hinges on race matters. Race matters in the United States shape our economic future. Ignoring race matters essentially erodes our eminence and shrinks our footprint in the world because our demographics are changing. If we don't deal with race, we can't deal with the future of America. But race matters are rarely discussed when we consider our nation's future. What do we see when we look in the mirror and when we look down the road? What do we see when we look at our country, and what do we see when we look at the future? What kind of decisions are we making now that will impact us in the future?

We could answer those questions by simply looking at the aftermath of the tragedy of August 29, 2005. The aftermath of the natural disaster of Hurricane Katrina, and the subsequent manmade disaster of the broken levees, gave living expression to numbing statistics about the quality of life for African American people, particularly in New Orleans. According to the African American Leadership Project,[1] a group of African American leaders who were personally affected by the hurricane and serve as watchdogs and advocates, between 35 and 40 percent of African American people in New Orleans lived in poverty. Between 40 and 50 percent of them were underemployed. Of African American households, 62 percent had annual incomes of less than $25,000 a year. Thirty-one thousand children were undereducated. Only 14 percent of New Orleans businesses were owned by African Americans, although the African American population exceeded 65 percent.

New Orleans had abject poverty pre-Katrina. Then the hurricanes hit and the levees broke, and the city ended up with 80 percent of its housing under water. People living on high ground were much less affected than the mostly black folks in the city's "Lower Ninth," but in many ways the flood waters did not know race or class. However, relief and recovery efforts revealed striking disparities that followed race and class lines. Still, many people, especially some in the African American Leadership Project, thought that the aftermath of the tragedy might be a practical and historical opportunity to promote racial justice, equity, and healing.

Prior to Katrina, New Orleans was a city where injustice was rife. It was a tourist economy where the success of the gaming industry was tied to business owners' ability to pay people minimum wage or only a dollar or two above it. When Harrah's and the other casinos came to New Orleans, there was a notable increase in every social indicator that you expect from gambling, including heightened levels of spousal abuse and alcoholism, as well as an understandable increase in crime. How do you take a glittering Harrah's and place it six blocks from a housing project where jobless-ness is rife and expect everything to be OK? The juxtaposition of glitter and grit is a slap in the face to economic justice. It's an in-your-face manifestation of inequality.

Does the post-Katrina rebuilding process offer New Orleanians an opportunity to humanize the city? Have people embraced the international concept of a right to return for displaced people? What should happen to the displaced population? Could New Orleans become a model of a sustainable and just city in a global era? After working for about six months, the African American Leadership Project issued a statement in which they asked that some of the federal dollars be used to improve human development and capacity and issues like literacy, to rebuild the physical infrastructure, and to rebuild institutional services and systems. Even as their ambi-tious and hopeful vision was being put forth, several members of the Rebuild New Orleans Committee were speaking openly about a shrunken footprint.

Some civil leaders appeared content with the mass displacement of New Orleanians and openly delighted at the notion that the city would be different. An industrialist and "business turnaround specialist" quoted in the September 8, 2005, issue of the Wall Street Journal, James J. Reiss, commented that post-Katrina New Orleans would be different "demographically, geographically, and politically."[2] Although he did not explicitly discuss racial difference, his comments and subse-quent actions and behaviors certainly implied as much. Reiss led the New Orleans Regional Transit Authority, which he resigned from following the rejection of pro-posals to reduce the city's already limited bus service and lay off half of the system's 700 employees. In June 2006, Reiss advocated for eliminating the public transporta-tion system on the grounds that the city could not afford it. What did we learn in the wake of the levee breaking? Of African Americans in New Orleans, 35 percent had no cars, compared to 15 percent of whites. In part, this explains why many did not leave. In this context, Reiss' efforts to quash public transportation in the name of affordability suggests that only those who can afford automobiles need return to New Orleans, effectively excluding over a third of the city's African American pop-ulation. What would the international community think of our position in the global economy if they knew of our inability to maintain domestic needs and prior-ities such as public transportation? What happens when the footprint shrinks?

Reiss was not alone in his myopic, Machiavellian ignorance. House Speaker Dennis Hastert said that New Orleans should simply be "bulldozed." Republican Congressman Richard Baker of Baton Rouge, Louisiana, responded to the devasta-tion from Katrina by saying, "We finally cleaned up public housing in New Orleans. We couldn't do it, but God did." How would we respond if an international neigh-bor made comparable statements in the wake of a disaster that impacted not a mostly black community such as New Orleans but a mostly white one such as Maine? We would decry the comment, condemn it, and insist that others do the same. What is happening in New Orleans? While the African American Leadership Project would push the city toward striving for social and economic justice, other forces would align to work for a whiter version of the status quo. Politics, chicanery, and racism are pulling the city in a direction to shrink the footprint. Those forces, and the people that support them, are making conscious decisions to remove and exclude African American people permanently from New Orleans. I would also suggest that what is happening in New Orleans is happening in cities across the country, but in a more languid, less crisis-driven way.

The first thing we would have to look at in the wake of Katrina is what happened in those five days when people were abandoned. Spike Lee's painful and moving HBO film, *When the Levees Broke: A Requiem in Four Acts*, describes it best, but the written record is equally powerful. What we know is that there was an attempt to sear racial stereotypes into the American consciousness. If you were black, you were looting, but if you were white, a loaf of bread merely came upon you through necessity. Miraculously, bread just came to you as if you had a bread magnet. Many of the people who were so-called looters were taking food, shoes, and other things back to their abandoned neighborhoods so that children, the elderly, and others could have those things needed to survive. That story was not told in the media, but people spread it word of mouth.

The other story that was not told concerned the mental health crisis that emerges when people lose everything and thus desperately seek to get their hands on something. Why do people with water up to their waists go get flat screen TVs? One would rightly presume that they were not intending to plug them in and risk electrocution. In the post-storm context, taking a flat screen television is an irrational decision that people in trauma make because they want to have something. In other words, one could literally view the city as being in the middle of a nervous breakdown. We did not see mental health professionals responding to the crisis or hear the issue of mental health being spoken of generally in the aftermath of the storm. Indeed, this is attributable, in part, to the fact that the health care industry was also impacted and many of these professionals left the city, too. Instead, pundocrats opined that African Americans were just crazed, stealing savages. More than a year later, a New Orleans *Times-Picayune* reporter has written about his depression movingly, and the New Orleans City Council released information that the city's suicide rate, at 26 per 100,000 residents has nearly tripled in a year. The city has lost many of its mental health workers, and depression is compounded significantly by post-traumatic stress. While candid conversations about mental health have begun to occur, few have been willing to reassess the harsh indictment of African American New Orleanians and consider that some of the behavior witnessed in the storm's aftermath may have been a function of mental health breakdowns. However, people were committed to the stereotypes and the lies. The media had seared this notion of black people as criminals into our collective consciousness. We didn't want to consider other explanations. Those were "the days of looters and lies" because they were all lies. The stories we heard about children being raped turned out to be untrue. Yes, there were a couple of rapes, but the evidence showed that the number of rapes that occurred were no greater than what usually occurred in New Orleans or, frankly, in any other American city.

Of course, even one rape is one too many, but the lies about rapists roaming "out of control" contributed to the stereotypes that led the governor to focus the efforts of the National Guard on protecting property instead of saving people. Indeed, there were no murders committed inside the Superdome. In my opinion, the United States of America committed murders by deciding not to go and rescue Americans who subsisted without food and water for five days. There were diabetic people without access to medicine; there were people who were immobile and were abandoned. The Red Cross collected money for hurricane relief but did not enter New Orleans because of "safety" issues. Newspapers reported that Mr. Bush took a helicopter over New Orleans to assess the damage while Condoleezza Rice shopped for designer shoes in New York City. Who really committed crimes after the levees broke?

Thousands of children were separated from their parents with many not reunited until months after their separation. The callous manner in which families were separated is reminiscent of the separations that took place during slavery.

Another important aspect of the Katrina disaster concerned the Federal Emergency Management Agency (FEMA) and its then-director Michael Brown. While people were dying, newspapers reported an aide telling Mr. Brown to "Roll up your sleeves so it looks like you're working hard." Another paper quoted Brown in New Orleans asking, "Where can I get something to eat that's not fried?" While people were dying, Brown cared more about his appearance and appetite than the loss of life. After that, we saw the uneven, arbitrary way in which FEMA assistance was administered and distributed. While some got ATM cards worth $2,500, others received checks for $800 although they didn't have banks. Different people got different forms of assistance without rationale. Although some have focused on alleged fraud that existed, few have focused on the fraud was perpetrated by FEMA. It would be expected that those who had nothing would willingly accept a $2,500 debit card from FEMA.

Essentially, FEMA gave people money to leave. They eventually helped get people out of the city but is not working effectively to bring them back. Despite the impact of that displacement, there was abject insensitivity on the part of some. For example, Barbara Bush, the former first lady of the United States, visited Katrina evacuees at the Houston Astrodome and commented, "These people are doing very well for themselves." Her comments help make it possible to comprehend the abhorrent inattention to poverty that emanates from the current Bush administration. Barbara Bush's repulsive condescension was only dwarfed by her son's failure to respond to the Americans who were disadvantaged by the broken levees.

During my time at the Superdome on September 11, 2005, with Minister Louis Farrakhan and a delegation who came to address the needs of evacuees, I saw children curled up into fetal positions and old people sitting in their wheelchairs with vacant stares. The floor was concrete. We saw many people in need of a hot meal or immediate medical attention. The political structure is one that exploited the lack of organization in New Orleans and the over-organization in Mississippi. Ray Nagin and Kathleen Blanco and George W. Bush do not get along. Louisiana is a Democratic state, marginally. New Orleans is most emphatically a Democratic city, despite the fact that Ray Nagin is a neophyte reform Republican elected in the Democratic city. The fissures in intergovernmental cooperation were more than apparent. Mr. Bush held money for political advantage even as he promised Mississippi Senator Trent Lott (R) a new house. Although Lott is able to and will rebuild his house, how many others will? Examinations of federal appropriations in the wake of Katrina revealed that Mississippi got $5.2 billion compared to $6.2 billion for Louisiana. Mississippi households will get 4 times as much for rebuilding as Louisiana households. Indeed, Mississippi Republicans Trent Lott and Todd Cochran were able to take full advantage of the political system. However, infighting between Kathleen Blanco and Ray Nagin produced little results for Louisiana. Blanco and Nagin could have benefited from the assistance of the former mayor and Urban League president, the politically savvy Marc Morial, but did not seek it. Morial has strongly rejected the notion of the shrinking footprint, but Nagin has ignored him.

From a political perspective, what has also been developed is a set of quasi-legal bodies that both overlap and seep away authority from elected officials. There is the Rebuild New Orleans Commission, the Rebuild Louisiana Commission, and other entities that claim responsibility for determining the reconstruction vision for the city. As of this writing, in October 2006, there are several rebuilding and recovery plans, and the governor's housing rebuilding initiative, the Road Home Homeowner Assistance Program, among others. That many of these programs are administered on the state level and not city level has caused tension among some politicians. However, while politicians jockey for positions of leadership and power, people are suffering.

Hurricane Katrina and the broken levees destroyed much of the New Orleans public education system. According to an Urban Institute report, fewer than 20 of about 115 public school buildings remained usable in the storm's aftermath. By November, the state had taken over the New Orleans public education system. "We see an opportunity to do something incredible," Governor Blanco said when she signed the authorizing legislation. What benefits resulted from the take-over remain questions in light of the fact that 7500 teachers and employees who worked in the public school system and more than 400 other school employees including truck drives, custodians, and cafeteria workers were fired. Many learned that they were terminated not through formal notice but rather through reports on the local evening news. The takeover, and the subsequent firings proved to be a blow to the African American middle class in New Orleans, as the majority of the New Orleans Public School workers were African American.

The New Orleans Public Schools (NOPS) system currently receives funding for just 9,300 students in the four district schools and twelve charter schools that the school board operates. The NOPS has inherited all of the previous bond debt and operating deficits from the system but exercises little control over the schools themselves. Much of the revenue, controlled by the state, is being diverted to charter schools. In some cases, charter schools have been allowed to take over buildings that once belonged to public schools. Many of the charter schools have admissions requirements that are exceedingly high for many prior public school students. The takeover of NOPS exacerbated Katrina's devastation of the educational system, at least for some students. For example, the children of Tulane faculty have access to a charter school that was once a fairly strong but open public school. The students who used to go to this public school prior to Katrina now have to compete for a limited number of slots. While some are excited about Tulane's institutional involvement in public education, there is something of a class bias that results from the transfer of a public school into private hands, with many former students unable to experience the benefits of the new arrangement.

Mayor Nagin says he wants people to come back to New Orleans. For working-class people who want to return and live in the city, jobs are an absolute prerequisite. Today, many African Americans have expressed anger at the Latino immigrants, especially Mexican and Nicaraguan, who have come to New Orleans in search of jobs. While much of the immigrant population lives in single-male housing, African American men who have families can only return to the city and secure low-wage jobs if they are willing to live in dormitory-style housing.

The same companies that made billions in Iraq have also profited tremendously in New Orleans by exploiting low-wage and undocumented labor. The way to prevent that kind of exploitation is to make sure that undocumented workers belong to unions that will fight for secure job conditions and a living wage. The politics of immigration are part of the conversation that must take place regarding the reconstruction of New Orleans, particularly as some have attempted to pit black against brown. Fairness would dictate that we support a living wage for all workers, documented or undocumented, and institute preferences for those who are native New Orleanians.

A law passed by the New Orleans City Council in May 2006 said if you did not gut your house by August 28, 2006, the city could use its eminent domain powers to claim the house. On average, gutting of a home costs at least $5,000, with higher rates charges based on the severity of the damage to the home. However, analysis of the labor market reveals that there were not enough contractors present in the region to carry out the necessary gutting. The Association of Community Organizations for Reform Now (ACORN), a national grassroots organization that focuses on economic justice issues, mobilized volunteers to gut more than 1500 homes at no cost by the one-year anniversary of the storm. The organization has also been on the cutting edge in organizing for displaced people in New Orleans. Despite their work, though, the die is cast against poor people, especially poor black in New Orleans.

The Congressional Black Caucus offered a piece of omnibus legislation, H.R. 4197, in late 2005. Congress has not yet acted on it and apparently refuses to. One of the bill's ten provisions mounts a powerful challenge. That provision requires the President to chart how we might eliminate poverty in the United States. The bill also seeks to address a number of challenges concerning rebuilding, homeownership, and education. However, requiring executive-branch solutions that would eliminate poverty gets to the heart of the matter. After all, the national conversation about poverty was fairly dormant until the levees broke.

Loyola Law School Professor Bill Quigley, an economic justice advocate and activist, considers the problems in New Orleans comparable to the problems happening elsewhere in the country. The poor are being pushed out. Quigley believes that officials might as well place a sign up at the border that says, "poor people don't come back."

Although there is a housing shortage in New Orleans, there is also unoccupied, available housing. Perfectly livable public housing projects are boarded up and have steel plates screwed into the windows. These places were not destroyed in the storm, and displaced persons could come back to them. Interestingly, some of these places are just a few blocks from the French Quarter. However, the Department of Housing and Urban Development (HUD) does not want public housing anymore; there is a national movement towards mixed-use housing. While luxury condos are receiving subsidies, people who could return to New Orleans and live in sturdy, empty low-income public housing are being kept away. Currently, residents of the Lafitte public housing project are planning to sue HUD to prevent the demolition of their homes.

With the changing face of cities, one of the things we see is a political shift. Already in New Orleans a white attorney, Stacy Head, replaced an African American female city councilor, Renee Gill Pratt. Pratt's politics are distinctly to the left of Head's, but many of her constituents remain displaced and were thus unable to get back to vote. The urban demographic and political shift is happening in many cities—Washington DC; New York; Boston; and Oakland, among many others. The changing face of cities means the changing face of urban politics. In the long run, it means the diminution of African American political power.

WHAT DOES THE FATE OF NEW ORLEANS SAY TO THE WORLD?

What did the United States reveal about itself in the treatment of New Orleans? How does this further erode our moral authority? What will happen when others treat us the way we've been trying to treat New Orleans? What happens when our footprint shrinks?

Fifteen months after Hurricane Katrina hit New Orleans, the city is still unsettled. Billions of federal dollars have been directed to the area, but the people who need money most for resettlement haven't received it. Instead, the same profiteers who have exploited the Iraq situation have gained in New Orleans. By late October 2006, the debate about the shrinking footprint and neighborhood development continues. The world's eyes have been on the United States as we have dealt with our citizens, and in the world's eyes we are lacking. Our nation's treatment of New Orleans belies the notion of our greatness and of our eminence.

Venezuelan President Hugo Chavez's September 2006 speech at the United Nations was provocative and on point. Indeed, there is only one letter between "evil" and "devil." In his 2004 State of the Union Address, Mr. Bush described Libya, Syria, and Iran as axis of evil. People get upset about name-calling, but Mr. Chavez is not the only one who engages in the act of political rhetoric. When Pat Robertson calls for his assassination, no one in his church says anything about supporting murder. Hugo Chavez's remarks regarding America's hypocrisy and reduced eminence shed light on our government's poor response to the Katrina crisis.

In *The Paradox of Loyalty: An African American Response to the War on Terrorism*, University of Maryland political scientist Ron Walters defined foreign policy equity. He asked why we can't have something called foreign policy justice. We fought for China to get the Olympics, despite their repressive human rights policies, but we argue against other countries playing a role in world economic development because their human rights policies are not correct. We talk about human rights in Iraq, but we know this conflict is not about human rights. First it was about weapons of mass distraction, then oil, and now ego. At one point, First Lady Laura Bush incredibly claimed that our invasion of Iraq was about women's rights.

Iowa Senator Tom Harkin said he understood the frustration that Chavez and others in the world harbor towards our nation's foreign policy arrogance. While calling the Chavez rhetoric "incendiary," he also recognized it as the rhetoric of frustration. However, when the President of the United States uses that same kind of invective, he is met with little criticism or condemnation.

We had enormous moral authority after September 11. The world saw what we saw. Many people believed that our actions were aimed at stamping out terrorism in the world. If we had simply gone to Afghanistan, we would still have enormous support from all kinds of countries. However, we chose to invade Iraq. In September 2002, before the invasion of Iraq, Mr. Bush described Saddam Hussein as "the guy who tried to kill my dad." In other words, his actions were clearly personally motivated and thus irrational. Should we squander our nation's collective future for a family feud?

It's not just squandering esteem and moral authority; the war in Iraq is also eroding our economy. We are spending so much money on the war that we have eliminated the surplus that remained from President Clinton's tenure and amassed one of the largest deficits we have ever seen. Because of that deficit, we have systematically cut social programs. We had so little money that we did not fulfill the request of the Army Corps of Engineers to repair the levees before they broke in 2005.

Is the United States at the end of its eminence? Certainly, we have communicated our weakness to the world by our failure to respond to New Orleans in an effective way. But several factors suggest that unless we make major changes, we are coming to the end of our era of world dominance. The education situation is perhaps the most important. Our ability to compete has been impaired by the eroding state of our educational system and our country's self-divestment of that system. Although we once we had the best educational system in the world, now the poor quality of inner city K-12 education raises pipeline issues, and our self-divestment of public higher education raises other issues. Our system still has strengths, especially in the humanities and sometimes the social sciences, but we are losing our edge in engineering and science. These fields are ones that dictate whether we have sufficient capital to build and maintain effective levee systems to protect cities such as New Orleans, among other things.

The second issue is China, a nation on a fast track to full industrialization. They fully intend to compete with the United States and Western Europe in the next twenty years. Right now, one in twenty households in China has a car. Most people get around by bicycle. But in ten years, we should expect that there will be increasing reliance on cars and perhaps one in ten households will have a car. What pressure will this place on gasoline prices? China represents one quarter of the world's population and consumption possibility. If they are as fuel inefficient as we are, what does that mean for the rest of the world, and what does that mean for us? Are we prepared to compete with this new superpower?

The third sign that our eminence is ending is the arrogance that we bring to the table on foreign policy. It has had reverberations in terms of the way global nations respond to us. Many think the economic consequence of our own arrogance is that some oil-producing countries will shift from the dollar to the euro. When they do that, it essentially weakens the United States' world position. The dollar has been the primary world currency since after World War II. If we have to compete with the euro, which has become more stable, we may end up in a situation where we will be talking about balance-of-payment problems that weaken the ways we do business day to day.

Our nation's failure to deal with our diversity, and especially to deal with race matters, also signals the end of our eminence. Part of this is connected to the education issue. While our nation is divesting education, that divestment clearly does not affect all Americans equally. Upper-middle-class Americans are likely to continue to be educated at our nation's best universities and institutions. It is the students who need help, particularly those who need financial assistance, who are finding education less attainable. We can talk about affirmative action as much as we like, but the real challenges in the twenty-first century are educational access and affordability. Public colleges and universities are going to have less money available. Will the quality of education suffer, and if it does, will it erode our global position? The strength of our educational system and student access to that system are directly tied to our country's ability, or lack thereof, to deal with and have smart discussion about ongoing racial crises in the United States. Some may argue that this is a post-racial era. However, the data certainly do not suggest so. The average white household has an income of $51,000 a year, and the average black household has $31,000. The black unemployment rate is twice the white unemployment rate. These disparities are particularly heightened in New Orleans. Every socioeconomic indicator shows that race matters, but there is a reluctance to deal with race problems in a real way.

We are like ostriches, burying our heads in the sand. Demographically, the fastest-growing population is the Latino population. The slowest-growing population is the white population. Rates of African American growth fall somewhere in between that of Latinos and whites. If you are of a certain age and thinking about who will be your nurse in your golden years, she is likely to be a woman of color. Many of our workforce issues could be solved if we talked about educating people of color and those whites who can't afford education.

Our future hinges on the development of a competitive and able workforce. We are not technologically superior to the rest of the world. However, what we have that other countries don't is the diversity that we've generally ignored. When we pay attention to this diversity, it strengthens us, but when we ignore it, it erodes our ability to move into a future with strength and moral authority. The world looked at what happened post-Katrina, and what they saw was a great nation that did not care about black people and poor people. Our treatment of our very own citizens raised questions about the empty rhetoric we announce before the United Nations about how much we care about a thriving democracy. New Orleans showed us at our very worst—venal, uncaring, and inefficient.

What happens when our footprint shrinks? What happens when the United States goes the way of Great Britain, that once-great superpower that is now teetering at the edge, with high poverty, unemployment, crime, and racial problems? We are poised to be the next dinosaur, the next Britain, the next once-great nation. The way we treat New Orleans is, in my opinion, a metaphor for what we can and must do with our nation. If we can fix New Orleans, we can fix our nation. But if we cannot fix New

Orleans, we are sending a clear signal that we lack the will to help our own. If we cannot help our own, how can we lead in the transformation of the rest of the world? Our failures in New Orleans lead me to believe that we have reached the end of our nation's eminence.

BIBLIOGRAPHY

Much of this information comes from interviews I conducted July 24–25, 2006, for *The Salteaux Report*, a television show in development. There are more than fifteen hours of taped interviews that deal with Katrina, broken levees, and the recovery, especially around housing, economic development, and education. This material will be cited as Malveaux Report, 2006.

Bucks, Bruce, Arthur Kennickell and Kevin Moore. "Recent Changes in US Family Finances: Evidence from the 2001 and 2004 Survey of Consumer Finances." *Federal Reserve Bulletin*, January 2006.
Columbus, Danae. News for June and July, New Orleans City Council, July 26, 2006.
Danielson, Darwin. "Harkin Defends Venezuelan President's UN Speech Against Bush." *Radio Iowa*, September 21, 2006.
Democracy Now. "Interview with Joe DeRose, Communications Director, United Teachers of New Orleans." http://www.democracynow.org.
DeNavas-Walt, Carmen, Bernadette D. Proctor, and Cheryl Hill Lee. *Income, Poverty, and Health Insurance Coverage in the United States: 2004*. U.S. Census Bureau, Current Population Reports, P60–229. Washington, DC: U.S. Government Printing Office, 2005.
DeNavas-Walt, Carmen, Bernadette De. Proctor and Cheryl Hill Lee. *Income, Poverty and Health Insurance Coverage in the United States: 2005*. U.S. Census Bureau, Current Population Reports, P60–231, U.S. Government Printing Office, Washington, DC, 2006.
Hill, Paul, and Jane Hannaway. *The Future of Public Education in New Orleans*, Washington, DC: The Urban Institute, 2006.
Kamenetz, Anya. *Generation Debt: Why Now is a Terrible Time to Be Young*. New York: Riverhead Press, 2006.
King, John. "Bush Calls Saddam 'The Guy Who Tried to Kill My Dad.'" *CNN*, September 27, 2002.
Malveaux Report interview with Bill Quigley, July 2006.
Message from the School Board President, New Orleans Public Schools, October 17, 2006. http://www.nops.k12.la.us (accessed [date]).
Reckdahl, Kathy. "Like A Ton of Bricks." *Gambit Weekly*, New Orleans, October 2006.
Rose, Chris. "Hell and Back." New Orleans *Times-Picayune*, October 22, 2006.
Walters, Ronald. "The U.S. War on Terrorism and Foreign Policy Justice." In Julianne Malveaux and Reginna Green, eds., *The Paradox of Loyalty: An African American Response to the War on Terrorism* (New York: Third World Press, 2003).

NOTES

1. The African American Leadership Project has been a voice and force of resistance to the shrinking footprint, as well as a force for economic justice in New Orleans. Their Web site is at http://www.hurricane-katrina.org/2006/07/africanamerican.html.
2. Christopher Cooper, "Old Line Families Escape Worst of Flood and Plot the Future," *Wall Street Journal*, September 8, 2005.

CHAPTER 17

"THE CITY I USED TO. . .VISIT": TOURIST NEW ORLEANS AND THE RACIALIZED RESPONSE TO HURRICANE KATRINA

LYNELL THOMAS

Late last August, like so many others around the world, I sat glued to Hurricane Katrina news coverage and the now-infamous images of African Americans and their corporeal indictment of American politics and policies. In my case, I was desperate for information about my husband and family members who had remained in our New Orleans home during the storm. In those critical days of anxiety and outrage, I recognized in a new and painful way how representations of race define and delimit citizenship. I wondered in disbelief, first, how news reporters who were intractably anchored in the French Quarter had repeatedly and falsely reported that the city had escaped major damage as flood waters engulfed most of the city; second, how even as levees remained breached, rescue efforts were suspended because of alleged violence (later revealed to be largely exaggerated), leaving tens of thousands of people stranded and unprotected from the rising flood waters; and finally, how local, state, and national public officials and media collaborated in promoting destitution, debauchery, and complete degradation as the most predominant and enduring image of black New Orleans.[1]

A year later, my initial disbelief has given way to the realization that Hurricane Katrina revealed the public level of invisibility of New Orleans's black population, the limited categories and conventions available for identifying that population even after the storm victims demanded recognition, and the public costs of this distortion of black people's lives. I might have foreseen this outcome if I had given more weight to my own research on representations of race in contemporary New Orleans tourism. In fact, the responses to Hurricane Katrina eerily emulate New Orleans's tourism narrative, which limits New Orleans to the French Quarter and the city's European identity, labels historically and predominantly black areas of the city as dangerous, and obscures and distorts the African presence and participation in the development and sustenance of the city. Given that tourism is New Orleans's most lucrative industry and that most outsiders' perceptions of the city are mediated through its tourist identity, it should come as little surprise that even in the midst of unprecedented crisis, the representation of New Orleans during Hurricane Katrina continued to be dominated by the troubling images and ideas of the city's all-too-familiar and overdetermined tourism narrative.

The city's contemporary tourism industry invites white visitors to participate in a glorified Southern past. Black residents, if they appear at all in this narrative, appear as secondary characters who are either servile or exotic—always inferior to whites and never possessing agency over their own lives. Hence, New Orleans's pre-dominant tourism narrative is predicated on the propagation, albeit distorted, and the negation of New Orleans's black history. White tourists are encouraged to view the Old South with a sense of loss and nostalgia as they play the part of wealthy planters by visiting plantations and touring historic home that display the wealth of a bygone era, by shopping for the ubiquitous "contemptible collectibles" in French Quarter shops and hotel lobbies, and by themselves being cared for by the highly vis-ible African American service workers often "performing" to garner a larger tip.[2] Visiting tourist New Orleans promises historical authenticity by bringing visitors to the past itself, not just a scale reproduction of the past.[3] Unlike many other Southern cities, where black cultural traces have been largely erased from the mainstream nar-ratives, New Orleans tourists are invited to consume black cultural production by listening to black music and hearing anecdotes of the secret rites of quadroon balls and voodoo ceremonies. Yet these purportedly black experiences are limited to the safe tourist space of the heavily police-patrolled French Quarter. Outside these boundaries, the city's majority African American population is portrayed as physi-cally and socially threatening. These multiple representations sustain the stereotypes of black servility and inferiority, and in the process, they ignore the historical and contemporary realities of the city's African Americans.

New Orleans tourism perpetuates these representations by focusing almost exclu-sively on two periods in history: the colonial period under French and Spanish rule and the antebellum period following the Louisiana Purchase. The emphasis on these periods (to the exclusion of more recent history) and the depiction of these periods together create a very particular racial image of the city. Even when the tourism nar-rative represents African American culture and history, it does so through the lens of New Orleans exceptionalism, thereby exempting New Orleans from any real racial problems in the past or present. Consequently, assertions of French and Spanish leniency during the colonial period; depictions of a romantic, paternal slave system during the antebellum period; and an oversimplified insistence on a constantly self-sustaining, vibrant class of free people of color who themselves were aligned with white interests than with black interest obscure the fact that enslaved and free blacks in both colonial and antebellum New Orleans fought in a myriad of ways to obtain their freedom, assert their rights, and demand equitable treatment.

The representation of New Orleans's colonial history exemplifies this distortion with numerous references to the enduring legacy of the French and Spanish. The official New Orleans tourism Web site announces, "New Orleans, with its richly mottled old buildings, its sly, sophisticated—sometimes almost disreputable—air, and its Hispanic-Gallic traditions, has more the flavor of an old European capital than an American city."[4] Another tourism article refers to New Orleans as "the most European American city," although historian Gwendoyn Midlo Hall's designation of New Orleans as "the most African city in the United States" is far more accurate.[5] The colonial construction of New Orleans accredits the city's white, European heritage with most of what makes the city unique—from its architecture to its zydeco. This construction privileges a particular episode in New Orleans history and a particular racial and cultural group. New Orleans's tourism literature exalts European and particularly French culture as the most predominant and influential culture in New Orleans despite the tremendous presence and influence of Africans and Native Americans throughout the city's colonial history.[6] The tourism narrative employs New Orleans's colonial history to emphasize the racial distinctions that made New Orleans a world apart to the exclusion of the characteristics that bind New Orleans to the rest of the United States, particularly the Southern United States. The city's portrayal as racially unique is girded by the notion that the racism and racial conflict that continue to haunt other U.S. cities is largely lacking or lessened in New Orleans because of its unique historical and cultural legacy.

The tourist industry's homage to the French and Spanish extends to those nations' purported humane and even charitable treatment of enslaved Africans. Generally, tour guides and tourism literature focus almost exclusively on slavery during the colonial period, emphasizing a supposed unique and favored position of New Orleans slaves who were protected by the Code Noir, were given Sundays off to congregate in Congo Square, and were allowed to purchase their own freedom. One Web site's assessment is typical:

> There were. . .important differences between Louisiana and other slave-driven Southern states. Here slavery was more in the West Indian mould [sic] than the Anglo-American. The Black Code of Louisiana, established by the French, upheld by the Spanish and then effectively broken by the Americans, gave slaves rights unparalleled elsewhere, including permission to marry, meet socially and take Sundays off. The black population of New Orleans in particular was renowned as exceptionally literate and cosmopolitan.[7]

Despite acknowledging a slave past, this account privileges slavery in the colonial period as a time of European beneficence, an idea refuted in recent historiography. Historian Thomas Ingersoll argues, "The principal intent of the Black Code of 1724. . .was evident in the many measures aimed at ensuring the subjugation of all blacks and separating the races so as to limit the numbers of mixed-race and free black people."[8] Even Ingersoll's detractors who argue for distinctions between the British and French or Spanish systems of slavery generally attribute gains made by Africans and their descendants primarily to their own agency, aptitude, and creativity, not to their enslavers' compassion or generosity. Despite contrary historical evidence, New

Orleans emerges in its colonial tourism narrative as a city whose attitudes about race and treatment of its residents of African descent—in both historical and contemporary representations—are exceptionally benign.

Just as New Orleans's tourism narrative diminishes the brutality and degradation of the slave system under the French and Spanish, it also erases completely the more rigid system of slavery that evolved in the nineteenth century, marked by a more tenuous position for the city's free people of color and the near impossibility for bondsmen and women to purchase their own freedom.[9] While the emphasis on New Orleans's European or French heritage highlights the city's distinctiveness, an emphasis on the antebellum period in the tourism literature blurs those distinctions by situating New Orleans in the clichéd construction of a romantic antebellum South. The Old South is evoked countless times in descriptions of contemporary New Orleans. Visitors are invited to experience firsthand the Old South, although this experience rarely incorporates the experiences of those bondsmen, women, and children who supported the mansions and sustained, through great sacrifice and hardship, the fabled "good old days."

Most Web sites and tourism brochures advertise or feature nearby plantations and plantation tours, which are represented as authentic recreations of New Orleans's antebellum past.[10] With few exceptions, however, this authenticity is reserved solely for decorative and architectural replications of the antebellum South. Advertisements and descriptions of the plantations and tours are interspersed with language that almost literally reconstructs an antebellum mythology. "Grand winding staircase, original slave cabins, magnificent oaks"; "tours by guides in [a]ntebellum dress"; "Belter & Mallard furnishings, five hand-painted ceilings, faux marbling and wood graining"; and "elaborate wrought iron trimmed galleries, ornate friezes and medallions" conjure up hackneyed notions of "romance, history, and beauty" associated with the "grandeur and elegance of 19th century southern living."[11] Although River Road and many of its plantations are listed in the National Register of Historic Places, very few of the plantation tours interrogate the history that the sites represent, namely the history of slavery, which enabled the wealth and accompanying lifestyles that the plantation tours celebrate.

Even when slave dwellings are present on these properties, they form part of the backdrop to the romanticized Southern mystique, thereby enhancing, not countering, the mythology of a glorious Southern past.[12] Many of these plantations and their outbuildings, including the slave cabins, have been renovated into bed and breakfasts, exploiting the performative possibilities of a white antebellum Southern mythology. Examples include Oak Alley Plantation, Restaurant & Inn and Nottoway Plantation Restaurant & Inn, which advertise their sites as places to "enjoy elegant accommodations overlooking the Mississippi River or the charming surroundings of the overseer's cottage."[13] The Southern past is reduced to a checklist of objects and images—period furnishings, oak and fruit trees, Greek Revival or Italianate architecture, slave quarters—that provide ideal settings for film and literary portrayals of the antebellum South.[14] Implicit in the invitation for visitors to "feel the gentle breeze of southern hospitality on a tour that takes you back to a time of mint juleps, gracious living and the glory of the Old South," is either the erasure of the South's slave past or appropriation of that history to suit a more romantic, idyllic narrative of slavery.[15]

The portrayal of a relatively benevolent, paternal slave system relies on some of the very images and arguments of nineteenth-century pro-slavery ideology. What results is a re-writing of the historical narrative that erases the violence, exploitation, and depravation of slavery and the accompanying culpability of whites.

This rewriting of slavery is not limited to outlying plantation tours. New Orleansonline.com includes a page on historic homes within the city limits of New Orleans. By visiting selected "architectural treasures"—mansions and "palatial antebellum residences"—the page invites visitors to "step into the past" and "learn about the culture and history of the city's colorful and hospitable past."[16] While descriptions of these historic homes do not allude to the city's slave history (yet another erasure of that history), the romanticizing of slavery for local and tourist consumption abounds in the city. The designation of former slave quarters as chic, quaint, and marketable by area tour guides, hotels, realtors, and restaurants illustrates how the history of slavery is removed from its historical context in New Orleans's tourism industry. Consequently, these venues can offer potential visitors and buyers the "sheer romance" and the opportunity to "go native" that coincide with being entertained, fed, or housed in "quite attractive" or "liberated, languorous cottages," which once housed the city's enslaved, black population.[17]

In these topsy-turvy appropriations of slavery, the remnants of New Orleans's slave system are beautified and decontextualized. These descriptions imaginatively wrest slavery from its historical and ideological contexts and repackage it for tourist consumption; visitors to the city—for a price—are able to reap the material and psychological rewards of others' labor. Beyond these imaginative reenactments of slavery, several New Orleans French Quarter restaurants and hotels literally bear the mark of the city's slave past.[18] The façade of the upscale Omni Royal Hotel still carries the ironic inscription "Change"—the remaining words from "Slave Exchange" following the building's renovation.[19] In addition to the promotion of former slave quarters, the sale, display, and dispersal of an array of slavery images and reproduced artifacts create a disturbing tourist landscape. French Quarter businesses greet customers with mammy- or Sambo-adorned signs, packaging, and products. These images appear on everything from store signage to souvenirs and perpetuate an idea of the antebellum South that caricatures and trivializes black contributions to and contests over Southern history. Despite the omnipresence of these markers of slavery, the tourism narrative generally fails to include a discussion of those who were enslaved. Even when antebellum slavery is discussed, tourist narratives focus primarily on New Orleans's wealth and prosperity during the nineteenth century, not on the human cost exacted because of this wealth.

Usually, however, slavery is not discussed, despite the wealth of firsthand accounts of the slave system in nineteenth-century New Orleans. The city was, after all, the leading North American slave-trading city by the 1850s, with numerous slave auction houses and regular auctions.[20] De Tocqueville and Beaumont observed that slavery in Louisiana was equally as brutal as elsewhere in the South, and numerous nineteenth-century slave narratives record the harrowing experiences that bondsmen and women faced in New Orleans's slave market.[21] The exclusion of these accounts to promote New Orleans's racial exceptionalism further perpetuates an idea of a city that is devoid of any real black presence or significance.

This omission is compounded by the fact that New Orleans's tourism narrative does not give any sustained attention to the historical period following the Civil War and continuing through the post-civil rights era, the very period when most African Americans obtained their freedom and made more concerted efforts to demand and exercise their civil rights. The dismissal of this history continues to shape tourist accounts of New Orleans's more recent past. In the rare instance when recent history is presented to tourists, it is a history that does little to challenge the mythology of New Orleans's racial, cultural, and social harmony. For instance, NewOrleans.com explains that New Orleans's 1960 school desegregation "was not marked by the racial strife found in other Southern cities."[22] This rather innocuous account, of course, disclaims the years of struggle preceding and following the 1960s desegregation movement.[23] Because so much effort goes into sustaining New Orleans's image as a historic city through a reverence for the city's past, careful preservation efforts, and attention to genealogies and historical figures, the history of New Orleans presented in tours and tourism literature gains an air of credibility and comprehensiveness it may not always deserve, particularly as it relates to the city's multiracial heritage.

On the other hand, multicultural and black heritage tours do provide competing narratives that acknowledge New Orleans's multiracial heritage and the contributions of black people to the city. However, these tours generally reinforce the racial ambivalence and anxiety of the mainstream narrative or remain otherwise ephemeral or marginal. Increasingly, as New Orleans tourism attempts to respond to public and scholarly demands for a more inclusive representation of history, both mainstream and black heritage tours incorporate the language and symbols of diversity, multiculturalism, and black history.[24] Despite this seeming embrace of different cultural, linguistic, and social influences, the city's multiracial tourism narrative is cloaked in the conventions of a problematic multiculturalism that in the end promotes a clichéd, trivialized understanding of race and aggrandizes the structures of oppression that it purports to disrupt. In other words, the language may have changed, but the mythology of racial exceptionalism remains the same. In many ways, the message of multicultural New Orleans is often as conservative and regressive as more racially exclusive representations of the city.

The New Orleans melting pot—or more accurately "gumbo" pot—motif demonstrates the limitations of the multicultural narrative. The success of New Orleans's culinary assimilation of foods from throughout the world is used to symbolize an equally successful and satisfying racial and cultural assimilation in the city. The 2002–3 New Orleans Official Visitors Guide describes the people of New Orleans as being "as diverse and unique as the ingredients in gumbo, and each group has brought its own influences to the city."[25] Although much of this gumbo pot rhetoric originated as a direct challenge to previously dominant tourism narratives that ignored or distorted the contributions by non-Europeans, even the city's so-called multicultural tours offer divergent narratives of New Orleans's multicultural heritage and suggest the problem of associating terms, such as "melting pot," "diversity," or "multiculturalism" exclusively with the progressive agendas of social change.[26] Often these terms create and sustain an image of social transformation, particularly in reference to racial equality, that is not substantiated by the realities of New Orleans's historical and contemporary experiences. What results is yet another mythology of

New Orleans's racial history, perhaps a more palatable romance than the one portrayed by the plantation and slave cabin stories, but one equally as inaccurate. In this construction of the city, New Orleans is (and has always been) a place of racial harmony. This notion of unity that transcends racial, class, religious, and other divisions is echoed in former mayor Marc Morial's introduction to the 2000 New Orleans Official Visitors Guide. Morial—son of the city's first African American mayor Ernest "Dutch" Morial—boasts, "Proud of our heritage, we have combined the influences of our European, African, Caribbean and, of course, American forefathers into how we live, what we eat and how we celebrate"[27] The present mayor Ray Nagin takes this idea even further in his welcome in the official multicultural visitor guide, *Soul of New Orleans*. He invokes the customary "cultural gumbo" and proffers the city's diversity as a commodity "within easy reach of visitors and locals alike." Most tellingly, Nagin attributes the origin of jazz to the "waves of immigration from Africa, Europe, the Caribbean, Latin America and Asia." In an effort to portray present-day New Orleanians as equal participants in the city's culture, this assessment obscures the history of jazz, whose origins are generally attributed to the united efforts of Afro-Creoles and African Americans, as well as obscures the history of slavery, which Nagin euphemistically substitutes with New Orleans's "immigration from Africa."[28]

Other tourism sites likewise portray historical and contemporary New Orleans through the lens of a noncontentious multiculturalism. This "history" has become so diluted as to be rendered ineffectual in challenging white privilege and the continuing legacy of racism. Instead, New Orleans's construction as a multicultural city works to reinforce the idea that racism in the city is either nonexistent or aberrant. Generally overlooked in this notion of the gumbo pot is the potential and actual divisiveness of the languages, cultures, ethnicities, races, and economic classes of the city's inhabitants since the city's founding. Just as de Crevecoeur's "melting pot" referred to the amalgamation of many races into one American identity, the New Orleans gumbo pot motif suggests that all people, regardless of national, racial, social, or cultural background forge a new Creole identity once they become New Orleanians. However, just as the melting pot has been criticized for its underlying assumption that its designation of a new American race was in fact a Western identity available only to European immigrants, so, too, the gumbo pot's Creole identity often excludes the city's non-European population or mandates a very particular profile for people of color that reifies the divisions a Creole identity is supposed to mitigate.

The term *creole* has had a very circuitous and contentious evolution. Historians acknowledge that the term's meaning has shifted over time and in relation to different political, historical, and economic developments. Throughout Louisiana's history, Creole has been used at different times and for different reasons to describe all native-born inhabitants of the early colony, irrespective of race; those native-born colonists and their descendants of French or Spanish heritage; native-born colonists and descendants of interracial unions among French, Spanish, African, and American Indian settlers; and numerous combinations thereof. Different groups continue to wrangle for entitlement to an exclusive Creole identity. Despite these continued contests over the term, many definitions of the term advanced in the tourist literature leave little room for Afro-Creoles, whether poor or enslaved, or even the much-touted free people of color. Instead, tourist literature often uses the term to distinguish between the wealthy, aristocratic, Catholic, in-grown (white) Creoles and the more industrious, hard-working, Protestant, though less interesting, (white) Americans who arrived in the city following the Louisiana Purchase. In most cases, the term is stripped of its racial, cultural, and historical meanings. Even on SoulofAmerica.com, a national black travel Web site, the term's technical meaning is said to refer to "a city resident who claims to be of French descent and speaks French," which would eliminate many New Orleanians who consider themselves Creole. Now, according to the site, the term refers to "practically all residents who enjoy life and cooking and music of New Orleans. They truly believe in the art of sophisticated living, no matter how short their average life span."[29] This attempt to neatly obliterate the contests over meaning and power conceals the serious political, economic, legal, and social connotations of the term.

Although *creole* is a term that remains contested in New Orleans's tourism and in the historiography, the tourist literature problematically uses and portrays *creole* both as a metaphor for New Orleans's all-inclusive society and as a way to signify divisions and racial exclusions.[30] This anxiety over the designation of Creole identity is most clearly reflected in the widespread claim that Creoles of Color, because of their economic and political wherewithal and racial anxiety, identified more with upper-class whites than with enslaved or poor blacks throughout the city's history. The tourist literature's insistence on a racially and economically exclusive denotation of Creole sets up an impossible stratification of society and insists even now on the purity of race as a legitimate means of classification.

In this system, it somehow goes without question that—despite a history of both inter- and intraracial unions among enslaved and free blacks; shared language and culture; familial and social bonds between enslaved and free blacks; and economic, educational, and professional diversity within the free black population, as well as a range of phenotypes among the black population, both enslaved and free—New Orleans's communities of free people of color and enslaved blacks can be neatly and categorically distinguished by skin color, class, education, profession, and/or racial identity.[31] Historian Jerah Johnson counters this myth, basing his argument on the recent proliferation of scholarship on Afro-Creoles in Louisiana:

> Acutely conscious of their legal rights and their group's interests as well as the tenuous and fragile nature of their position, [free Creoles of Color] tended to act with an exceptionally high degree of cohesiveness. At the same time, individual members of the group freely associated with the European colonials, the African slaves, and the Indians, both free and slave. Work, service, trade, and *plaçage*, the developing institution of formalized mistress-keeping, brought them into close contact with the European community, while close cultural and family bonds tied them to both slave and Indian communities. Except for recently arrived islanders, there were few free people of color who did not have relatives, often immediate family members, among the African slaves and not infrequently among the Indians.[32]

Unlike the tourist narrative's fixed Creole identity, recent historiography recognizes the complexity and variety of Creole society, a society that included free blacks immigrating from other parts of the United States, free dark-skinned blacks, enslaved light-skinned French-speaking blacks, poor free blacks, inter-class and inter-color unions, and any number of combinations thereof. These encounters and interconnections illustrate the inexact, ambiguous uses and meanings of the term *creole* throughout the city's history and belie the simplistic definitions included in tourist representations. The contention over the term continues among different groups that claim to be Creole and fight for a Creole identity rooted in history, culture, and genealogy.

In the end, New Orleans's multicultural narrative promotes a vision that limits and distorts the actual black presence and experience in New Orleans. One reason for this distortion is a dearth of tours and sites that reference the city's deeply rooted African culture. Many potential sites and areas of the city that would serve as ideal settings for black heritage sites have either been destroyed, neglected, or undeveloped by the city of New Orleans. A relevant example has been the city's neglect of its jazz history. The creation of jazz in New Orleans by classically trained and self-taught black musicians influenced by African musical traditions is inextricably linked to the development of a unique African American culture. Yet critics point out that these early jazz musicians, including the much-touted Louis Armstrong, are shamefully underrepresented by historical monuments. In fact, in her article on New Orleans's fledgling jazz tourism, one reporter notes that the former homes and hangouts of the city's early black jazz musicians are "now abandoned buildings, dilapidated Creole cottages, and in some cases, parking lots."[33] New Orleans's physical landscape, with its deteriorating and demolished jazz structures—even prior to the destruction leveled by Hurricane Katrina—is suggestive of its tourism landscape that also neglects the city's black heritage.

Another reason for the omission of potential black heritage sites and spaces is their designation as "unsafe." Although New Orleans's crime and murder rates have been consistently high for the past twenty years, additional resources and police patrols are dispatched to certain tourist areas of the city, particularly the French Quarter and central business district, which attract the largest volume of tourists. Consequently, tourism brochures, Web sites, and safety notices reassure visitors about their safety in these areas while tacitly or overtly discouraging ventures into other areas of the city. Many of these other areas of the city are either predominantly or historically black neighborhoods; they are almost always excluded from the accepted demarcation of tourist New Orleans. In fact, despite the recent trend to encourage visitors to explore beyond the city's French Quarter, the sanctioned spaces for these excursions rarely include black neighborhoods or black heritage sites.[34]

The failure of even the city's multicultural tourism narrative to present a viable counternarrative to the myth of New Orleans's racial exceptionalism points to the difficulty of overcoming a deeply entrenched narrative that provides economic benefit to the city and psychological rewards to many of its locals and visitors. After all, to fully revise this narrative would necessitate not simply an inclusion of characters of color or a commitment to more accurate or authentic cultural representations but a revision of the idea of New Orleans itself. Such a revision is a costly one, for the idea of New Orleans—a city of uncontested contradictions that include its simultaneous claim of uniqueness from the rest of the South and its Old South romance, its celebration of racial exoticism and mixture and its insistence on racial boundaries and exclusiveness, and its allure of decadence and danger and its continual demarcation of safe spaces—is one that appeals to and validates a broad spectrum of locals and visitors. For many whites, this idea of New Orleans provides a safe, sanctioned space to indulge in black culture and unite with black bodies, if only vicariously. In the context of New Orleans tourism, this racialized consumption is afforded without censure and with the added benefit of absolving whites of guilt and culpability for a racist past or present. The idea of New Orleans has also provided psychological and economic benefit for blacks in the city, who are not only able to subsist on tourism dollars, but for whom a glorified past of beneficent slavery, wealthy and independent free people of color, and racial harmony counters a far less redeeming and dignified history of slavery, racism, and degradation.

Despite these psychological and material benefits, Hurricane Katrina epitomized the devastating human cost of New Orleans's racial mythology. Ultimately, those who were most responsible for representing the city and the citizens of New Orleans in the aftermath of Hurricane Katrina did so from a very limited perspective that was mediated by the popular or tourist image of the city. Without the proper terms and conventions to represent New Orleans's rich and varied black experience, the coverage of and response to the hurricane grotesquely mimicked the distortions and stereotypes of the tourism narrative. Because of the limitations of the New Orleans tourism narrative, which misrepresents the city's black population, neglects and abandons New Orleans's black history, and clearly delimits the proper, safe New Orleans as nonblack, representations of the city in the aftermath of Hurricane Katrina ignored the complexity and diversity of New Orleans's black experience and continued to entrench the boundaries around the New Orleans French Quarter, abandoning the rest of the city as too dangerous and menacing to rescue or rebuild. In an article detailing the widespread exaggeration and false reporting in the aftermath of Hurricane Katrina, local journalists surmise, "The picture that emerged was one of the impoverished, masses of flood victims resorting to utter depravity, randomly attacking each other, as well as the police trying to protect them and the rescue workers trying to save them. [New Orleans mayor Ray] Nagin told [Oprah] Winfrey the crowd has descended to an 'almost animalistic state.'"[35]

More frequently, Hurricane Katrina survivors forged a community of mutual support and protection in the days following the storm instead of waiting for the uncoordinated and ineffective national response to materialize. Residents pooled resources by sharing food, water, generators, working cell phones, and necessary supplies; they coordinated and staffed makeshift shelters in abandoned homes, schools, and churches; they patrolled neighborhoods to search for elderly and infirm relatives and neighbors; they held continuous vigils to discourage looters; and they saved lives using their own fishing boats, rafts, and whatever else would float.[36] This type of community effort reflects black New Orleans's history of grassroots organizing, strong kinship ties, and multigenerational cooperation, a portrayal absent from the wake coverage and public portrayals of the city in the aftermath of Katrina.

Instead, the vilification of black New Orleans as an incorrigible drain on financial and civic resources continues to shape the debate about whether to proceed with rebuilding efforts and illustrates the limited terms and conventions available to policy-makers and news media to imagine who New Orleans residents are and what they might need now. We see the public policy impact of the city's tourism narrative in the insufficiency of federal relief, redlining of black neighborhoods, lack of a coordinated rebuilding effort, disregard of the interests of primarily black renters and public housing tenants, silence around the decimation of the black middle class, and exclusion of minority contractors, community leaders, and black residents from the rebuilding efforts.[37]

That is why it is not surprising that during a brief visit to New Orleans less than five months after the storm, President George W. Bush bypassed the worst-hit, predominantly black areas of the city. Instead, his comments to Mayor Nagin and a group of business owners reflected the degree to which New Orleans's tourism image continues to obfuscate the reality of racial dignity and racial inequality in the city. Bush announced to the group,

> I will tell you, the contrast between when I was last here and today, is pretty dramatic. It may be hard for you to see, but from when I first came here to today, New Orleans is reminding me of the city I used to come to visit. It's a heck of a place to bring your family. It's a great place to find some of the greatest food in the world and some wonderful fun. And I'm glad you got your infrastructure back on its feet. I know you're beginning to welcome citizens from all around the country here to New Orleans. And for folks around the country who are looking for a great place to have a convention, or a great place to visit, I'd suggest coming here to the great—New Orleans.[38]

That the president returned to the familiar image of the "city [he] used to visit"—one of frivolity and flavorful food—in the midst of a national catastrophe speaks volumes about the impact and enduring legacy of New Orleans's problematic representations of race. Even in the midst of heart-wrenching devastation and potentially new understandings about race and poverty in the city and the nation, the president circulated a construction of a New Orleans without racial problems at best, or worse, without a black population worthy of attention.

Given the tourism narrative's influence on political and economic support, we must demand and create a more truthful, thoughtful, and positive historical assessment, interpretation, and reimagining of New Orleans on the part of cultural institutions, the tourism industry, and public officials. What is at stake is not only a guide for remembering the past but a blueprint for re-envisioning New Orleans's future. Henceforth, we can ignore the tourism narrative and its exclusions and influence only at our own peril. The limitations of the responses to Hurricane Katrina uncovered the ways that racial representations within popular culture profoundly impact the way that we live and die.

NOTES

1. For an example of media coverage in the aftermath of the hurricane, see "Official: Astrodome Can't Take More Refugees," *FoxNews.com*, September 2, 2005, http://www.foxnews.com/story/0,2933,168112,00.html
2. For a similar analysis of Jamaican tourism and its reliance on the plantation economy and the social relations of slavery, see Frank Fonda Taylor, *To Hell with Paradise: A History of the Jamaican Tourist Industry* (Pittsburgh: University of Pittsburgh Press, 1993).

3. New Orleans Tourism Marketing Corporation's 1995 travel survey indicates that an exceptionally high percentage of visitors to New Orleans consider their experience in the city historically notable. According to the survey, New Orleans tourists visited historic sites more than any other activity in the city and more than anyplace else in the country. Sixty-four percent of New Orleans tourists visited historic sites. Eighty-eight percent of those tourists who engaged in sightseeing thought that their experience was historically notable, which exceeds the national average. "New Orleans 1995 Travel Year: Final Report" (New Orleans: Longwoods International, 1996), 166.

4. Jonathan Fricker, "Uncommon Character," *New Orleans Metropolitan Convention and Visitors Bureau*, June 2000, http://www.neworleanscvb.com/ (accessed March 1, 2003).

5. Will Coviello, "Style Points: French Quarter Architecture," *Visitor Magazine* (January 2003): 10; Gwendolyn Midlo Hall, "The Formation of Afro-Creole Culture," in *Creole New Orleans: Race and Americanization*, ed. Arnold R. Hirsch and Joseph Logsdon (Baton Rouge: Louisiana State University Press, 1992), 59.

6. For detailed arguments about how Africans' language, customs, political traditions, labor and technology influenced and sustained the new colony, see Raphael Cassimere Jr., *African Americans in New Orleans before the Civil War* (New Orleans: University of New Orleans, 1995); Gwendolyn Midlo Hall, *Africans in Colonial Louisiana: The Development of Afro-Creole Culture in the Eighteenth Century* (Baton Rouge: Louisiana State University Press, 1992); Kimberly S. Hanger, *Bounded Lives; Bounded Places: Free Black Society in Colonial New Orleans, 1769–1803* (Durham, NC: Duke University Press, 1997); Hirsch and Logsdon, eds., *Creole New Orleans*; Daniel Usner, *Indians, Settlers, and Slaves in a Frontier Exchange Economy: The Lower Mississippi River Valley before 1803* (Chapel Hill: University of North Carolina Press, 1992); Charles Vincent, *The African American Experience in Louisiana* (Lafayette: University of Southwestern Louisiana, 1999).

7. "Louisiana Basics," *Baton Rouge City Guide*, http://www.baton-rouge.com/Baton Rouge/labasics.htm.

8. Thomas Ingersoll, "Free Blacks in a Slave Society: New Orleans, 1718–1812," *The William and Mary Quarterly* 48, no. 2 (1991): 176.

9. See Ingersoll, *Mammon and Manon in Early New Orleans: The First Slave Society in the Deep South, 1718–1819* (Knoxville: University of Tennessee Press, 1999); Walter Johnson, *Soul by Soul: Life Inside the Antebellum Slave Market* (Cambridge, MA: Harvard University Press, 1999).

10. Most New Orleans tourism guides and Web sites feature plantations located on River Road, listed on the National Register of Historic Places and described in the 2002 *Louisiana Official Tour Guide* as "plantation alley." River Road stretches for about 70 miles between New Orleans and Baton Rouge on both sides of the Mississippi River. The road is noted for the elaborate plantations that were constructed by exorbitantly wealthy sugar cane planters in the years preceding the Civil War. See also, "The River Road," *National Park Service*, http://www.cr.nps.gov/nr/travel/louisiana/riverroad .htm. In addition to charter companies that customize tours, Cajun Pride Swamp Tours, Charilie's Tours, Good Old Days, Gray Line, Machu Picchu, Tours by Isabelle, Old River Road Plantation Adventure, Steppin' Out Tours, and New Orleans Tours all offered tours from New Orleans to River Road plantations prior to Hurricane Katrina.

11. "Les Grandes Dames of the River Road," Oak Alley and Nottoway Plantations, *New Orleans Official Visitors Guide* 1 (2000): 127; "Evergreen Plantation—New Orleans Tours," *New Orleans Official Visitors Guide* 1 (2000): 116; "Houmas House Plantation and Garden," *New Orleans Official Visitors Guide* 1 (2000): 116; "San Francisco Plantation," *New Orleans Official Visitors Guide* (Winter/Spring 1999): 107;

"Romance, History & Beauty on the Great River Road: Oak Alley Plantation," *New Orleans Official Visitors Guide* (2002–3), 131; and "Tezcuco Plantation," *New Orleans Official Visitors Guide* (Winter/Spring 1999): 107.

12. San Francisco Plantation has even transplanted slave cabins and a school room on the property to lend more verisimilitude to the plantation. See "San Francisco Plantation," *New Orleans Official Visitors Guide* (Winter/Spring 1999): 127.

13. *New Orleans Official Visitors Guide* (2002–3): 42.

14. Films, such as *Hush, Hush, Sweet Charlotte, Interview with a Vampire, Primary Colors*, and *Gone with the Wind* were either filmed at or inspired by Louisiana plantations, including Houmas House, Oak Alley, Chretien Point, and Greenwood Plantations. See *Louisiana Official Tour Guide*, 152, 194, 208; "Media," *Oak Alley Plantation*, http://www.oakalleyplantation.com/media/.

15. See "Oak Alley Plantation," *Gray Line Tours*, http://www.graylineneworleans.com/plantation.shtml.

16. "Historic Homes: Step into the Past," *New Orleans Online*, http://www.neworleansonline.com (accessed March 6, 2003). The list includes the Hermann-Grima Home, The Williams Residence, Beauregard-Keyes House, Longue Vue House and Gardens, and the Women's Opera Guild Home.

17. Jolene Bouchon, "Big Easy Dining: Feelings Café," *Epicurious*, http://www.epicurious.com/e_eating/e03_restaurants/neworleans/ (accessed October 7, 2003); "Past Perfect Reservations" http://www.ppreservations.com/ (accessed October 7, 2003); "New Orleans Real Estate: Elegant Victorian Home," *Nolarealtor.com*, http://www.mathildenelson.com/vt_3727_coliseum.htm (accessed October 7, 2003); "Past Perfect Reservations," http://www.ppreservations.com/ (accessed 7 October, 2003).

18. Examples include the Original Pierre Maspero's Restaurant, which is advertised as the site of the Old Slave Exchange, and Feelings Café (promoted for its "charming piano bar"), which once served as the slave quarters for D'Aunoy Plantation.

19. The Omni Royal is on the site of the former St. Louis Hotel, which held the slave auctions in the nineteenth century. This account is given on several New Orleans tours, including Le Monde Creole, African American Legacy Heritage Tour (in association with the Tennessee Williams Festival), and Eclectic Tours' "Freedom's Journey: An African American Perspective" walking tours.

20. See Johnson, *Soul by Soul.*

21. Introduction to Part 1, in Hirsch and Logsdon, eds., *Creole New Orleans*, 9. New Orleans functions literally and symbolically in nineteenth-century slave narratives as "down river," or the epitome of the slaveholding Deep South. New Orleans figures prominently in nineteenth-century slave narratives as the ultimate punishment for bondsmen and women. In these narratives, New Orleans is equated to its slave market, the center of slave trading in the United States. Its market is associated with the lowest elements of slave society, namely, unscrupulous slave traders and harsh or negligent slave masters. New Orleans is the constant threat for runaways, unwilling concubines, and other recalcitrant bondsmen and women. Even the threat of being sent to New Orleans serves as an act of correction because the city/slave market personified for bondsmen and women the evils of the slave system—heart-wrenching separation of families, the moral undermining and abuse of black womanhood, and inhumane conditions, marked by unrelenting labor and inadequate provisions. Although numerous nineteenth-century narratives include references to New Orleans and its feared slave market, three narratives, in particular, give sustained accounts: Henry Bibb's 1849 *Narrative of the Life and Adventures of Henry Bibb*; William Wells Brown's 1847 *Narrative of William Wells Brown, a Fugitive Slave*; and Solomon Northup's 1853

Twelve Years a Slave. For a brief discussion of New Orleans' association with the evils of slavery, see John Cleman, *George Washington Cable Revisited* (New York: Twayne, 1996), 5–6.

22. Jeff Crouere, "History of New Orleans" *NewOrleans.com*, http://www.neworleans.com (accessed 2 April 2003).

23. For a detailed history of the struggle for civil rights in New Orleans and Louisiana, see Adam Fairclough, *Race and Democracy: The Civil Rights Struggle in Louisiana, 1915–1972* (Athens: University of Georgia Press, 1995).

24. For different analyses of the heritage industry in the US, see Richard Gable and Eric Handler, *The New History in an Old Museum: Creating the Past at Colonial Williamsburg* (Durham, NC: Duke University Press, 1997); Barbara Kirshenblatt-Gimblett, *Destination Culture: Tourism, Museums, and Heritage* (Berkeley: University of California Press, 1998); Hal Rothman, *The Culture of Tourism, the Tourism of Culture: Selling the Past to the Present in the American Southwest* (Albuquerque: University of New Mexico Press, 2003); Fath Ruffins, "Mythos, Memory, and History: African American Preservation Efforts, 1820–1990," in *Museums and Communities: The Politics of Public Culture*, ed. Ivan Karp, Christine Mullen Kreamer, and Steven D. Lavine, 506–611 (Washington, DC: Smithsonian Institution Press, 1992).

25. "Explore the City of Mystery," *New Orleans Official Visitors Guide* (2002–3), 14.

26. For instance, Tour with Chris and Tours by Isabelle, two of the four multilingual tour companies included on the NewOrleansOnline.com's multicultural pages before Hurricane Katrina, offer typical plantation tours romanticizing the Old South. In another publication, I critique a third tour, Le Monde Creole French Quarter walking tour, for its inability to fully extricate itself from New Orleans's racial mythology, despite its inclusion of black characters and storylines. See "Race and Erasure in New Orleans Tourism" (PhD diss., Emory University, 2005), chs. 1 and 3. The African American heritage tours listed on the Web site, in contrast, propose a revisionist approach to New Orleans's typical tourism narrative by focusing on the "voice of the slave" in plantation tours and/or by providing a black heritage city tour that incorporates a detailed history of African Americans in New Orleans. See "Heritage and Multilingual Tours," *New Orleans Online*, http://www.neworleansonline.com (accessed March 6, 2003).

27. Marc H. Morial, "Dear Friends," *The New Orleans Official Visitors Guide* 1 (2000), 3.

28. C. Ray Nagin, "Welcome to New Orleans," *The Soul of New Orleans* (New Orleans: New Orleans Multicultural Tourism Network, 2002), 5.

29. "Social Traditions and Cemeteries," *Soul of America*, http://www.soulofamerica.com (accessed 5 June 2003).

30. For scholarly discussions of Creoles of Color, see Caryn Cosse Bell, *Revolution, Romantics, and the Afro-Creole Protest Tradition in Louisiana, 1716–1868* (Westport, CT: Greenwood, 1996); Carl A. Brasseaux, *Creoles of Color in the Bayou Country* (Jackson: University Press of Mississippi, 1994); Rodolphe Desdunes, *Our People and Our History* (Baton Rouge: Louisiana State University Press, 1973) [*Nos Hommes et Notre Histoire*, 1910]; Virginia Dominguez, *White by Definition: Social Classification in Creole Louisiana* (New Brunswick, NJ: Rutgers University Press, 1986); James H. Dormon, ed., *Creoles of Color in the Gulf South* (Knoxville: University of Tennessee Press, 1996); Hall, *Africans in Colonial Louisiana*; idem., "Myths About Creole Culture in Louisiana: Slaves, Africans, Blacks, Mixed Bloods, and Caribbeans," *Cultural Vistas* 12, no. 2 (Summer 2001): 78–89; Kimberly S. Hanger, *Bounded Lives, Bounded Places: Free Black Society in Colonial New Orleans, 1769–1803* (Durham, NC: Duke University Press, 1997); idem., "Conflicting Loyalties: The French Revolution

and Free People of Color in Spanish New Orleans," *Louisiana History* 34, no. 1 (1993): 5–33; Hirsch and Logsdon, eds., *Creole New Orleans*; Thomas Ingersoll, "The Slave Trade and the Ethnic Diversity of Louisiana's Slave Community," *Louisiana History* 37, no. 2 (1996): 133; Jerah Johnson, "New Orleans' Congo Square: An Urban Setting for Early Afro-American Culture Formation," *Louisiana History* 32, no. 2 (1991): 117–57; Paul F. Lachance, "The Formation of a Three-Caste Society: Evidence from Wills in Antebellum New Orleans," *Social Science History* 18 (Summer 1994): 211; Joseph Logsdon, "Americans and Creoles in New Orleans: The Origins of Black Citizenship in the United States," *Americastudien/America Studies* (Germany) 34, no. 2 (1989): 187–202.

31. The historiography on Creoles of Color has until recently advanced this same idea of Creole conservatism. Most notably, historian David C. Rankin has argued that Creoles of Color unwaveringly upheld elite, white interests through slave ownership, endogamy, and their insistence on a cultural, political, and social divide between them and enslaved blacks. See, David C. Rankin, "The Forgotten People: Free People of Color in New Orleans, 1850–1870" (PhD diss., University of Chicago, 1976); idem., "The Impact of the Civil War on the Free Colored Community of New Orleans," *Perspectives in American History* (1977): 377–416; idem., "The Politics of Caste: Free Colored Leadership in New Orleans During the Civil War," in *Louisiana's Black Heritage*, ed. Robert R. MacDonald, John R. Kemp, and Edward F. Haas (New Orleans: Lousiana State Museum, 1979), 107–46. For a similar argument, see Laura Foner, "The Free People of Color in Lousiana and St. Domingue: A Comparative Portrait of Two Three-Caste Slave Societies," *Journal of Social History* 3 (1970): 415–22; Hanger, *Bounded Lives, Bounded Places*. For recent scholarship that argues for Creole radicalism and alliances between enslaved and free blacks, see Bell, *Revolution*; Hall, *Africans in Colonial Louisiana*; and Hirsch and Logsdon, *Creole New Orleans*.

32. Jerah Johnson, "Colonial New Orleans: A Fragment of the Eighteenth-Century French Ethos," in Hirsch and Logsdon, eds., *Creole New Orleans*, 53.

33. Ronette King, "Blowing Life Back into the Birthplace of Jazz," *Newhouse News Service*, http://www.newhousenews.com (accessed June 7, 2003).

34. For instance, while NewOrleans.com does include a fairly extensive section on neighborhoods, the site provides a warning for only predominantly black New Orleans East: "Neighborhoods vary in this area, so be aware. Take taxis at night." Ironically, this warning seems appropriate to most, if not all, neighborhoods in the city, yet it is applied to only one. Furthermore, the only areas of New Orleans East that are highlighted in the section include the white neighborhoods of the Holy Cross community and the suburb of St. Bernard Parish, neither of which are truly part of New Orleans East. See "Areas of the City: New Orleans East," http://www.neworleans.com (accessed April 2, 2003). Another Web site joins the prevailing demarcation of the French Quarter as safe, with this warning: "The area along the **N. Rampart St**. side of the Quarter, and some parts near **Esplanade** are probably *not safe any time*" (emphases original). The site lists the "iffy-to-dangerous" areas of the city as "anywhere across Esplanade Avenue into Elysian Fields" and "anyplace approaching N. Rampart Street, from Canal to Esplanade, with an additional note that the public housing development at the former site of the historic Storyville neighborhood "look[s] rough to me." See "Safety Tips," *Dave and Susie's Guide to Romance in New Orleans*, http://www.angelfire.com/jazz/davensusie/safety.htm.

35. Brian Thevenot and Gordon Russell, "Rumors of Deaths Greatly Exaggerated," New Orleans *Times-Picayune*, September 26, 2005.

36. For examples of Weblogs that corroborate this type of community effort, see Quixote, "New Orleans, Hurricane Katrina, Update #3," *Acid Test Weblog*, September 1, 2005, http://acid-test.blogspot.com (accessed August 1, 2006); "St. Mike's Hardware," *Katrina New Orleans Weblog*, October 24, 2005, http://katrinola.typepad.com/katrina _new_orleans/2005/week43/index.html.

37. Charles Babington, "Some GOP Legislators Hit Jarring Notes in Addressing Katrina," *Washington Post*, September 10, 2005; Jennifer Barrett, "'A Right to Rebuild,'" *Newsweek*, January 13, 2006, http://www.msnbc.msn.com/id/10841718/site/ newsweek; Julia Cass, "Notable Mardi Gras Absences Reflect Loss of Black Middle Class," special to the *Washington Post*, February 25, 2006; James Dao, "Study Says 80% of New Orleans Blacks May Not Return," *The New York Times*, January 27, 2006; Mike Davis, "Who is Killing New Orleans?" *The Nation*, March 23, 2006, http://www.thenation.com/doc/20060410/davis.

38. "President Participates in Roundtable with Small Business Owners and Community Leaders in New Orleans," *The White House*, January 12, 2006, http://www.whitehouse .gov/news/releases/2006/01/20060112-1.html.

CHAPTER 18

THE SOCIAL CONSTRUCTION OF DISASTER: NEW ORLEANS AS THE PARADIGMATIC AMERICAN CITY*

CHESTER HARTMAN AND GREGORY D. SQUIRES

The water- and wind-driven devastation that wracked New Orleans and the entire Gulf Coast region during and after the 2005 hurricane season is virtually without parallel in recent U.S. history. A staggering two million people were displaced (Hsu 2006). In the wake of Katrina and Rita came a series of striking events, most of which were also without parallel. Illustrations include:

- Credible accusations of dereliction, even financial improprieties, on the part of national sacred cows such as the Red Cross and the Humane Society, leading to firings, resignation of the Red Cross president, criticisms by international Red Cross organizations, and official state and federal investigations (Strom 2005, 2006a, 2006b; Salmon 2006a, 2006b; Nossiter 2006a). Workers for the Federal Emergency Management Agency (FEMA) also have been accused of bribery (Lipton 2006c).
- Foreign aid coming to us, not from us—and then mishandled. The United Arab Emirates was the leader, contributing $100 million (Lipton 2006b). (Cuba, on the other hand, proposed to send medical personnel and equipment— as it does to many countries—which the State Department ignored.)
- Pets taking center stage: many flood-endangered folks, the elderly in particular, refused evacuation when they learned from government workers that they

---

* Parts of this essay are drawn from chapters in our edited collection, *There Is No Such Thing as a Natural Disaster: Race, Class and Hurricane Katrina*, with considerable updating.

could not take their pets—in effect, their family—with them. In response, full-page ads, with maudlin photos, later appeared in the *New York Times* and the *Washington Post*. The Humane Society of the United States cleverly entitled them "No Pet Left Behind" in their attempt to generate support for the Pets Evacuation and Transportation Standards (PETS) Act, signed into law in late 2006.

- A mandating of moratoriums (albeit ephemeral) on mortgage foreclosures and evictions—an intervention in the private housing market virtually unheard of since Depression days.
- Absentee voting problems (only partially solved) on an unprecedented scale for the April 22 New Orleans primary and May 20, 2006, runoff elections.
- The limitations of insurance coverage, as insurance companies flee coastal areas and the federal flood insurance program went into the red to the tune of $23 billion as a result of post-Katrina claims (Treaster 2006a; Treaster and Dean 2006).
- The Iraq war comes home, as a strapped U.S. military, bogged down in the Middle East, was unable to send National Guard, reservists, or active personnel and equipment to help out in the many ways the federal government has supplemented state and local resources in the past.

But of course the most salient and ongoing story in New Orleans and cities throughout the United States is one of poverty and racism—all those dramatic, pathetic shots, on television and in the papers, showing who the prime victims were and their helplessness, suffering, and abandonment —reflecting the current realities of race and poverty in America. Correspondent Wolf Blitzer, on September 1, 2005, lamented on CNN: "You simply get chills every time you see these poor individuals. . .so many of these people. . .are so poor and are so black, and this is going to raise lots of questions for people who are watching this story unfold" (CNN 2005).

While such images were no surprise to the community organizers, journalists, academics, and others who deal with these issues every day, for all too many segments of America this was regarded as a "wake-up call" (Turner and Zedlewski 2006; Broookings Institution 2005; Pastor et al. 2006; Dyson 2006). Speaking from New Orleans just a few weeks after the storms, President Bush asserted: "Poverty has its roots in racial discrimination, which cut off generations from the opportunity of America. . . .We have a duty to confront this poverty with bold action. . . .Let us rise above the legacy of inequality" (Bush 2005). Conservative *New York Times* columnist David Brooks, while mislabeling the events as merely a "natural disaster," recognized that they "interrupted a social disaster" (Brooks 2005).

And so the real questions are these: What created the pre-Katrina world? What should the post-Katrina world be? How might we get from here to there? And since Katrina is really a shorthand for a set of economic, social, and political conditions that characterize most of metropolitan America, what lessons and models does this provide for the nation as a whole? Ironically, we might say that in some sense we are fortunate that iconic New Orleans is the focus: one can hardly imagine equivalent national attention, had the locus been Spokane, Toledo, or Utica.

PRE-KATRINA NEW ORLEANS

Pre-Katrina New Orleans, like most major U.S. cities, was characterized by extreme levels of poverty and racial segregation. The local poverty rate has long been high, and poor residents have been heavily concentrated. New Orleans's poverty rate in 2000 was 28 percent, compared to 12 percent for the nation. The number of high-poverty census tracts (tracts where 40 percent or more of the residents are poor) grew from thirty in 1980 to forty-nine in 2000. The number of people living in these tracts increased from 96,417 to 108,419. Consequently, among U.S. cities, New Orleans had the second-highest share of its poor citizens (38 percent) living in such neighborhoods in 2000. In addition, the black poverty rate of 35 percent was more than three times the white rate of 11 percent, and 43 percent of poor blacks lived in poor neighborhoods (Jargowsky 1996, 2003; Brookings Institution 2005; Wagner and Edwards 2006). And New Orleans has long been highly segregated. According to two common indicators of racial segregation—the Index of Dissimilarity and the Isolation Index—New Orleans is one of the ten or fifteen most racially segregated among the nation's fifty largest metropolitan areas. As a Brookings Institution report summed up the situation: "By 2000, the city of New Orleans had become highly segregated by race and had developed high concentrations of poverty. . . .[B]lacks and whites were living in quite literally different worlds before the storm hit" (Brookings Institution 2005).

But where New Orleans stands relative to other metropolitan areas is almost beside the point. Big cities throughout the United States all contain large numbers of poor people, many neighborhoods of concentrated poverty, and highly segregated housing patterns. Why is this? What are the consequences? Oddly enough—and even though nothing from his personal history or his follow-up actions reflects such understanding and analysis (in his January 2007 State of the Union address, there was not a single mention of Katrina)—George W. Bush got it right: We need to understand the history and the legacy of inequality and why generations of Americans have been and continue to be cut off from opportunity (Massey and Denton 1993; Briggs 2005).

STRUCTURAL RACISM

Racial disparities and poverty are not the result primarily of individual actions or character traits, as the culture-of-poverty theory asserts (Thernstrom and Thernstrom 1997; McWhorter 2000). They are the cumulative result of a long history of institutional arrangements and structures that have produced current realities. We can start with the 250 years of African American slavery and the longer-term effects that status has had on wealth creation, family life, and white attitudes toward—as well as treatment of—blacks. Eleven Southern states (including Louisiana) seceded; a bloody civil war followed; and the defeated states (selectively) asserted a claim of "states' rights" as a means of limiting national intervention. Afterwards, a century of legal segregation throughout the South—overturned by the civil rights movement and several court rulings—ensued, with less formal barriers at work in other parts of the country. Even progressive national policies—in particular, those introduced in the New Deal period—were racially discriminatory (Katznelson 2005).

The Social Security system, when introduced, categorically excluded two occupations, courtesy of Southern members of Congress: farmworkers and domestics. Not coincidentally, these were occupations dominated by racial minorities, particularly African Americans. Federal housing programs provided minimal home ownership assistance to minority households and reinforced patterns of residential segregation. The GI Bill following World War II similarly provided relatively little education and housing assistance to minorities, compared to the massive benefits whites secured from this program. Even when African Americans were offered these federal benefits, all too often they were still effectively denied by educational institutions, housing providers, and employers.

But, of course, this is not "just history"—and in any case, history has clear and powerful continuing impacts (Williams 2003; Brown et al. 2003). "Redlining" by lending institutions and insurance companies is still all too common. School conditions for black and white students are very different, and, *Brown* notwithstanding, K-12 schools are resegregating all over the country, providing minorities with inferior education, which in turn perpetuates intergenerational disadvantage. Housing and employment discrimination is rife, as demonstrated by reams of scholarly literature. Exclusionary zoning regulations, racial steering by real estate agents, federally subsidized highways, and tax breaks for homeowners as well as suburban business development prop up the system. Racial health disparities abound. The criminal justice system—incarceration rates, sentencing patterns, the laws themselves—reflects extreme racial disparities (Massey and Denton 1993; Mauer 2006; Smelser et al. 2001; O'Connor et al. 2001; Squires and Kubrin 2006). Need we go on? Concentrated poverty and racial segregation severely reduce opportunity of all types. As sociologist Douglas Massey (2001), co-author of the classic *American Apartheid*, observes: "Any process that concentrates poverty within racially isolated neighborhoods will simultaneously increase the odds of socioeconomic failure."

The racial segregation and concentration of poverty resulting from these forces have shaped development in New Orleans and metropolitan areas around the country. One consequence is that in New Orleans those with means left when they knew the storm was coming: They had access to personal transportation or plane and train fare, money for temporary housing, in some cases second homes. Guests trapped in one luxury New Orleans hotel were saved when that chain hired a fleet of buses to get them out. Patients in one hospital were saved when a doctor who knew Al Gore contacted the former vice president, who was able to cut through government red tape and charter two planes that took them to safety. This is what is meant by the catch phrase "social capital"—a resource most unevenly distributed by class and race. Various processes of racial segregation have resulted in middle- and upper-income whites being concentrated in the outlying (and in New Orleans, literally higher) suburban communities, while blacks have been concentrated in the low-lying central city, where the flooding was most severe. And they had difficulty escaping: most notoriously, on August 31, 2005, police in the West Bank city of Gretna blocked a bridge from New Orleans, preventing large numbers of African American evacuees from leaving the deluged city (Hamilton 2006).

INFRASTRUCTURE AND UNEVEN DEVELOPMENT

A related key issue is the failure to maintain critical public services, including the infrastructure (e.g., levees in flood-prone areas). In the Katrina case, officials long knew the protective levees surrounding the city were inadequate, leaving it vulnerable to precisely the type of disaster that occurred on August 29. But whether it is the levees in New Orleans, the bridges in the San Francisco Bay Area and the Twin Cities, or the public schools in almost every city, such public services are generally viewed as expenses that need to be minimized rather than essential investments to be maximized for the purpose of enhancing the quality of life in the nation's cities. In its *2005 Report Card for America's Infrastructure*, the American Society of Civil Engineers concluded: "Congested highways, overflowing sewers and corroding bridges are constant reminders of the looming crisis that jeopardizes our nation's prosperity and quality of life." Assessing twelve infrastructure categories, the Society gave the nation a "D" for its maintenance efforts, noting there had been little improvement in recent years and asserting an as-yet unfunded $1.6 trillion investment need over the next five years (American Society of Civil Engineers 2005). The consequences have not been and will not be race- or class-neutral. Low-income people and people of color are disproportionately dependent on public transportation to get to work and to shop, on local police to keep their neighborhoods safe, and on emergency services of all types. They have fewer private resources to serve as cushions in times of stress caused not only by outside forces like hurricanes but by personal disasters such as sudden unemployment, unexpected illness or injury, or other vagaries of modern life. As James Carr observed, if the city of New Orleans had been a more diverse community, it may well have had the political clout to secure the levees long ago (Carr 2005).

The clear "bottom line" is that, while there still is plenty of racist behavior by individuals, incompetence by FEMA and other public and private bureaucracies, corruption on the part of government contractors and their partners in the public sector (Eaton 2006), and other widely reported forms of malfeasance and misfeasance, by far the most potent force in creating these extreme disparities is institutional racism—"color-blind racism," as it is often termed (Bonilla-Silva 2003)—something that most black people understand and experience but most white people do not. Consequently, it should have been no surprise when Katrina hit New Orleans that the areas damaged were 45.8 percent black, compared to 26.4 percent in undamaged areas, and that 20.9 percent of the households in damaged areas were poor, compared to 15.3 percent in undamaged areas. And if nobody is allowed to return to damaged areas, New Orleans will lose 80 percent of its black population, compared to just 50 percent of its white population (Logan 2006).

New Orleans and the Gulf Coast region generally, like virtually all metropolitan areas in the United States, experience many costs of racism, concentrated poverty, and uneven development. These forces may well shape, and hinder, redevelopment efforts in and around both New Orleans and other communities seeking paths to prosperity for their citizens (Mann 2006). Inequities associated with race, class, gender, and other socially constructed markers are not inevitable. They reflect the conscious choices made by political and economic decision-makers and implemented by public and private institutions. Different choices are available in a post-Katrina world.

SPECIFIC COSTS AND SPECIFIC LESSONS OF KATRINA

Not all past U.S. disasters were so poorly handled by government. While there were both mistakes and positive lessons to be learned, a look at the Chicago Fire of 1871, the 1906 San Francisco Earthquake, the 1927 Mississippi flood, the 1930s Dustbowl, and Hurricane Andrew in 1992 is highly instructive, showing the importance of a comprehensive revitalization approach to recovery, rather than simple rebuilding; involvement of the affected persons in their own recovery; the importance of oversight and accountability; the need for ecological balance; and the appropriate division between private- and public-sector responsibilities (Powers 2006). Recovery in some instances focused on restoring the status quo, in others on true reform—depending on who was in the decision-making role. Recovery in New Orleans, no doubt, has been and will be a contested process. A brief examination of key areas illustrates these dynamics.

HOUSING

Housing and rehousing (temporary and permanent) are of course critical issues for family life, access to jobs, schools and other community facilities, and household finances. The extent of destruction (of both privately-owned as well as public and assisted housing) was unprecedented—a National Low Income Housing Coalition September 2005 analysis showed 302,405 housing units seriously damaged or destroyed by Hurricane Katrina, 73 percent of all units in the jurisdictions studied. Slightly under half (47 percent) were rental units, 71 percent affordable to low-income households (Crowley 2006). Subsequent HUD data showed that 932,944 homes were damaged in Alabama, Louisiana, Mississippi, Texas, and Florida, 30 percent of which sustained severe or major damage (U.S. Department of Housing and Urban Development 2006). Of the 103,019 occupied public or assisted units (including privately-owned units rented by Section 8 voucher holders) in the Katrina-affected areas of Louisiana, Mississippi, and Alabama, 41,161 were damaged, 15,199 so severely damaged as to render them uninhabitable (Jackson 2006).

And the government—national, state, and local—largely botched efforts at relocation and replacement housing (Fischer and Sard 2006; Frank and Waters 2005; Hsu and Connolly 2005; Lipton 2006a; Steinhauer and Lipton 2006; Torpy 2005; Weisman 2005). In the scramble to find shelter for displaced people as they were dispersed to cities across the county, HUD encouraged local public housing authorities to give admission priority to evacuees over people on the waiting lists—thereby pitting one needy group against another, a predictable situation at a time when there are 4.5 million more extremely low-income households in the United States than there are affordable rental units (Pelletiere 2006).

FEMA's initial response after the disaster was its standard disaster response: it ordered trailers, providing them rent-free for up to eighteen months. Other FEMA housing relief programs provide funding for emergency shelters and cash grants to individuals for rental assistance, home repairs, and other personal costs. FEMA ordered 300,000 travel trailers and mobile homes but placed them mostly in trailer camps ("FEMAvilles") they set up, isolated from transportation, jobs, schools, health care, and shopping (Cohn 2005). The agency was able to install only 500 a day, with a waiting list of 40,000 in Louisiana alone (Collins and Lieberman 2006). News images of thousands of trailers awaiting delivery and installation sinking into the mud in a field near Hope, Arkansas became a symbol of FEMA's incompetence (Brand 2005; Neuman 2006). Siting of trailers has been met with "Not in my backyard!" resistance by some parish governments in Louisiana and by neighbors in New Orleans (Hustmyre 2005; Jensen 2005; Tizon 2005; Reuters 2006).

Perhaps the most bizarre problem with trailers is that they are structurally unsuitable for hurricane-prone areas (Barbour 2006). Added to that are the serious health perils to occupants from high-formaldehyde-emitting particle board and composite wood, causing serious eye, lung and nose irritation; the Gulf's hot, humid climate increases the rate at which these toxic, caracinogenic vapors are released. The vast majority of the trailers FEMA ordered were built very quickly, likely with poor quality control (Spake 2007). More than 500 hurricane survivors living in these trailers and mobile homes have sued 14 manufacturers in federal court, accusing them of using inferior materials in a profit-driven rush to build the temporary homes (Storm Victims Sue Over Trailers, Associated Press, Aug. 9, 2007). Most disturbingly, since early 2006 FEMA knew and suppressed warnings from its own health workers about these trailer hazards—the agency stopped testing occupied trailers after discovering formaldehyde levels 75 times the U.S.-recommended safety threshold for workplaces, and more stringent standards likely are appropriate for tight living spaces where occupants (many of whom are children) spend more time than at their workplace (Hsu 2007b, 2007c; Palank 2007).

As of late April 2007, some 86,000 families were still living in government trailers, under unstable conditions (Eaton 2007a). A typical account of one such FEMA trailer park, in Hammond, Lousiana—a rural area an hour north of New Orleans—was headed, quoting one resident, "We Called It Hurricane FEMA" (Whoriskey 2007a). Five trailer parks in Baton Rouge, on airport land, are closing as the airport has declined to renew the lease. "What trauma victims need most, stability, is just what has proved most elusive," noted one *New York Times* story (Dewan 2006a). Nor is the agency's profound dereliction limited to the housing area or the 2005 events: as one account noted, "As many as six million prepared meals stockpiled near potential victims of the 2006 hurricane season spoiled in the Gulf Coast heat last summer when the Federal Emergency Management Agency ran short of warehouse and refrigeration space" (Hsu 2007a). In July 2007, FEMA announced it was throwing away the last of 42,000 tons of ice purchased at a price of $24 million for the total 112,000 ton buy when Katrina struck—after paying $12.5 million in storage fees, plus $3.4 million to melt what's left (*Washington Post* 2007b).

Yet another expensive and contentious form of temporary housing was hotel/motel rooms, furnished first by the Red Cross, then taken over by FEMA. At the peak of hotel usage, FEMA reported paying for 85,000 rooms a night (U.S. Department of Homeland Security 2006). FEMA's repeated attempts to end the hotel program and compel the evacuees to move elsewhere, when there were no alternative quarters available, resulted in widespread public outcry and lawsuits (*McWaters v. FEMA* 2005; Blumenthal and Lipton 2005; Driver 2005; *San Francisco*

Sentinel 2005; *Austin-American Statesman* 2005; Lipton 2006d). A Louisiana-based federal judge issued a temporary restraining order enjoining FEMA from proceeding with its December 15, 2005 hotel assistance deadline, calling FEMA's actions "numbingly insensitive" and "unduly callous" (*McWaters v. FEMA* 2005). And in late November 2006, a federal judge ordered FEMA to restore housing assistance and pay back rent to at least 11,000 families, calling the agency's cut-off unconstitutional and "Kafkaesque" (*New York Times* 2006; Dewan 2006d)—a decision FEMA appealed. In other lawsuits against FEMA, other federal judges used such terms as "incomprehensible" and "a legal disaster" (Apuzzo 2006). The details of other, mostly inadequate government rehousing efforts are catalogued in Crowley 2006 and Quigley 2007.

Insurance has been a further and quite serious problem. Insurance companies have been insisting that their policies do not cover water damage caused by flooding—and proving that there was wind damage (which is covered) rather than water damage has been highly contentious (Treaster 2006c). Some 6,600 insurance-related lawsuits have landed in Federal District Court in New Orleans alone, and as a recent summary of popular sentiment put it: "Every neighborhood [in New Orleans] is full of horror stories about companies that reneged on their promises, offered only pennies on the dollar in settlements, dribbled out payments, deliberately underestimated the costs of repairs, dropped longtime customers and sharply increased the price of coverage" (Eaton and Treaster 2007). And yet, as an October 2006 *New York Times* headline reported, "Earnings for Insurers are Soaring"—with the jump page headlined, "Record Profits Expected for Insurers in '06" (Treaster 2006b). The Mississippi Attorney General has sued one of the major carriers, State Farm and Casualty, and a Mississippi Grand Jury heard testimony on possible criminal charges against State Farm (*New York Times* 2007).

A further dimension of the housing problem has been salvaging, rebuilding, and reoccupying New Orleans's damaged homes—deciding what houses or parts of houses are salvageable, how to carry out the salvage and rebuilding work expeditiously and at modest cost, who will do the work, and of course how to do it in ways that do not endanger the health and safety of those carrying out these tasks. "Deconstructing" homes, rather than bulldozing them—careful dismantling in order to re-use the building materials, providing skilled employment, and reducing landfill dumping—has been suggested (Zdenek et al. 2006).

A related problem has been the overwhelming focus on homeowners, with wholly inadequate attention paid to the rental housing supply, both private and public—which, of course, is disproportionately where the region's poor families live (Carr 2005). Despite the huge hit the region's rental stock took, few replacement units have been built or restored for renters; consequently, rents have risen markedly: HUD's Fair Market Rent (a measure used for the agency's Section 8 rent subsidy program) for a two-bedroom apartment in the New Orleans metropolitan area was $978 in mid-2007, up from $676 in pre-Katrina 2005 (Brookings Institution 2007).

But neither is Louisiana's Road Home program—to assist homeowners in repairing or replacing their damaged homes—anywhere near successful: as of June 2007, it was estimated that the federally funded program may be as much as $5 billion short, and as of May 2007, only 16,000 of 130,000 applicants had received money (*Washington Post* 2007a; Whoriskey 2007b).

New Orleans's public housing program and projects—a major resource of low-rent housing for the city's poor—were in poor shape before the storms. HUD went so far as to place the city's Housing Authority in receivership in 2002. There is considerable pressure not to replace or renovate damaged and destroyed units; instead, to redevelop those projects located in potentially upscale neighborhoods—the city's public housing population was nearly 100 percent black—for entirely different uses and users (Nelson and Varney 2005; Filosa 2006b; Wilgorin 2005). Louisiana Congressman Richard Baker was heard to say, a few days after the storm, "We finally cleaned up public housing in New Orleans. We couldn't do it, but God did it" (Babington 2005).

In December 2006, HUD vastly added to the city's housing problems with its decision to demolish more than 4,500 public housing apartments—a move characterized as "the most prominent skirmish in the larger battle over the post-Katrina balance of whites and blacks in New Orleans and how decisions on rebuilding shape the city's demographic future" (Cass and Whoriskey 2006). Noteworthy is what the *New York Times* architecture critic had to say about New Orleans housing projects, in a plea for restoration: "Built at the height of the New Deal, the city's public housing projects have little in common with the dehumanizing superblocks and grim plazas that have long been an emblem of urban poverty. Modestly scaled, they include some of the best public housing built in the United States. . . .[T]he notion [of dynamiting the projects] is stupefying" (Ouroussoff 2006). And, in a declaration submitted as part of a lawsuit to prevent HUD from going ahead with its plan to demolish these projects, MIT Architecture Professor John Fernandez, following a five-day survey of 140 units in four projects (Lafitte, C. J. Pete, B. W. Cooper, and St. Bernard), asserted: "My inspection and assessment found that no structural or nonstructural damage was found that would reasonably warrant any cost-effective building demolitions. . . .I did not find any conditions in which the. . .residential units themselves could not be brought to safe and livable conditions with relatively minor investment. . . .[R]eplacement of these buildings with contemporary construction would yield buildings of lower quality and shorter lifetime duration, the original construction methods and materials of these projects are far superior in their resistance to hurricane conditions than typical of new construction, and with renovation and regular maintenance, the lifetimes of the buildings in all four projects promise decades of continued service that may be extended indefinitely" (Fernandez 2006).

The centrality of housing to racial issues has been vividly illustrated in many ways following the storms. Racial discrimination has been demonstrated in the following ways:

First, studies by the National Fair Housing Alliance (2006a, 2006b) found that black evacuees were treated less favorably than white evacuees in their attempts to obtain housing. Using standard "paired tester" techniques, the Alliance revealed the litany of standard discriminatory housing market practices in two-thirds of their tests: Some landlords represented to black home seekers that vacant livable units were unavailable or unlivable, while showing several homes to whites; black home seekers were charged more rent and higher deposits than their white counterparts; rental agents failed to return messages to African American home seekers while returning the calls of their white counterparts; rental agents offered special inducements like lower security deposits to white home seekers, while failing to offer the same to their black counterparts. Parallel racial discrimination in the French Quarter's tourist establishments was revealed in similar testing by the Greater New Orleans Fair Housing Action Center (2006a). To wit: African American customers at Bourbon Street bars and nightclubs received less favorable treatment than their Caucasian counterparts and were charged more for drinks; house rules concerning minimum number of drinks and dress codes were more frequently and stringently enforced against black testers than white testers (Korosec 2005; Greater New Orleans Fair Housing Action Center 2006).

The Greater New Orleans Fair Housing Action Center (2006b) also examined listserv postings for rehousing offers (including one sponsored by FEMA)—most of which reflected admirable charitable instincts, but some of which demonstrated far less admirable, but widely prevalent (and self-deluding), racism in housing patterns (Filosa 2006a). Among the postings (Perry 2006a; Greater New Orleans Fair Housing Action Center 2007; Korosec 2005):

- "I would love to house a single mom, with one child, but white only."
- "Not to sound racist, but because we want to make things more understandable [sic] for our younger child, we would like to house white children."
- "Provider would provide room and board for $400, prefers 2 white females."

Second, in September 2006, St. Bernard Parish, right outside New Orleans, passed an ordinance barring single-family homeowners from renting their home to anyone except a blood relative without special permission from the parish council. Given St. Bernard's history and reputation as a segregated, predominantly white community, the motives for this extraordinary measure were not hard to decipher. (Nearly 93 percent of the parish's owner-occupied housing is white-occupied, and potentially thousands of homeowners who left the parish after Katrina may wish to rent out their otherwise-empty homes.) The council cited the need to "maintain the integrity and stability of established neighborhoods as centers of family values and activities," and one of the supportive councilors remarked, "We don't want to change the aesthetics of the neighborhood." The Greater New Orleans Fair Housing Action Center, represented by attorneys from the Lawyers Committee for Civil Rights Under Law and John Relman Associates, immediately filed a motion in Federal District Court for a temporary injunction, claiming a clear violation of the Fair Housing Act, leading the council to announce it would suspend enforcement of the ordinance (but not repeal it). A second, similarly motivated council ordinance imposed a year-long moratorium on the redevelopment of multifamily housing, prohibiting renovation without strict screening and pre-approval by the council itself (Chen 2006; Bazille 2006; National Fair Housing Alliance 2006).

But the overall problem is massive. As Sheila Crowley writes: "Solving the levee problem may pale in comparison to solving the housing problem. It is impossible to disentangle the housing problems in the Gulf Coast from its other major institutional crises. The challenge for public officials and private enterprise is to restore housing, health care, schools, jobs and commercial establishments in concert with one another. Most people cannot return in the absence of any one of these elements of community life. Moreover, rebuilding has stalled because there is not enough housing for the workforce needed for the task" (Crowley 2006; see also Rivlin 2006; Sayre 2006.) To this list of linked needs should be added child care—"without child care, parents miss days of work or lose jobs" (Reckdahl 2007). And the broader context is that as a nation we do not take housing problems seriously enough. We are light years away from establishing and implementing a right to decent, affordable housing (Hartman 2006). And, of course, fundamental to the future of housing in the Gulf Coast is the question of who has the right to return. Sheila Crowley again: "While policymakers have been dithering, the modern-day carpetbaggers have moved in. Speculators are buying up property at bargain prices, and multinational corporations are getting richer off of FEMA contracts. Longtime residents sometimes do not have clear title to properties handed down through the generations. Evacuees in far-off places are tempted to take what looks like a windfall of needed immediate cash—usually well below what a knowledgeable on-site seller could command—for their homes" (Crowley 2006).

EDUCATION

Almost as important as housing—for families with K-12 children—is schools. Prior to the storm, the system was one of worst in the country, according to Michael Casserly of the Council of the Great City Schools, who observed: "Before Katrina's onslaught, the children of New Orleans were isolated racially, economically, academically and politically in public schools that were financed inadequately, maintained poorly, and governed ineptly" (Casserly 2006). The damage to the educational enterprise on August 29 was enormous, as over two-fifths of the system's schools (disproportionately those in low-income, African American neighborhoods) sustained severe wind and flooding damage, many beyond repair, and almost as many suffered moderate damage. And so tens of thousands of K-12 students wound up in different school districts and different states, disrupting their curriculum and teacher/student as well as peer relationships. A great many of them missed months of formal education (Dewan 2006c). New Orleans Charter Science and Math High School (2006) produced a collection of personal narratives by the school's students chronicling their evacuations from Katrina. Dewan (2007) describes a moving art therapy program at the largest trailer park for Katrina evacuees, Renaissance [sic] Village, in Baker, Louisiana, which helps children deal with their ongoing posttraumatic stress. Their evocative drawings will be displayed at the New Orleans Museum of Art in an exhibit entitled "Katrina Through the Eyes of Children."

The U.S. Department of Education estimated that some 372,000 students—preschool through college—were displaced from the states hit by the storms. Louisiana alone estimated that some 105,000 of its students were dislocated and not attending their home schools—creating sudden and severe burdens on the receiving school systems, such as overcrowded classrooms. Texas indicated it received some 40,200 out-of-state students as a result of Katrina and Rita; Georgia was accommodating 10,300 students; Florida, 5,600 (Jacobson 2006). These newly arriving students, of course, arrived without academic, discipline, immunization, or health records. As of the Summer of 2007, just 45 percent of New Orleans's public schools had opened

(Brookings Institution 2007). A state school board member is quoted as saying: "The teacher shortage is real. The book shortage is real. We have a labor shortage. There is a shortage of bus drivers. The whole food-service industry is short of workers" (Nossiter 2006b).

A major change in the New Orleans public school system has been state takeover of a large portion of the system and a shift to charter schools, a controversial move and one that has been pushed as a more general goal by market-oriented advocates in the education reform field (Saulny 2006a; Center for Community Change 2006; Adamo 2007). An ancillary result (some claim, a goal) is weakening of the teachers' union; the district's teachers were furloughed in the weeks immediately following Katrina, and their right to return to the system on a seniority basis was replaced by state authority to hire and place teachers in the schools it had seized (Maggi 2005). Teachers in most public charter schools are on year-to-year contracts without collective bargaining leverage. United Teachers of New Orleans has filed suit challenging these arrangements (United Teachers of New Orleans, Louisiana Federation of Teachers, and American Federation of Teachers 2007).

A more recent phenomenon has been a steep rise in violence and misbehavior in the reopened schools, due in large part to the return of teenagers without accompanying parents. Some parents, for a variety of reasons—many job-related—have chosen to remain, at least temporarily, but possibly permanently, in the cities to which they were evacuated, but they gave in to their children's entreaties to return to friends and a familiar environment, making do as best they can with respect to sleeping and eating arrangements. One New Orleans high school, the largest one still functioning, where up to a fifth of the 775 students live without parents, is described as having "at least 25 security guards, at the entrance, up the stairs, and outside classes. The school has a metal detector, four police officers and four police cruisers on the sidewalk." A student observed, "We have a lot of security guards and not enough teachers." Another student added, "It's like you're in jail. You have people watching you all the time" (Nossiter 2006b). One can only imagine the long-range impact of this schooling crisis on the future lives of thousands of angry, lonely, deracinated teens. The new focus on "the school-to-prison pipeline" around the country certainly has taken root here.

The future of the city's school system remains murky, in part because no one can reliably predict how many families will eventually return, who they will be, and to which parts of the city they will return. The experience of better schooling for their children in another city may be a key decision-making factor for some families. But, as Michael Casserly (2006, 211–12) observes: "One can also see flashing yellow lights on the horizon. The preliminary plans for remaking the city's public schools were designed by people who were substantially different in hue from educational decision-makers before the storm, a touchy issue in these still racially-sensitive times. The proposal to replace the currently elected school board with an appointed school board, for instance, is bound to exacerbate concerns among community activists and parents that the schools, along with other city agencies, are being highjacked by alien forces." And Casserly (2006, 213) appropriately extends this warning: "New Orleans is not the only city. . .in which our poorest children are concentrated and isolated in such a way. . . .And it is not the only one that embodies the nation's neglect of its poor. One can see the same pattern in many other cities across the country—if one is only willing to open one's eyes. And, in other cities, we run the same risk. . .whatever the next storm, wherever the next levees."

HEALTH

Another critical service is public health. The storms created their own public health problems—notably, toxins from damage to buildings and vehicles; brackish, sewage-contaminated flood water; decomposing bodies; vermin; and many other sources.

The immediate crisis was described by Dr. Evangeline Franklin (2006), Director of Clinical Services and Employee Health for the City of New Orleans:

> Aside from immediate and long-term mold and the lack of water, food, shelter and sanitation facilities, there was concern that the prolonged flooding might lead to an outbreak of health problems in many neighborhoods among the remaining population. In addition to dehydration and food poisoning, there was the potential for communicable disease outbreaks of diarrhea and respiratory illness, all related to the growing contamination of food and drinking water supplies in the area. After dewatering and house-to-house searches for dead bodies and animals, public health concerns centered around acute environmental hazards related to houses standing in water for several weeks (mold, bacteria, concealed rodents, snakes and alligators). This was combined with ruptured sewage lines, refuse, structural instability, debris, the lack of sanitary water (what water they had was usable only for flushing toilets), as well as a lack of gas and electricity. . . . There was concern that the chemical plants and refineries in the area could have released pollutants into the floodwaters. People who suffer from allergies or chronic respiratory disorders, such as asthma, were susceptible to what some health officials have dubbed "Katrina Cough." On September 6, it was reported that *Escherichia coli* (*E. coli*) had been detected at unsafe levels in the water that flooded the city. The CDC [Center for Disease Control and Prevention] reported on September 7 that five people had died of bacterial infection from drinking water contaminated with *Vibrio vulnificus*, a bacterium from the Gulf of Mexico. (Franklin 2006)

As was the case with the schools, New Orleans was a community at risk well before August 29. An Urban Institute report (Zuckerman and Coughlin 2006, cited in Franklin 2006, 185) noted that before the storms hit: "According to the United Health Foundation's 2004 State Health Rankings. . .Louisiana ranked lowest overall in the country. It numbered among the five worst states for infant mortality, cancer deaths, prevalence of smoking, and premature deaths. . . .Louisianans also had among the nation's highest rates of cardiovascular deaths, motor vehicle deaths, occupational fatalities, infectious diseases, and violent crime." This majority-black city with extreme levels of poverty produced a de facto caste system of health care, providing unequal, lesser treatment for the poor, the uneducated, the homeless, the immigrant, the uninsured, and others who are disenfranchised (Institute of Medicine 2003; Franklin, Hall, and Burris 2005). State funding cutbacks beginning in 2004 led to dramatic reductions by more than two-thirds in ambulatory care services for the poorest residents and reduced access to nonemergency care for the uninsured—leading to increasingly congested emergency rooms.

Most of the city's hospitals, located in the center of the city, were damaged or flooded, leading to loss of over half of the state's hospital beds and one of the state's two Level 1 trauma centers (Barringer 2006). As of the Summer of 2007, only 13 of New Orleans's 23 pre-Katrina hospitals were open (Eaton 2007b; Brookings Institution 2007). Doctors' offices were rendered unusable, as were pharmacies. As part of general layoffs of city workers, Health Department staff for its clinics went from 250 to 72. Health workers at all levels evacuated, and a great many likely will not return. As of spring 2006, there were virtually no dentists practicing in the city. Absence of former patient loads of course is a factor influencing health providers' return plans, and physicians need hospitals, support facilities, labs, x-ray units and pharmacies to provide the proper level of health care. One major problem is the loss of health records—leading to strong recommendations for development of a system of portable, electronic, transmittable records (Franklin 2006).

Higher rates of illness among evacuees have been reported, especially among children. Mental health problems are rife, with a notable rise in suicides (Turner 2006). The Regional Administrator for Health Unit 2 in Baton Rouge Parish, where some 30,000 persons relocated following the hurricanes, noted that the parish's mental health clinic has seen a 30 percent increase in post-Katrina office visits. "Many experience depression, anxiety, Post-Traumatic Stress Disorder and other mental illness due to hurricane displacement and trauma. . . . Transitional housing areas are reporting problems with child abuse, sexual abuse, domestic violence and substance abuse as people struggle to cope with the loss of their community/support system while living in close proximity to each other in tiny FEMA trailers" (Roques 2007). Decomposed bodies are still being found, adding to the still incomplete toll of storm-related deaths. And toxic waste problems doubtless will persist and show up—Agent Orange–like—in years to come (Dewan 2006a, 2006b; Redlener 2006; Nossiter 2005; Connolly 2005; Cass 2006; Hsu and Eilperin 2006; Barringer 2005, 2006; Cole and Woelfle-Erskine 2006; Saulny 2006b; Turner 2006).

A range of political, economic, and social forces have contributed to health care challenges in New Orleans. The city's historic absence of a manufacturing sector prevented the development of a strong labor movement with its demands for health care benefits. Cultural patterns ("*Laissez les bons temps rouler*"), and the high-calorie, high-cholesterol local cuisine were contributing factors to the local health picture (Turner 2006b; Franklin 2006).

In sum, the medical problems and unequal access to health care were predictable, just one of the many areas in which poverty and race in our society compound vulnerability (Franklin 2006).

ECONOMIC DEVELOPMENT

The issue of future (and past) economic development is prominent as well. Pre-Katrina, as Robert K. Whelan points out, "the local economy was highly polarized, with some professional people doing very well and a much larger number of people employed in low-wage jobs." (Whelan 2006). Employment growth was stagnant during the 80s and 90s, and the city registered substantial job losses in port-related industries and manufacturing while gaining jobs in health care, tourism, legal services, social services, and education (Whelan, Gladstone, and Hirth 2002). The metropolitan area lost more than 200,000 jobs in the wake of the storms; in November 2005, the unemployment rate in the New Orleans area was a staggering 17.5 percent.

Rebuilding economically has been hampered by multiple failures on the part of the Small Business Administration and FEMA (Nixon 2007), as well as the failure of insurance companies to make prompt and full payment, followed by the withdrawal of several key companies from the area with respect to writing new policies. Government cleanup and construction contracts went to firms from Texas and Arkansas, rather than to local firms, which would have had the benefit of shoring up the local economy. Labor contractors and outside firms imported large numbers of immigrant workers, a substantial portion of whose wages did not circulate in the local economy but were remitted to their home countries. And predictably, many were not paid promptly, were paid less than what they were promised, or were not paid at all. MacDonald (2007) offers a case study of Maryland day laborers who traveled to Louisiana and Mississippi, working long hours, for a promised $10/hour, shoveling mud and debris out of Gulf Coast gambling casinos and other dirty, dangerous jobs—only to have to wait two years to get their paychecks—and then only by virtue of a lawsuit on their behalf brought by a Maryland advocacy organization that assists Latinos (See also Beutler 2007; Browne-Dianis et al. 2006). For a constructive approach to employment generation and economic development in reconstruction projects—framed around conflicts but applicable as well to other disasters, see Mendelson-Forman and Mashatt 2007. See also Browne-Dianis et al. 2006.

Unequal access to credit—residential as well as commercial—has been a serious problem in the past and likely will continue to be. John Taylor and Josh Silver (2006) have documented the extensive redlining that existed pre-Katrina and its relation to unemployment and poverty conditions. In 2004, African Americans received just 15 percent of all market-rate home loans written in the New Orleans metropolitan statistical area, where they comprised 34 percent of the population—leaving them to rely on high-cost, often predatory, lending. Had they received market-rate mortgages in proportion to their population, they would have received 4,269 additional loans worth $458 million. In the area of small business lending, during 2004, Community Reinvestment Act–covered lenders made loans to 38 percent of the small businesses in Mississippi's minority neighborhoods, compared to 51 percent of the small businesses in white neighborhoods. Equalizing the 51 percent rate would have increased the number of such loans in Mississippi's minority neighborhoods by 6,588, worth $307 million. While income and creditworthiness obviously play a role, differences in unequal access to credit are rooted in the structural dimensions of race and space, as noted above.

A further important element in the lending picture is the location of bank branches: such local branches are generally found to boost small business lending, but there are far fewer bank branches in low-income and minority neighborhoods—as an extreme example, the predominantly minority, lower-income Lower Ninth Ward and St. Claude neighborhoods of New Orleans have just one branch, while bank branches are clustered around the French Quarter tourist hub and predominantly white neighborhoods across the city. As Taylor and Silver write, "the financial sector. . .has yet to address the inequalities between minorities and whites that were magnified by the hurricanes' damage" (Taylor and Silves 2006).

WHAT DOES THE FUTURE LOOK LIKE?

What, then, is the future of New Orleans—and by extension, America's metropolitan areas?

A story in the September 8, 2005, *Wall Street Journal* offers one vision:

> Despite the disaster that has overwhelmed New Orleans, the city's monied, mostly white elite is hanging on and maneuvering to play a role in the recovery when the floodwaters of Katrina are gone. . . .The power elite of New Orleans—whether they are still in the city or have moved temporarily to enclaves such as Destin, Fla., and Vail, Colo.—insist the remade city won't simply restore the old order. . . .The new city must be something very different, Mr. Reiss [chairman of the city's Regional Transit Authority, who helicoptered in an Israeli security company to guard his Audubon Place house and those of his neighbors] says, with better services and fewer poor people. . .,"Those who want to see the city rebuilt want to see it done in a completely different way: demographically, geographically and politically."

This speaks to the replanning process and who is involved in it. Is there to be a true right of return (Wellington 2006)—especially for those who lived in the most run-down and vulnerable parts of the city, such as the infamous Lower 9th Ward? (See ACORN 2007; Nossiter 2007; Solnit 2007) How can we avoid placing people in danger of future storms and levee failures? Who is responsible for carrying out plans? (An interesting international dimension to these issues occurred in November 2006, when a delegation of Katrina/Rita survivor advocates traveled to Thailand to meet with representatives of those impacted by the 2004 tsunami and the more recent earthquake in Yogyakarta, Indonesia, "to claim their human rights and dignity, while leading the process of rebuilding their communities with full participation and empowerment." Reference was made to government responsibility for such survivors as Internally Displaced Persons, who have special protections under international human rights standards. (See National Economic & Social Rights Initiative 2006.)

Peter Marcuse offers a counter to the *Wall Street Journal* vision: "the principle guiding the planning efforts should not simply be subservience to the desires of the 'monied, mostly white elite,' sweeping the area's past problems under the table and its poorer residents out the door" (Marcuse 2006). Rather the process should provide true democratic participation of all those affected by the disaster (involving evacuees who have not returned, as well as those now in New Orleans) and equitable distribution of costs and benefits. That latter goal speaks to economic development for the poor, a true safety net, fair compensation for what has been lost. Beyond New Orleans, "the goals should. . . be. . . moving towards making the cities and region affected a model of what American communities should and could be," and the federal government needs to play the key financing role. (See also Birch and Wachter 2007.)

The most detailed, comprehensive approach—combining democracy and equity, backed by the needed implementation resources—is the Congressional Black Caucus'

Hurricane Katrina Recovery, Reclamation, Restoration, Reconstruction and Reunion Act of 2005 (H.R. 4197 of the 109th Congress, with ninety-one cosponsors.) It did not pass, and it was not reintroduced in the 110th Congress, but elements of it—dealing with grants to states to respond to disasters and improvements to the Small Business Administration loan guarantee program—have been introduced as free-standing bills.

In December 2006, an elaborate "community engagement" process was held, involving evacuees in New Orleans, Baton Rouge, Atlanta, Houston and Dallas (with satellite gatherings in sixteen other diaspora cities attended by numerous, but fewer, evacuees), produced by America Speaks, designed to produce a comprehensive "bottom-up" rebuilding plan (Unified New Orleans Plan 2006). And in the same month, New Orleans Mayor C, Ray Nagin appointed Edward Blakely, a highly regarded urban planner, former Chair of the University of California–Berkeley Urban Planning Department, as "executive director for recovery management" for his city (Nossiter 2006c).

john powell and his colleagues at the Kirwan Institute for the Study of Race and Ethnicity (2006, 60-61) offer the following guidance on what is the most critical issue--the role of race in America:

> Questions about *why* African Americans are more likely than whites to be poor, and why poor African Americans are more likely to live in areas of concentrated poverty, are questions that were neither asked nor answered. . . .There was little critical discussion of how historical patterns of segregation contributed to the racial layout of the city, and how structures worked together to produce racial disparities and economic inequality (Muhammed et al. 2004). . . .[B]roadening how we think and talk about race is critically important for making sense of today's world. Doing so also raises critical questions about the shrinking middle class, our anemic investment in public space, the meaning of merit in a purported meritocracy, and the promises and failures of the American experiment—all of which concern every American. Once we are able to discuss race and racism in these broad terms, we will be able to construct a response not only to the damage wrought by Katrina, but also to that which occurs across the country every day. . . .[Katrina] created an opportunity for reexamining the connections between race and class, and deciphering precisely how race has been inscribed spatially into our metropolitan areas. In short, it has provided a rare chance to discuss the links between race, equity, justice and democracy.
>
> Race, as a transformative tool, can and should be applied to more than just the rebuilding effort in New Orleans. Racialized poverty, segregation, and the decaying infrastructure of our central cities are common problems plaguing urban areas nationwide. Used properly, race allows us to examine how institutional failings affect everyone, and enables us to re-imagine a society where democracy and democratic ideals are not constricted and undermined by structural arrangements.

We end with this observation by three academics (Frymer et al. 2006):

> "The experience of African Americans in New Orleans can serve as the 'miner's canary,' as Lani Guinier and Gerald Torres argue. Similar to the way in which canaries alerted miners to the specter of poisonous air, the fates that befall people who are disadvantaged by inequalities based on, for example, race, class, and gender are signifiers

of society-wide inequalities. If policymakers and the public heed the lessons of Katrina and make efforts to address the structural and institutional sources of American inequality, perhaps the brunt of future disasters will not be borne by those who are least able to endure their costs."

Sober words.

REFERENCE LIST

ACORN. 2007. A people's plan for overcoming the Hurricane Katrina Blues—A comprehensive strategy for building a more vibrant, sustainable, and equitable 9th Ward. http://www.rebuildingtheninth.org.

Adamo, Ralph. 2007. NOLA's failed education experiment: Privatization runs amok in the post-Katrina New Orleans school system. *The American Prospect.* August 15. http://www.prospect.org/cs/articles?article=nolas_failed_education_experiment_.

American Society of Civil Engineers. 2005. *2005 report card for America's infrastructure.* Reston, VA: American Society of Civil Engineers.

Apuzzo, Matt. 2006. Judge tells FEMA to "get going" on aid. *Washington Post,* December 14.

Austin-American Statesman. 2005. With holidays coming, evacuees will be sent packing. *Austin-American Statesman,* November 22.

Babington, Charles. 2005. Some GOP legislators hit jarring note in addressing Katrina. *Washington Post,* September 10.

Barbour, Governor Haley. 2006. Testimony before Committee on Appropriations, U.S. Senate. March 6. http://www.appropriations.senate.gov/hearmarkups/gov'stestimony-fullapprops.march06.htm.

Barringer, Felicity. 2005. Toxic residue of hurricane stirs debate on safety. *New York Times,* December 2.

———. 2006. Long after storm, shortages overwhelm N.O.'s few hospitals. *New York Times,* January 23.

Beck, Chris, and Preston Browning 2006. Fables of the deconstruction. Op-ed. *New York Times,* November 30.

Beutler, Brian. 2007. In the lawless post-Katrina cleanup, construction companies are preying on workers. July 16. http://www.alternet.org/story/56958.

Birch, Eugenie L., and Susan M. Wachter, eds. 2007. *Rebuilding urban places after disaster: Lessons from Hurricane Katrina.* Philadelphia: PennPress.

Blumenthal, Ralph, and Eric Lipton. 2005. FEMA broke its promise on housing, Houston mayor says. *New York Times,* November 17.

Bonilla-Silva, Eduardo. 2003. *Racism without racists: Color-blind racism and the persistence of racial inequality in the United States.* New York: Rowman & Littlefield.

Brand, Aaron. 2005. Rep. Mike Ross critical of FEMA mobile homes delay. *Texarkana Gazette,* December 15.

Briggs, Xavier de Souza, ed. 2005. *The geography of opportunity: Race and housing choice in met ropolitan America.* Washington, DC: Brookings Institution Press.

Brookings Institution. 2005. *New Orleans after the storm: Lessons from the past, a plan for the future.* Washington, DC: Brookings Institution.

———. 2007. *The New Orleans Index—Second Anniversary Special Edition: A review of key indicators of recovery two years after Katrina.* Washington, DC: Brookings Institution.

Brooks, David. 2005. "Katrina's silver lining." *New York Times,* September 8.

Brown, Michael K., Martin Carnoy, Elliott Currie, Troy Duster, David B. Oppenheimer, Marjorie M. Shultz, and David Wellman. 2003. *White-washing race: The myth of a color-blind society.* Berkeley: University of California Press.

Browne-Dianis, Judith, Jennifer Lai, Marielena Hincapie, and Saket Soni. 2006. *And injustice for all: Workers' lives in the reconstruction of New Orleans.* Washington, DC: Advancement Project.

Bush, President George W. 2005. Address to the Nation from Jackson Square on Hurricane Relief. http://www.whitehouse.gov.

Carr, James. 2005. Comments at conference on "Predatory home lending: Moving toward legal and policy solutions." John Marshall School of Law, Chicago, September 9.

Carr, Martha. 2005. Housing shortage hinders rebound. New Orleans *Times-Picayune*, October 16.

Cass, Julia. 2006. For many of Katrina's young victims, the scars are more than skin deep. *Washington Post*, June 13.

Cass, Julia, and Peter Whoriskey. 2006. New Orleans to raze public housing. *Washington Post*, December 8.

Casserly, Michael. 2006. Double jeopardy: Public education in New Orleans before *and* after the storm. In *There is no such thing as a natural disaster: Race, class and Hurricane Katrina*, ed. Chester Hartman and Gregory D. Squires. New York: Routledge.

Center for Community Change. 2006. Dismantling a community. Washington, DC: Center for Community Change. http://www.communitychange.org/issues/education/publications.

Chen, Michelle. 2006. Housing watchdogs call post-Katrina ordinance "racist." *New Standard*, October 6. http://www.newstandardnews.net/content/index.cfm/items/3731.

Christensen, Congresswoman Donna M., Brian Weinstock, and Natasha H. Williams. 2006. From despair to hope: Rebuilding the health care infrastructure of New Orleans after the storm. *Harvard Journal of African American Public Policy* 12:17–22.

CNN. 2005. "The situation room." http://transcripts.cnn.com.

Cohn, Jonathan. 2005. Trailer trash: Katrina victims need real housing. *The New Republic*, September 26.

Cole, Oskar, and Cleo Woelfle-Erskine. 2006. Rebuilding on poisoned ground. *ColorLines* (Spring): 26–28.

Collins, Senator Susan, and Senator Joseph Lieberman. 2006. Letter to Secretary of U.S. Department of Homeland Security Michael Chertoff. http://www.nlihc.org/news/031406collinsletter.pdf.

Connolly, Ceci. 2005. Katrina's emotional damage lingers. *Washington Post*, December 7.

Crowley, Sheila. 2006. Where is home? Housing for low-income people after the 2005 hurricanes. In *There is no such thing as a natural disaster: Race, class and Hurricane Katrina*, ed. Chester Hartman and Gregory D. Squires. New York: Routledge.

Dewan, Shaila. 2006a. Storm evacuees found to suffer health setbacks. *New York Times*, April 18.

———. 2006b. In attics and rubble: More bodies and questions. *New York Times*, April 11.

———. 2006c. For many, education is another storm victim. *New York Times*, June 1.

———. 2006d. FEMA ordered to restore evacuees' housing aid. *New York Times*, November 30.

———. 2006e. Storm evacuees remain in grip of uncertainty. *New York Times*, December 6.

———. 2007. Using crayons to exorcise Katrina. *New York Times*, September 17.

Driver, Anna. 2005. Some Katrina victims face NY homeless shelters. *Reuters*, November 18.

Dyson, Michael Eric. 2006. *Come hell or high water: Hurricane Katrina and the color of disaster.* New York: Perseus Books.

Eaton, Leslie. 2006. In Louisiana, graft inquiries are increasing. *New York Times*, March 18.

————. 2007a. Hurricane survivors to buy U.S. trailers or pay rental fee. *New York Times,* April 27.

————. 2007b. Shuttered hospitals ensure slow recovery in New Orleans. *New York Times,* July 24.

Eaton, Leslie and Joseph R. Treaster. 2007. Insurers bear brunt of anger in New Orleans. *New York Times,* September 3.

FEMA's latest fumble. 2005. *Northeast Mississippi Daily Journal,* November 18.

Fernandez, John E. 2006. Declaration of John F. Fernandez, U.S. District Court for the Eastern District of Louisiana. *Yolanda Anderson et al. v. Alphonso Jackson et al.* Civil Action No. 06-3296. October 23.

Filosa, Gwen. 2006a. Housing discrimination hits the Web: Post-Katrina ads cited in federal complaints. New Orleans *Times-Picayune,* January 3.

————. 2006b. Displaced residents demand access to public housing. New Orleans *Times-Picayune,* June 3.

Fischer, Will, and Barbara Sard. 2006. *Housing needs of many low-income hurricane evacuees are not being adequately addressed.* Washington, DC: Center on Budget and Policy Priorities.

Frank, Representative Barney, and Representative Maxine Waters. 2005. Reps. Frank and Waters: Where's the Katrina Housing Plan? Press Release. November 8. http://www.house.gov/banking_democrats/pr11082005a.html.

Franklin, Evangeline. 2006. A new kind of medical disaster in the United States. In *There is no such thing as a natural disaster: Race, class and Hurricane Katrina,* ed. Chester Hartman and Gregory D. Squires. New York: Routledge.

Franklin, Evangeline, Patrick Hall, and Nancy Burris. 2005. *Getting people healthy in New Orleans.* 2 Vols. New Orleans Health Department. Unpublished.

Frymer, Paul, Dara Z. Strolovitch, and Dorian T. Warren. 2006. Katrina's political roots and divisions: Race, class, and federalism in American politics. *Understanding Katrina: Perspectives from the social sciences.* http://understandingkatrina.ssrc.org/FrymerStrolovitchWarren.

Greater New Orleans Fair Housing Action Center. 2006. Audit of Bourbon Street discrimination against African-American males. Unpublished manuscript.

————. 2007. For rent, unless you're black: An audit report and study on race discrimina-tioin in the greater New Orleans metropolitan rental housing market. New Orleans: The Center.

Hamilton, Bruce. 2006. Bridge standoff still under scope: Gretna faces lawsuit for stopping evacuees. New Orleans *Times-Picayune,* January 4.

Hartman, Chester. 2006. The case for a right to housing. In *A right to housing: Foundation for a new social agenda,* ed. Rachel G. Bratt, Michael E. Stone, and Chester Hartman. Philadelphia: Temple University Press.

Hattiesburg American. 2005. "What next" for victims of Katrina? *Hattiesburg American* (Hattiesburg, MS), November 16.

Hsu, Spencer S. 2006. 2 million displaced by storms. *Washington Post,* January 13.

————. 2007a. Tons of food spoiled as FEMA ran out of storage space. *Washington Post,* April 13.

————. 2007b. FEMA knew of toxic gas in trailers. *Washington Post,* July 20.

————. 2007c. FEMA to let Katrina victims move from trailers into hotels. *Washington Post,* September 5.

Hsu, Spencer S., and Juliet Eilperin. 2006. Safety of post-hurricane sludge is disputed. *Washington Post,* February 23.

Hsu, Spencer S., and Ceci Connolly. 2005. Housing the displaced is rife with delays. *Washington Post,* September 23.

Hustmyre, Chuck. 2005. Residents, evacuees jam meeting. *Baton Rouge Advocate*, November 9.

Institute of Medicine. 2003. *Unequal treatment: Confronting racial and ethnic disparities in health care*, ed. Brian D. Smedley, Adrienne Y. Stith, and Alan R. Nelson. Washington, DC: National Academy Press.

Jackson, Secretary Alphonso. 2006. Testimony before Committee on Banking, Housing, and Urban Affairs, U.S. Senate. http://www.nlihc.org.

Jacobson, Linda. 2006. Hurricane's aftermath is ongoing. *Education Week*, February 1.

Jargowsky, Paul. 1996. *Poverty and place: Ghettos, barrios, and the American city*. New York: Russell Sage Foundation.

———. 2003. *Stunning progress, hidden problems: The dramatic decline of concentrated poverty in the 1990s*. Washington, DC: Brookings Institution.

Jensen, Lynne. 2005. FEMA trailer plans hit stoplight in New Orleans. New Orleans *Times-Picayune*, December 1.

Katznelson, Ira. 2005. *When affirmative action was white: An untold history of racial inequality in twentieth-century America*. New York: Norton.

Korosec, Thomas. 2005. Survey finds bias in evacuee housing: 66 percent of white callers got better deals in Houston and 16 other cities. *Houston Chronicle*, December 27.

Lipton, Eric. 2006a. Trailers, vital after hurricane, now pose own risks on Gulf. *New York Times*, March 16.

———. 2006b. Hurricane relief from abroad was mishandled. *New York Times*, April 7.

———. 2006c. FEMA workers accused of bribery. *New York Times*, January 26.

———. 2006d. FEMA is set to stop paying hotel cost for storm victims. *New York Times*, November 16.

Logan, John R. 2006. The impact of Katrina: Race and class in storm-damaged neighborhoods. Working paper, Spatial Structures in the Social Sciences, Brown University.

MacDonald, Christine. 2007. The Big Sleazy: Area day laborers helped a local contractor with the Katrina cleanup. Two years later, they're getting their paychecks. *Washington City Paper*, August 31.

Maggi, Laura. 2005. State to run New Orleans schools. New Orleans *Times-Picayune*, November 23.

Mann, Eric. 2006. *Katrina's legacy: White racism and black reconstruction in New Orleans and the Gulf Coast*. Los Angeles: Frontlines Press.

Marcuse, Peter. 2006. Rebuilding a tortured past or creating a model future: The limits and potentials of planning. In *There is no such thing as a natural disaster: Race, class and Hurricane Katrina*, ed. Chester Hartman and Gregory D. Squires. New York: Routledge.

Massey, Douglas S. 2001. Residential segregation and neighborhood conditions in U.S. metropolitan areas. In *America becoming: Racial trends and their consequences*, ed. Neil J. Smelser, William Julius Wilson, and Faith Mitchell. Washington, DC: National Academy Press.

Massey, Douglas S., and Nancy Denton. 1993. *American apartheid: Segregation and the making of the underclass*. Cambridge, MA: Harvard University Press.

Mauer, Marc. 2006. *Race to incarcerate*. New York: New Press.

McWaters v. FEMA. 2005. Case 2:05–cv–05488–SRD–DEK, Document 38.2005. E.D.La 2005. December 12.

McWhorter, John. 2000. *Losing the race: Self-sabotage in black and white*. New York: Free Press.

Mendelson-Forman, Johanna, and Merriam Mashatt. 2007. Employment Generation and Economic Development in Stabilization and Reconstruction Operations. Washington, DC: United States Institute of Peace. http://www.usip.org/pubs/specialreports/srs/srs6.pdf.

Muhammed, Dedrick, Attieno Davis, Meizhu Lui, and Betsy Leondar-Wright, 2004. The state of the Dream 2004: Enduring disparities in black and white. *United for a Fair Economy*. http://www.faireconomy.org/press/2004/StateoftheDream2004.pdf.

National Economic & Social Rights Initiative. 2006. Media Advisory. December 1.

National Fair Housing Alliance. 2006. *"Still No Home for the Holidays" for Katrina Survivors*. Washington, DC: The Alliance.

National Low Income Housing Coalition. 2005. National housing advocates call on administration and Congress for action on housing hurricane victims. Press Release. http://www.nlihc.org/press/113005pr.html.

Nelson, Rob, and James Varney. 2005. "Not in my back yard" cry holding up FEMA trailers: Emotional tone of opposition hints at role and stereotypes of race, class. New Orleans *Times-Picayune*, December 26.

Neuman, Johanna. 2006. The land of 10,770 empty FEMA trailers. *Los Angeles Times*, February 10.

New Orleans Charter Science and Math High School. 2006. *From the Second Line*. Houston: Katrina Writing Project.

New York Times. 2006. Kafka and Katrina. *New York Times*, December 2.

———. 2007. Mississippi Attorney General sues State Farm over Katrina claims. *New York Times*, June 12.

Nixon, Ron. 2007. House committee to examine recent performance of S.B.A. *New York Times*, February 8.

Nossiter, Adam. 2005. Hurricane takes a further toll: Suicides up in New Orleans. *New York Times*, December 27.

———. 2006a. F.B.I. to investigate Red Cross over accusations of wrongdoing. *New York Times*, March 31.

———. 2006b. Students after the storm, left alone and angry. *New York Times*, November 1.

———. 2006c. New Orleans picks professor to lead efforts on rebuilding. *New York Times*, December 6.

———. 2007. In New Orleans, progress at last in the Lower Ninth Ward. *New York Times*, February 23.

O'Connor, Alice, Chris Tilly, and Lawrence D. Bobo, eds. 2001. *Urban inequality: Evidence from four cities*. New York: Russell Sage Foundation.

Ouroussoff, Nicolai. 2006. All fall down. *New York Times*, November 19.

Palank, Jacqueline. 2007. FEMA faulted on response to risks in storm trailers. *New York Times*, July 20.

Pastor, Manuel, Robert D. Bullard, James K. Boyce, Alice Fothergill, Rachel Morello-Frosch, and Beverly Wright. 2006. *In the wake of the storm: Environment, disaster, and race after Katrina*. New York: Russell Sage Foundation.

Pelletiere. Danilo. 2006. *The rental housing affordability gap: Comparison of 2001 and 2003 American Housing Surveys*. Washington, DC: National Low Income Housing Coalition.

Perry, James. 2006a. Testimony before Subcommittee on Housing and Community Opportunity, U.S. House of Representatives, February 28. http://financial services.house.gov/media/pdf/022806.pdf.

———. 2006b. Fair Housing and Hurricane Katrina. Presentation at National Fair Housing Alliance Fair Lending Conference, Washington, DC, October 12–13.

powell, john a., Hasan Kwame Jeffries, Daniel W. Newhart, and Eric Stiens. 2006. Towards a transformative view of race: The crisis and opportunity of Katrina. In *There is no such thing as a natural disaster: Race, class and Hurricane Katrina*, ed. Chester Hartman and Gregory D. Squires. New York: Routledge.

Powers, Michael P. 2006. A matter of choice: Historical lessons for disaster recovery. In *There is no such thing as a natural disaster: Race, class and Hurricane Katrina*, ed. Chester Hartman and Gregory D. Squires. New York: Routledge.

Quigley, William P. 2007. Obstacle to opportunity: Housing that working and poor people can afford in New Orleans since Katrina. *Wake Forest Law Review* 42:393–418.

Reckdahl, Katy. 2007. Crisis in child care: The extreme shortage strains local families and the economy. New Orleans *Times-Picayune*, May 14.

Redlener, Irwin. 2006. Orphans of the storm. *New York Times*, May 9.

Reuters. 2006. Trailer Trouble Chills New Orleans-FEMA Relations. *Reuters*, April 3.

Rivlin, Gary. 2006. Patchy recovery in New Orleans. *New York Times*, April 5.

Roques, Jamie M. 2007. Urban MCH leaders assess local urban mental health services (E-Roundtable). *City Lights* 16 (1): 1, 4–5. http://webmedia.unmc.edu/community/city match/CityLights/CLsum07.pdf.

Salmon, Jacqueline L. 2006a. Red Cross, Humane Society under investigation. *Washington Post*, March 26.

———. 2006b. Counterparts excoriate Red Cross Katrina effort. *Washington Post*, April 5.

San Francisco Sentinel. 2005. Newsom slams Bush decision to cut off housing vouchers for Katrina victims at beginning of holidays. *San Francisco Sentinel*, November 17.

Sarbanes, Senator Paul. 2005. Letter to Senators Thad Cochran and Robert Byrd. December 8. http://www.nlihc.org/news/120905sarbanes.pdf.

Saulny, Susan. 2006a. U.S. gives charter schools a big push in New Orleans. *New York Times*, June 13.

———. 2006b. A legacy of storm: Depression and suicide. *New York Times*, June 21.

Sayre, Alan. 2006. Post-storm New Orleans economy a huger question mark: Housing shortages are on the minds of everyone. *Associated Press*, March 28.

Smelser, Neil J., William Julius Wilson, and Faith Mitchell. 2001. *America becoming: Racial trends and their consequences*. Washington, DC: National Academy Press.

Solnit, Rebecca. 2007. The Lower Ninth battles back. *The Nation*, September 10/17.

Spake, Amanda. 2007. Dying for a home. *The Nation*, February 26.

Squires, Gregory D., and Charis E. Kubrin. 2006. *Privileged places: Race, residence, and the structure of opportunity*. Boulder, CO: Lynne Rienner.

Steinhauer, Jennifer, and Eric Lipton. 2006. Storm victims face big delay to get trailers. *New York Times*, February 9.

Strom, Stephanie. 2005. President of Red Cross resigns; board woes, not Katrina, cited. *New York Times*, December 14.

———. 2006a. Red Cross fires administrators in New Orleans. *New York Times*, March 25.

———. 2006b. Bill would restructure Red Cross. *New York Times*, December 5.

Taylor, John, and Josh Silver. 2006. From poverty to prosperity: The critical role of financial institutions. In *There is no such thing as a natural disaster: Race, class and Hurricane Katrina*, ed. Chester Hartman and Gregory D. Squires. New York: Routledge.

Thernstrom, Stephan, and Abigail Thernstrom. 1997. *America in black and white: One nation indivisible*. New York: Simon & Schuster.

Tizon, Tomas Alex. 2005. La. trailer villages bring hope, some fear. *Los Angeles Times*, December 15.

Torpy, Bill. 2005. $11 million per night to house evacuees. *Atlanta Journal-Constitution*, October 15.

Treaster, Joseph B. 2006a. Home insurers embrace the heartland. *New York Times*, May 20.

———. 2006b. Earnings for insurers are soaring. *New York Times*, October 26.

———. 2006c. Judge upholds policyholders' Katrina claims. *New York Times*, November 29.

———. 2007. Grand jury looking at Katrina insurer. *New York Times*, January 18.

Treaster, Joseph B., and Cornelia Dean. 2006. Yet another victim of Katrina: Federal flood insurance program is itself under water. *New York Times*, January 6.

Turner, Dorie. 2006. Mental health problems from Katrina persist. *Washington Post*, November 9.

Turner, Margery Austin, and Shelia R. Zedlewski. 2006. *After Katrina: Rebuilding opportunity and equity into new New Orleans*. Washington, DC: The Urban Institute.

Unified New Orleans Plan. 2006. Preliminary Report, Community Congress II. December 2.

United Teachers of New Orleans, Louisiana Federation of Teachers, and American Federation of Teachers. 2007. *No experience necessary: How the New Orleans school takeover experiment devalues experienced teachers*. New Orleans: UTNO.

U.S. Department of Homeland Security. 2006. FEMA concludes short-term lodging program: Longer-term housing efforts continue. Press Release Number HQ-06–020. http://www.fema.gov/news/newsrelease.fema?id=23158.

U.S. Department of Housing and Urban Development, Office of Policy Development and Research. 2006. Current housing unit damage estimates: Hurricanes Katrina, Rita, and Wilma. February 12. Unpublished document.

Wagner, Peter, and Susan Edwards. 2006. New Orleans by the numbers. *dollars & sense* 264:54–55.

Warner, Coleman 2006. Rebuild sessions casting wide net. New Orleans *Times-Picayune*, December 1.

Washington Post. 2007a. Nation in brief. *Washington Post*, June 2.

———. 2007b. Nation in brief. *Washington Post*, July 18.

Weisman, Jonathan. 2005. Critics fear trailer "ghettoes"—Right, left target FEMA initiative. *Washington Post*, September 16.

Wellington, Darryl Lorenzo. 2006. New Orleans: A right to return? *Dissent* 3:23–35.

Whelan, Robert K. 2006. An old economy for the "new" New Orleans? Post-Hurricane Katina economic development efforts. In *There is no such thing as a natural disaster: Race, class and Hurricane Katrina*, ed. Chester Hartman and Gregory D. Squires. New York: Routledge.

Whelan, Robert K., David L. Gladstone, and Trisha Hirth. 2002. *Building a workforce development system in New Orleans*. New Orleans: College of Urban and Public Affairs, University of New Orleans.

Whoriskey, Peter. 2007a. "We called it Hurricane FEMA." *Washington Post*, March 12.

———. 2007b. $2.9 billion shortfall seen in Katrina aid. *Washington Post*, May 12.

———. 2007c. Victims of Katrina file rash of lawsuits. *Washington Post*, May 13.

Wilgoren, Jodi. 2005. Vouchers in their pockets, evacuees find it hard to get keys in hand. *New York Times*, October 28.

Williams, Linda Faye. 2003. *The constraint of race: Legacies of white skin privilege in America*. University Park: Pennsylvania State University Press.

Zdenek, Robert O., Ralph Scott, Jane Malone, and Brian Gumm. 2006. Reclaiming New Orleans' working-class communities. In *There is no such thing as a natural disaster: Race, class and Hurricane Katrina*, ed. Chester Hartman and Gregory D. Squires. New York: Routledge.

Zuckerman, Stephen, and Teresa Coughlin. 2006. *After Katrina: Rebuilding opportunity and equity into the* new *New Orleans: Initial policy responses to Hurricane Katrina and possible next steps*. Washington, DC: The Urban Institute.

CHAPTER 19

ARE THEY KATRINA'S KIDS OR OURS? THE EXPERIENCE OF DISPLACED NEW ORLEANS STUDENTS IN THEIR NEW SCHOOLS AND COMMUNITIES

KEVIN MICHAEL FOSTER

This chapter uses examples from the challenging year of teaching, learning, and healing that has occurred in Austin, Texas, to discuss the complex and varied impacts of forced dispersal upon African American schoolchildren who are now attending schools across the country. This chapter provides a picture of children, teachers, and community in adjustment, including examples of effective practices in working with children dispersed from New Orleans and examples of ineffective or racist responses to new students that have been present and that we must counter.

This chapter is grounded in the perspective and methods of a school- and community-engaged educational anthropologist. It reflects over 100 hours of work in a shelter immediately following Hurricane Katrina, and a subsequent school year of work in Austin public schools working on issues of black student achievement—including the achievement of close to 1,000 students and 10,000 families from New Orleans. It builds upon Katrina-related work published in *Transforming Anthropology* and forthcoming in *Perspectives on Urban Education*.

For most of us, the year starts on January 1 and goes from there. For the U.S. government, which maintains its own fiscal calendar, the year starts in October. In either case, the "New Year" is a time of new beginnings—new money to be spent, new ideas to be tried, new hopes for better days ahead. When things go well, the world of teachers, students, and schools is little different. In late August of each year, there are new teachers to meet, new students to teach, and new classroom communities to join or build.

So what did "New Year 2005" bring to our children from New Orleans and the Gulf Coast region? As we can imagine, for our children who were displaced by the aftermath of Hurricane Katrina, the new school year meant displacement, loss, and the realization of their worst nightmares. And what did the new school year bring to teachers whose cities took in those displaced following the Katrina aftermath? Our teachers and school administrators, along with many of the rest of us, began a journey into the surreal—where ideas about what was possible in this powerful nation were challenged, and where ideas about how and how much we care for *all of our* children, and about our lack of prejudice in this "colorblind society," were put to the test. It was also a year in which schools' utter lack of preparedness to teach and serve the diversity that is represented by our nation's students was on tragic display.

In short, and though an admitted generality, for those in schools across the country that were affected by Katrina, this past year has been a meeting of the traumatized (displaced children) with the insecure and ill-prepared (teachers, schools administrators, and schools)–sometimes with surprising and beautiful outcomes, but too often with predictable results that confirm our worst fears about the manner in which American schools routinely abuse poor black children.

This essay provides a glimpse into the ongoing circumstances and adjustments of children displaced from New Orleans[1] and of the schools and school districts in which they eventually enrolled following their relocation. It focuses on Austin, Texas (as one of the cities that has taken in thousands of new children and families) and is grounded in the observations and work of a community-engaged educational anthropologist. The essay reflects over 100 hours of work in a shelter immediately following Hurricane Katrina and a subsequent school year of work in Austin public schools working on issues of black student achievement—including the adjustment and achievement of close to 1000 black students and 10,000 black families from New Orleans. It builds upon Katrina-related writings published in *Transforming Anthropology*[2] and forthcoming in *Perspectives on Urban Education*.

FIRST DAYS OF SCHOOL

During the evacuation and immediately afterwards, Austin proved to be a decent and caring city, full of people who dropped everything to volunteer in any way they could. But what would happen in areas of ongoing and close interactions between displaced residents and those who were already established? Schools were one of the spaces of such ongoing interaction—as students met teachers and teachers met families. So what are some of the stories from those early days? Do they give us hope for what might come next for members of the New Orleans Diaspora?

Austin Independent School District (AISD) personnel spent several days in early September on-site at the Convention Center, enrolling students and assigning them to schools. As school and district personnel sat at computers and attended to those who lined up to enroll, some progressive volunteers (several of whom had themselves been displaced and had lost all of their material possessions) organized students and got them and their parents into line. As the first day of schooling *en masse* began for displaced children, busses lined up outside the Convention Center, and kids were sent to one of several schools chosen for their initial transition into AISD. As families transitioned out of the Convention Center, kids changed schools and ended up scattered across the District's 107 campuses—two or three here, none there, and in some cases, several dozen attending the same elementary, middle, or high school.

For one elementary school student things had started off well. As we sat at an arts and crafts center in the corner of the cafeteria area that had been set up in the Convention Center, he offered his perspective on his schooling circumstances. He shared a sentiment that I would hear from other young children—he hoped that he would get to stay in Austin. He liked his new school. It was clean, he explained; the people were nice; he liked his teacher.

Other students started out similarly well, even as the magnitude and difficulty of the circumstances meant that many approached school with little guidance or support. Once children were enrolled in schools, but before they left the Convention Center, most spent their afternoons at a large computer area, which had morphed into a video game area after some movie stars who live in Austin visited the Center and donated video game equipment. Kids gathered around large screen televisions and computers and watched one another play. Not too many seemed concerned about homework. Meanwhile, a small number of children went straight to their living areas that they and their families had set up. These were generally children whose parents had provided strict instructions to go straight to their cots after school and to do their homework. There they would quietly sit and work on math problems or craft stories that they had been instructed to write. A couple of hours later their parents would return from their day of job hunting, apartment hunting, or navigating the city or federal service bureaucracy.

For one high school, where several students from the shelter were initially enrolled, things hadn't started so well. It is sometimes impossible to think of everything, so perhaps no one is too blame for the placement of members of two rival gangs onto the same bus for the ride to school. The bus actually carried students from a variety of backgrounds and from different neighborhoods across New Orleans, so there were plenty of students who wanted nothing to do with the fights that broke out on the bus and compelled the bus driver to pull over and work to get everything settled down. When the bus finally pulled into the school lot, students who were already filled with trepidation and unsure about what to expect now felt the added burden of being associated with and talking about "the fights." This was their unfortunate welcome to high schooling in Austin.

Finally, in yet another case, things didn't go so well for a fifth grade teacher and one of her new students. Ms. B had introduced herself to her newest fifth graders in a manner that was unsurprising and routine. She asked students their names and about themselves and told them she would like to be called Ms. B. The response from Earl, one of her new students from New Orleans, was nothing she anticipated or had ever experienced. "I'd like to be called Ms. B," went her side of the conversation. "OK, Bitch" was his. The story of Ms. B and Earl, to which I'll return later in this chapter, provides a poignant glimpse of the challenges and possibilities inherent in teaching students who have experienced trauma—be that trauma of several hellish days (as during the Katrina aftermath) or of a lifelong nature (as in the case of thousands and thousands of black and poor children who had been systematically undereducated and abused in a broken public school system in New Orleans).

SETTLING IN

As the school year rolled on in Austin, the mobility rates of children from New Orleans was very high—students switched schools often as parents worked towards establishing stability. For many families, two immediate priorities emerged—finding jobs and finding affordable housing. The most pressing complication, of course, was making sure that both jobs and homes were accessible by bus. School placement was often dictated by those other pressing needs, such that as families continually worked to improve their circumstances, they were pressed to change schools as well. Things began to settle down after a time, but even at the end of the 2005–2006 school year, student mobility remained an important dynamic in the education of our children.

Throughout the 2005/6 school year, there were spaces of remarkable and dynamic teaching, examples of the inadequacy of schools to address the circumstances at hand, and some unfortunate cases of inexcusable behavior on the part of teachers and administrators. Citywide, one concern put forth by the city manager and the mayor was whether the City of Austin would be reimbursed by the Federal Government for the canceled Convention Center events, the overtime paid to city employees, and the expansion of the social service net to account for the new residents.[3] On campuses across the school district, one early concern was whether the "Katrina Kids" would "count" in the standardized testing accountability system (where schools are penalized for failing to adequately educate their students). In classrooms, one concern was how to handle "those bad Katrina Kids." Over the course of several months, some began to adopt the term "compassion fatigue" to capture their sense of having been asked to care too much, for too many, for too long.

Such discourses of self-concern, selfishness, privilege, and fear did not go uncontested, however, although they remain powerful and present. Discourses of despair, hardship, and defeat were answered head-on by competing discourses of compassion and (ironic) opportunity. Some nonprofit social service organizations, schoolteachers, community members, and others actively countered the emerging rhetoric of self-concern. Others simply did not care to notice them. Most notable was the resilience, strength, and resourcefulness of many of the families from New Orleans, as they saw a chance for a fresh start in a new place and were eager to make things work for them in their new environment.

Within the swirling context of competing ideas about our newest residents and their impact upon Austin, and amidst their lived realities of grieving and ongoing struggles regarding future plans, schooling went on. The story of Earl and Ms. B— the student who used profanity to address his teacher and the teacher who had to figure out how to handle the situation—captures some of the challenges and possibilities that some chose to acknowledge and work through in the midst of this whole confusing morass.

EARL AND MS. B

Teachers across the nation know the exercise well. It is captured by a statement like this: "Oh, she is so good with 'those children'; we should give them to 'her.'" "Those children" are children with special needs. The 'her' is deemed an effective teacher.[4]

In some schools and districts in Georgia, Illinois, or South Carolina, "those" may be migrant children. In California, Texas, New York, or Florida "those" may be English language learners. Often, "those" are the "discipline problems." All too often, "those children" are black boys. Regardless of the specifics, children are dehumanized by such labels, but in the cases of our best teachers—the ones who do the most and the best work—stigmatized children come into a classroom where their humanity is restored. In Austin 2005–2006, "those children" were "the Katrina Kids." At AISD's Slaughter Elementary the "her"—that effective teacher to whom the fifth-grade "Katrina Kids" were sent—was Ms. B.

After Ms. B's initial run-ins with Earl, she sent him to the principal's office. She did this with other kids as well. This backfired, however, and she quickly abandoned the strategy. The problem was that, as the children quickly figured out, her well-meaning administrators had trouble disciplining children who had been through so much trauma. As Ms. B relates the story, she figured this out one day when she over-head one student from New Orleans telling another in so many words, "If you get sent to this principal's office, just give your story—tell her all your stuff about what you've been through and then she won't do anything." In essence, the kids, through no fault of their own, were learning to be victims, to use their tragedy to their advantage to receive sympathy instead of punishment (and who can blame them—this is the type of strategy that smart ten-year-olds employ as they negotiate schools, parents, and authority). But this was a disaster for Ms. B's classroom. Kids came back as unruly as before, but now were emboldened in their misbehavior because there were no consequences for misdeeds.

As Earl continued to address Ms. B with curse words, she adapted her strategy (as good and thoughtful teachers do). Showing extraordinary patience (not over hours, days, or weeks but over months) she withheld attention from Earl. Unless a situation demanded intervention, she simply ignored Earl when he was uncivil. On the other hand, appropriate behavior was met with authentic and caring interest and engagement, which could have been difficult, given the many times when Earl had been so profoundly disrespectful.

Over the course of the year, an interesting dynamic emerged in Ms. B's classroom. Sometimes Earl wanted attention, and became tired of being ignored. At those moments, and with increasing frequency, Earl began to refer to the teacher by her name, and with appropriate respect. And then came what may have been the defining moment—just a moment—but one that Ms. B and Earl had been working towards. Just before the Christmas break, Ms. B did what she usually did at that time of year; she gave everyone a gift, along with a note about something she appreciated about them. Nothing big—a cute pencil, a pad, a trinket of appreciation. Earl received a gift, too. He took it with mild suspicion and surprise, but he took it.

And then came the break. And then back came Earl. By Ms. B's account, he was not suddenly perfect, but he was markedly different. He still cursed too much, but no longer at her, and not so much at other people. His classroom and academic engagement improved, and he generally behaved according to the expectations of the classroom community. Over a very difficult several months, Ms. B had earned the trust of a hurting little boy, and he had begun to respond. Earl's story is ongoing of course, and things are not perfect for him—in school or outside of it. But he is now a more integrated member of a school community because of Ms. B's work—he is taught, cared for, and held up to standards of behavior and academic behavior. This is a small story, of course, and it is a one of thousands of stories across Austin, and thousands of others across the country—some of which have turned out well, many others of which have gone horribly wrong.

There is an epilogue to the story of Earl and Ms. B. Once Earl and Ms. B had a reasonable student-teacher relationship, they were able to talk. Ms. B asked Earl what all the cursing was about. He told her in plain terms that cursing was what he was used to in his old school—his teachers cursed at him and he cursed back. It was the norm. In short, Earl's New Orleans schooling experience sounded like one of systematic abuse.[5]

Some would refer to Ms B. as an everyday hero. At the same time, we must recognize that what Ms. B offers is the spirit, perseverance, patience, and competence that is required of teachers if we are to have children endure the everyday challenges of life (let alone the horrific experience of Katrina) with their humanity and potential intact. Although not a trained counselor, Mrs. B and other teachers have had to maintain a difficult balance in working with so many traumatized children. On one hand, it is a teacher's responsibility to provide for the academic and moral development of all of the children in his or her classroom. According to a rich and historically rooted tradition of African American teachers teaching African American children, this has meant high academic and behavioral standards and expectations for students, along with adherence to an age-old adage that we need to hold our students to the highest standards because our children need to be twice as good to get half as far as their nonblack peers. On the other hand, the question that effective, caring and reflective teachers struggle with is whether this is all the more true when our kids suffer such trauma, or if under such absolutely horrific circumstances we better serve our children by backing off and "cutting some slack."

Trained counselors, let alone teachers, struggle with just how much to push and how much to ease up in such circumstances. What astute teachers and counselors in Austin reported and realized was that a range of inappropriate or difficult behaviors was to be expected of displaced children—not just cursing, but also fighting, crying, disengagement from the classroom, and sullenness, among other behaviors. In addition, poor behaviors were often wholly unavoidable, as they were fueled by circumstances completely beyond students' control—lack of sleep due to recurring nightmares; entire families in mourning for lost homes, lost family members, and lost stability; and additional home stress due to financial instability, overcrowding, and not knowing where the family would live next month. The reality that cannot be overstated is that our children from New Orleans have been through and continue to endure significant trauma, that this represents an ongoing challenge to those who take their charges as public servants seriously, and that in those cases where we are ineffective, children's tragedies are only compounded by our failures.

Our best teachers, heroic teachers like Ms. B, can, must, and do find a way to account for and even address students' realities as they teach, in fact subtly weaving healing processes into the classroom communities that they shepherd. Unfortunately, the fact that teachers like Ms. B have to be seen as heroes lets us know that she is not the current norm. Across Austin, many teachers and administrators have not been effective in addressing the needs of our children from New Orleans. In some cases, where administrators lacked the necessary socio-cultural awareness to build bridges of possibility out of the circumstances, kids ended up fighting with one another—kids from New Orleans trying to establish a sense of space and place, kids from Austin defending the same, and ineffective administrators not understanding what was happening and chalking it all up to the typical behavior of black people. In other cases, teachers, parents, and principals could barely hide their disdain for their new students. One administrator expressed that she "hated" the Katrina Kids. In classrooms across the district, teachers sought to have kids from New Orleans moved out of their classrooms and into someone else's classroom, into special education, or into the principal's office. In such cases, the trauma experienced by our displaced students was only deepened.

In such spaces of tragic ineffectiveness, we actually have schooling as usual in countless U.S. public schools. Educational anthropologist Michele Foster relates that she always knows when she is in a U.S. public school because of the line of black boys waiting to see the principal. To wit, I could have quickly exhausted my word count for this chapter by sharing horror stories from the past year. My hope, however, is that readers are at least generally aware of the unchecked racism and teacher ineffectiveness in our schools, so that while I can (and will) return to that topic in other writings, I have been able to focus here on those aspects of our current crisis that provide us (those in Austin, but also those in Houston, San Antonio, Baton Rouge and any other city, school or neighborhood with new residents from New Orleans) with positive examples that will best prepare us to move forward.

Some citizens from the Gulf Coast region have returned to places like New Orleans to rebuild, but many others have chosen to stay in their new locales. Of the thousand or so students that have entered AISD, close to seven hundred were still with us by the end of the 2005/6 school year. Issues for subsequent school years include (1) better accounting for the cultural and linguistic differences that our new children bring with them to school [this will mean anti-racist teacher training as well as training in cross socio-cultural competence], (2) more pro-action in our work to incorporate families from New Orleans into the fabric of everyday life in Austin schools, (3) the need for ongoing mental health monitoring and counseling, (4) better accounting (and less blaming) for behavioral issues that arise from students' ongoing posttraumatic adjustments, and (5) more and better support to facilitate the academic growth of the many students from New Orleans whose academic preparation is well below grade level.

In short, we have a lot of work to do. Austin schools have had a year of trial by fire, and I would be less than candid if I did not say that many of the kids we should care about the most have been burned on top of their burns, and that we have not done enough to heal wounds and promote health and recovery. But now that we have been through the year, there is no reason for us to not do better in the next. The lives of our children depend upon it. For my part, I will stay involved in the work in the schools for three reasons: (1) our kids are there, (2) their educational lives are at stake, and (3) the local district, while far from perfect, and while filled with many people who have no idea of just how ineffective they are, has shown a willingness to grow, learn, and improve. This is at least a foundation. It is a start.

NOTES

1. Children and families came from throughout the Gulf Coast region, but the overwhelming majority were from New Orleans.

2. Foster, Kevin Michael (2006) "Austin Shelter Notes," *Transforming Anthropology* 14 (1): 26–30.

3. The *Austin Business Journal* reported that the city was reimbursed 100% for its expenditures: "All together, Austin estimates it spent $22.65 million on its Katrina response—$5.47 million of that for emergency shelter operations. The total includes: $2.5 million in payments to vendors, $2.9 million for labor, use of city vehicles and related items, $17.1 million for interim housing for more than 1,200 households, covering everything from $12 million in rent to $500,000 in moving expenses. The Federal Emergency Management Agency fully reimbursed the city for those costs through public assistance grants from the Division of Emergency Management in the Texas Governor's Office" (*Austin Business Journal*, October 10, 2005). The city was not reimbursed for lost business in the Convention Center (*Austin Business Journal*, October 28, 2005).

4. For instance, teacher and researcher Jeff Duncan-Andrade notes the presence of this practice in a California school he has worked with for several years (2005).

5. In following up on this with others, several have shared that Earl's claim is not at all unreasonable. Former New Orleans teachers who now live in Austin, along with social workers and others from New Orleans, have all met this part of the story with no surprise and rather with sad shakes of affirmation to the possibility that some elementary school children (of whatever race) were subject to cursing by their teachers in their old schools (not to mention the readily evident, thorough under-education that is reflected by the abysmal academic levels of many of our newest students).

Chapter 20

Envisioning "Complete Recovery" as an Alternative to "Unmitigated Disaster"

Mindy Thompson Fullilove,

Fred Bosman, Henk Bakker, Pieter de Wit,

Eric Noorthoorn, and Robert Fullilove

Introduction

This essay represents a collaboration between two teams, the United States–based Community Research Group and the Dutch Mediant Psychiatric Group. The first part of the essay draws on the work of the American-based Community Research Group (CRG), which is co-led by Drs. Mindy and Robert Fullilove. For the past fifteen years, CRG has conducted studies of epidemics that have affected inner-city communities in the United States, including AIDS, violence, crack addiction, and obesity. The co-location of these epidemics in particular neighborhoods led CRG to study the contribution of spatial organization to illness. In particular, the human ecology studies of Drs. Rodrick and Deborah Wallace on forced displacement have provided important theoretical guidance to CRG's work. While studying forced displacement, and the seriousness of its consequences, CRG also looked for examples of complete recovery of ruptured communities. While there are many interesting examples of recovery, none are of the order of magnitude needed to achieve complete repair or to reverse the trends created by past upheaval. Rather, what CRG encountered was a hegemonic discourse that labeled forced displacement "progress," which ultimately justified scenarios of disaster without mitigation.

Thus, the Fullíloves were intrigued when they heard the story of Enschede from Dr. Pieter de Wit, one of many people who was engaged in recovery after a major disaster in the town of Enschede, The Netherlands. The second part of this essay draws on the work of the Mediant Aftercare Team, which provided a wide array of mental health services to the people affected by the Enschede disaster. Dr. de Wit and his colleagues, Fred Bosman, Henk Bakker and Eric Noorthoorn, invited the Fullíloves to visit Enschede, meet with key leaders of the recovery effort, visit the site of the disaster, and participate in a mini-symposium on the work of the Aftercare Team. Enschede benefited from massive national commitment to the "right to return," which proved to be a foundation for everything from individual treatment to neighborhood rebuilding. This work offers a model of complete recovery that challenges many of the hegemonic concepts surrounding U.S.-style forced displacement. The third part of this essay examines the implications of the Enschede model for complete recovery from the Gulf Disaster.

MASSIVE DISASTER IN THE GULF

In late summer 2005, Hurricanes Katrina and Rita hit the Gulf Coast of the United States, among the deadliest and most costly hurricanes ever experienced. Especially the first of these, Hurricane Katrina, mobilized wind force, water surge, and flooding to create havoc in small and large cities of the Gulf. The terrors of the storm were augmented by government failure at every level. Indeed, the actions of the Federal Emergency Management Agency created chaos, rather than calming it, with the result that hundreds of thousands of poor people, many black, were left in dangerous flood waters, without food, water to drink, medicine, or succor. When help arrived, it was often callous and random, dispersing the survivors of the flood throughout the forty-eight contiguous states without a plan for their arrival or their return home.

The American public, watching these events on television, was horrified that fellow citizens were treated in such a manner. As investigations of the disaster have collected the details, stories have accumulated of rescuers turned away or punished by federal, state, and local officials. This inefficiency and abandonment produced a death toll that stands at over 1,800 people, many of which were preventable deaths. A clear and persistent outrage and sorrow marks the public's reaction to this failure to rescue Americans in their hour of need.

This essay, however, addresses a problem that is unacknowledged in the hegemonic discourse: the problem of the secondary disaster attending the forced displacement of over a million people. In New Orleans alone, 80 percent of the city was flooded, destroying homes, businesses and urban infrastructure. Given the magnitude of the damage, a massive recovery plan was needed. Almost immediately, the shape of the plan began to emerge, characterized by three features: (1) poor blacks were congratulated on having moved to a "better place" where they might start over; (2) speculators began to purchase property from the displaced with the intention of rebuilding for the benefit of wealthier individuals; and (3) planners started to develop proposals for a "new" city, cleansed of the problems associated with poverty. What is striking is that the outrage that followed the failure of disaster management has been largely absent as this secondary disaster unfolds. In fact, many Americans, if queried, would support the new plans. We propose to challenge this acceptance by arguing that forced displacement is not "good luck" but rather an unmitigated disaster of a sort that has been a cyclical part of the African American experience.

REPEATED DISPLACEMENT IN AFRICAN AMERICAN HISTORY

Repetitive forced displacement is a major organizing process in the African American experience. The violence of the international and internal slave trade, which were accompanied by the development of a philosophy and social structure of racism, laid the groundwork for many forms of abuse. Among these abuses was involuntary and forcible removal of Africans from their homes. This trope of loss has been played out in the implantation of Jim Crow (1890–1910), the Great Migrations (1916–30 and 1940–70), the mechanization of agriculture (1940–70), urban renewal (1949–73), deindustrialization (1945–present), planned shrinkage (1974–present), HOPE VI (1992–present) and gentrification (1965–present).[1] During each of these important historical moments, African Americans have been pushed out of their homes or forced off the land they once occupied.

A large body of literature documents the many health and social costs of upheaval, particularly in refugee situations caused by war. In order to understand the costs of upheaval under peacetime conditions, M. Fullilove studied the long-term consequences of urban renewal, a federal program that destroyed urban neighborhoods in order to improve cities. She examined the costs associated with loss of a neighborhood. Interviews with displaced people in five American cities revealed that they had suffered from severe social, cultural, political, and economic losses that exposed them to further losses as time went on.[2] Based on these findings, Fullilove proposed that forced displacement be understood as causing a traumatic stress reaction among those who had lost all or part of their emotional ecosystem, a reaction she called "root shock."

The crucial issue in the concept of "root shock" is that it recognizes that the loss of a massive piece of human habitat causes harms that go far beyond individual emotional suffering and require united action on the part of social groups to ensure the recovery of the people and restoration of their way of life. Forced displacement has played a major role in social and health problems such as the overrepresentation of blacks in the AIDS epidemic and the surge of incarceration.[3] Thus the consequences of forced displacement are both severe and enduring.

A major reason for the heavy consequences of forced displacement is the failure of society to repair the damage caused by upheaval. Fullilove notes that, although displaced people suffered massive losses, increased rates of disease and earlier deaths, their upheaval was portrayed in mainstream media and described in political discourse as "progress."[4] Hence, these losses were neither acknowledged nor remedied. Even among African American people, the discourse surrounding issues of loss of home and place has been fractured and minimizing.

This distorted discourse on displacement was shaped by a number of additional factors that interacted with the assumptions of structural apartheid. These included upheaval among other populations related to economic disaster (the Great Depression and the accompanying drought), suburbanization, deindustrialization, and reorganization of the economy from the Rust Belt to the Sun Belt. These more general, but very dramatic, reorganizations of the American landscape translated into widespread relocation: approximately one in five Americans moves every year. The longstanding cultural ideal of chasing opportunity has both supported this movement and obscured its true costs for the population. Thus, massive movement of population is not restricted to African Americans, and its general integration into the popular imagination as "moving to the good life" has stood in the way of a deeper understanding of the traumatic effects of displacement. This is as true among mental health professionals as it is among the general public.

HURRICANES KATRINA AND RITA: "MOVING IS GOOD FOR YOU"

This disaster, but the latest in the series of unmitigated forced displacements, occurred in the context of a particular national discourse that seemed to both encourage and promote the idea of moving.[5] The displacement of poor black people from their homes was immediately reframed to fit this national discourse, Most notably, former First Lady Barbara Bush noted that "moving is good for you" while observing African Americans seeking shelter and relief in the Houston Astrodome.[6] News reports made it clear that poor black people would have a very difficult time getting back to their homes, but this was distorted as a real improvement in their situation. Indeed, one columnist called the concept of moving the "silver lining" of Katrina. This false depiction of the consequences of forced displacement has been the ideological underpinning for exclusionary planning for the future of New Orleans and other parts of the Gulf.

Given the very long litany of experiences of displacement without mitigation, there is a paucity of information about the processes and outcomes that accompany full recovery from disaster. In order to contribute to the development of a language of recovery that understands and repairs root shock, we pose the question: what should have happened after the disaster caused by Hurricanes Katrina and Rita in order to ensure full recovery of the people and their lifeworlds? This is a question of great importance, because in the absence of an ideal response, people will remain trapped in the "moving is good for you" paradigm.

In order to answer this crucial question, we report on the recovery process utilized in Enschede, The Netherlands, after a major explosion destroyed a substantial part of the city. While the scale of the disaster is admittedly smaller than that faced in the Gulf, the successful actions taken by national and local officials and other local leaders point towards the set of actions that we might call "tender reimplantation."

TENDER REIMPLANTATION IN ENSCHEDE

Enschede is a city with 150,000 inhabitants, situated in the eastern part of The Netherlands, in a region called Twente. This region was very stable from the Middle Ages to the beginning of the twentieth century, when the establishment of cotton and metalworking factories called for a massive influx of new workers. These workers came to the city from other parts of The Netherlands and other parts of Europe, as well as from Turkey, Morocco, and Syria, particularly members of the Syrian Orthodox religion.

As in many other parts of the developed world, much of the industry that helped Enschede grow had moved away by the end of the century. This left the Roombeck neighborhood with many abandoned or partially used factories that stood alongside residences. The neighborhood, despite its collection of derelict buildings, was well positioned near downtown, used for many purposes, and occupied by a wide variety of people. Redevelopment of the area was on the minds of city officials, but it was forced to the top of the agenda when, on May 13, 2000, at 3:35 PM, a fireworks factory exploded, killing 23 people, wounding 1,000, leaving 1,500 homeless, and destroying hundreds of acres of urban habitat.

THE RESPONSE

PHASE ONE: IMMEDIATE DISASTER RESPONSE

The first phase of disaster required the immediate intervention of firefighters, police, and medical personnel. As in any disaster of this magnitude, first responders came from great distances to provide help. These included firefighters from Munich, Germany, health professionals from many parts of the Netherlands, and others. In addition, there was an equally crucial political response by national leaders, who quickly committed to rebuilding the area in a safe and secure manner so that all former inhabitants could return to their homes. The "right to return" is a significant antidote to the profound feelings of loss that accompany the destruction of homes and neighborhoods.

A third part of the immediate disaster response was the rehousing of the homeless and the provision of funds for various necessities, including food and clothing, among other things. A part of this effort involved deepening the level of multicultural understanding of the ways in which each group hoped to receive help from others around them. A single government office was set up to help people manage all of the problems that they faced in getting on their feet again.

PHASE TWO: TREATMENT FOR PHYSICAL AND MENTAL WOUNDS

Medical and psychiatric care was immediately established for all of the people affected by the disaster. General practitioners handled the medical problems of the injured, while a special psychiatric unit handled the emotional problems related to what people had seen and lost. Part of the work conducted by the psychiatric team was ensuring cooperation among many sectors of the city government. This was essential to maintaining long-term cooperation and commitment to recovery.

PHASE THREE: REBUILDING THE AREA

Another special unit was established to develop a plan for rebuilding. Prior to the development of the plan, many groups of people, including former residents, were consulted about their memories of the neighborhood and their hopes for the future. This resulted in a series of principles for redevelopment that guided the architects in their design. This plan had eight elements:

- A district to return to (right to return)
- A lively district
- A familiar district
- A district with history
- A district with value for the future
- A district within boundaries
- A district within your own hands
- A safe district

The plan, once accepted by the people of the city, was implemented. In the first phase, much of the housing for low-income residents was rebuilt, allowing those most in need to move home again. In a next phase, people who wished to build their own homes began to buy plots of land, work with architects, and reconstruct their houses. The development of condominiums and cultural facilities followed. Six years after the disaster, a massive amount of rebuilding had been accomplished, and the intensity of work was obvious to those who visited the site.

PHASE FOUR: UNDERSTANDING THE DISASTER AND ITS CONSEQUENCES

As part of the national response, research in a number of areas was implemented immediately. In fact, the national group that studies environment pollution was in the area and arrived in Enschede hours after the disaster to begin studies of possible contamination of the site. Studies were also conducted to examine the use and effectiveness of mental health services. In addition, displaced residents and first responders were interviewed and surveyed about their firsthand knowledge of the disaster.

PHASE FIVE: HOUSEWARMING FOR THE NEW ROOMBECK NEIGHBORHOOD

As the new neighborhood emerged, the groups that helped to ensure its development shifted their relationship. As a marker of the end of the beginning, a "housewarming party" was held at which various parts of the city celebrated the repair and reconstruction of this central piece of Enschede.

UNDERSTANDING THE ENSCHEDE MODEL

The rapid recovery of Enschede is reminiscent of the recovery European cities experienced after World War II, due in no small part to the aid of the Marshall Plan. Recovery from a massive disaster requires an unambivalent commitment of resources. But the resources that are required often go far beyond the money supplied by major governments. The resources must include contributions from individual citizens, from the fire fighters who make the first stand against destruction to the social organizers who ensure that rebuilt neighborhoods develop a functional revitalization as well.

Some of the contributions that made the Enschede recovery so remarkable were

- the long-term maintenance of cooperation and commitment even after the postdisaster "honeymoon" phase ended,
- the engagement of many sectors of the city in developing a statement of principles about rebuilding, and
- the recurrent use of arts to help the recovery, ranging from children designing their own memorials to homeowners designing their own homes.

In the United States, we are familiar with the story of the little Dutch boy who understood that he had to tend to the leak in the dyke no matter how difficult it was for him. The actions in Enschede represent, on a larger scale, the commitment of Dutch citizens to protecting their cities and the human relationships they embody. Dutch society, like other societies, is challenged by rapid social changes from immigration to deindustrialization. Despite these changes, the people of Enschede, supported by the larger nation, worked together to rapidly repair a devastated neighborhood. This real story offers some benchmarks in recovery from disaster at any scale and in any place.

THE ENSCHEDE MODEL AND THE RECONSTRUCTION OF THE GULF

The Enschede Model is characterized by three features: (1) it is founded on a vision of complete recovery for the city and the affected individuals; (2) it was based on participation and inclusion; and (3) it embodied respect in interpersonal interactions. We have called this "tender reimplantation" to underscore the degree to which it addresses, in an appropriate manner, the traumatic stress of root shock. The Enschede Model is one in which people are aided to rebuild their life on the foundation of the world that was destroyed. This model respects culture, social ties, and urban form as the building blocks of a healthy society. In restoring each of these aspects of the lost neighborhood, the Enschede Model creates a stable foundation that helps ensure the future well-being of the affected people and of the city. The principles of the Enschede Model are not limited to a disaster of a particular level or scale; they can be easily adapted to a disaster of any size. That is because these are transferable principles of interpersonal respect and mutual responsibility.

Obviously, there is nothing about the exclusionary recovery process in the Gulf that resembles what happened in Enschede. The political fight, therefore, is to create a new national commitment to complete recovery that respects people's right to return home and their need to be treated with respect. The Enschede Model offers us many useful benchmarks for judging the degree to which we have succeeded in achieving these goals.

BIBLIOGRAPHY

Brooks, David. "Katrina's Silver Lining." *New York Times*, Opinion/Editorial, September 9, 2005.

Fullilove, Mindy. *Root Shock: How Tearing Up City Neighborhoods Hurts America and What We Can Do About It*. New York: Ballantine/One World, 2004.

Hanchett, Thomas. *Sorting Out the New South City: Race, Class, and Urban Development in Charlotte, 1875–1975*. Chapel Hill: University of North Carolina Press, 1998.

Hirsch, Arnold. *Making the Second Ghetto*. Cambridge: Cambridge University Press, 1983.

Powell, J. A., and M. L. Spencer, "Giving them the Old 'One-Two' Punch: Gentrification and the K.O. of Impoverished Urban Dwellers of Color." *Howard Law Journal* (Winter 2000).

Wallace, D., and R. Wallace, *A Plague on Your Houses: How New York Was Burned Down and National Public Health Crumbled*. London: Verso, 1998.

NOTES

1. These processes are not all equally well known to the American public, nor is their cumulative impact–what my colleague Rodrick Wallace has called "synergistic damage accumulation"—fully appreciated. The African slave trade, which dragged people from their homes in Africa and sold them into slavery in the Americas, took the liberty of 12 million who arrived alive. It is estimated that twice that number died on the journey within Africa and during the middle passage across the Atlantic. After the slave trade was banned in 1808, an internal slave market developed in the United States, which regularly sold slaves from Virginia and other more Northern states to the lower South. Emancipation restored people's liberty, but put them at the great disadvantage of owning no land and having no education. There was massive movement after the war, as people sought to reunite with family, go to school, find land or work, and begin their new lives as freedmen. This hopeful epoch came to a violent end with the institution of Jim Crow laws, which made African Americans second-class citizens, stripped of their right to vote or to be protected in the courts. The two Great Migrations represented people's efforts to make new homes in the city, where they might have more economic and political opportunity. This effort, too, was thwarted by the reification of segregation in the cities. Redlining, instituted in 1937, aggravated segregation by steering investment away from African American ghetto neighborhoods. Urban renewal then found these to be "blighted" and ordered them cleared for "higher uses." Catastrophic disinvestment in the 1970s and 1980s represented the active removal of assets—from fire stations to banks and supermarkets—from minority and poor neighborhoods. Many of those displaced by urban renewal and catastrophic disinvestment moved into housing projects and became vulnerable to a new "improvement" scheme, this one called HOPE VI. At the same time, poor and minority neighborhoods that had maintained some of their historic buildings and charm were targeted for gentrification, and the poor were forced to move again. In sum, the efforts of African Americans to free themselves and become first-class citizens have not only met with resistance but also have been actively undone by government programs operated in close cooperation with business leaders. See, especially, Hanchett, *Sorting Out the New South City* and Hirsch, *Making the Second Ghetto*, on the institution of segregation; Fullilove, *Root Shock*, on urban renewal; Wallace and Wallace, *A Plague on Your Houses*, on catastrophic disinvestment; and Powell and Spencer, "Giving Them the Old 'One-Two' Punch," on gentrification.

2. Fullilove introduced the concept of "root shock" in order to explain the social, cultural and emotional consequences of the loss of the near environment. Indeed, as many authors have argued, the near environment provides an external homeostatic system that maintains the individual's well-being. It has also been observed that complex

human habitat plays this role in supporting the organization of larger social groups. For example, a home helps a family function; a well-organized assemblage of housing, services, and businesses helps a neighborhood function; and so on. Because social organization is tightly linked to human health, we find that physical organization influences human health in its ability to support productive interpersonal relationships, among other factors.

3. Rodrick Wallace and Deborah Wallace have an extensive body of work on the effects of forced displacement on health. Their book, *A Plague on Your Houses*, offers an excellent introduction to their findings, as it describes the unintended consequences of forcing poor people out of their homes in the South Bronx in the 1970s. Among these consequences, they argue, was the widespread and rapid dissemination of AIDS in the South Bronx, New York City, and the region. Because of the region's dominating position in the U.S. urban hierarchy, the unleashing of an important epidemic in New York City had severe consequences for the nation's health.

4. During the course of the fieldwork on the long-term consequences of urban renewal, the CRG team heard a number of planners and politicians point out that "you have to break a few eggs to make an omelet."

5. The concept that "moving is good for you" is embedded in the thinking of US federal government. Two examples may suffice to make this point: (1) The HOPE VI project is breaking up so-called "distressed" federal housing projects, dispersing their residents to areas of less concentrated poverty. (2) HUD funded a large study, called "Move to Opportunity," in which residents of housing projects are given Section 8 so that they can move elsewhere.

6. Many people took up the theme that dispersal of the poor from New Orleans was a blessing in disguise. Among the first were Barbara Bush and David Brooks. Former First Lady Barbara Bush, during a visit to the Houston Astrodome shelter, said, "What I'm hearing, which is sort of scary, is they all want to stay in Texas. Everyone is so overwhelmed by the hospitality. And so many of the people in the arena here, you know, were underprivileged anyway, so this is working very well for them." In similar vein, David Brooks, columnist for the *New York Times*, wrote an op/ed piece called "Katrina's Silver Lining" (September 9, 2005), which argued that poor people had been separated from their dysfunctional neighborhoods and could be dispersed among middle class people, thereby creating more opportunity for their children.

ABOUT THE AUTHORS

Kathleen A. Bergin: Kathleen A. Begin, Associate Professor of Law, South Texas College of Law, BA, University of Massachusetts, 1994; JD, University of Baltimore School of Law, 1997; LLM, New York University Law School, 1999.

Professor Bergin teaches and writes in the areas of Constitutional Law, Critical Race Theory, Feminist Jurisprudence, and International Human Rights. Her writings have appeared in the *Yale Journal of Law and Feminism, University of Miami Law Review, Pace Law Review,* and other academic journals. She has spoken nationally and internationally on the risks of human rights violations following natural disaster and was recognized in 2005 as a "Local Hero" by the Houston Bar Association for her work with evacuees from Hurricane Katrina. Prior to joining academia, Professor Bergin served as the Derrick Bell Fellow at New York University Law School, completed a judicial clerkship at the Connecticut Supreme Court, and worked as a litigation associate in Boston, where her practice focused on commercial disputes, employment discrimination, and civil rights.

Darwin Bond Graham: Darwin Bond Graham is a graduate student in sociology and a teaching assistant in the department of Black Studies at the University of California Santa Barbara. In New Orleans he has worked with many post-Katrina social justice organizations including the Survivor's Village, Hands Off Iberville, and Common Ground. He is currently working on an ethnographic account of postdisaster social movements in the city, specifically the movement to reopen and reoccupy public housing.

Kristen Clarke: Kristen Clarke Co-Director of the Political Participation Group at the NAACP Legal Defense and Educational Fund, Inc. Prior to joining LDF, Ms. Clarke worked for several years in the Civil Rights Division of the U.S. Department of Justice through the Attorney General's Honors Program. Between 2000 and 2003, Ms. Clarke served as a trial attorney in the Voting Section of the division where she handled enforcement efforts under the Voting Rights Act of 1965. Between 2003 and 2006, Ms. Clarke worked as a federal prosecutor in the Criminal Section of the division, where she helped promote criminal civil rights enforcement efforts. She handled a range of police misconduct, police brutality, hate crimes, human trafficking, and obstruction matters. The vast majority of her work at the Justice Department was concentrated in the Deep South. Ms. Clarke received her AB from Harvard University and a JD from Columbia Law School, where she served as editor in chief of the *National Black Law Journal.* She has also served as an Editorial Assistant for *Souls.*

Erica M. Czaja: Erica M. Czaja is a survey coordinator in the survey research division of the RAND Corporation. Czaja received her BA in psychology from the University of Michigan, Ann Arbor, and her MA in the social sciences from the University of Chicago. Her research interests include American public opinion and political behavior, public discourse, racial and ethnic politics, social movements, public policy, and the politics of Hurricane Katrina.

K. Animashaun Ducre: K. Animashaun Ducre is an assistant professor in the Department of African American Studies at Syracuse University. She received her PhD in environmental justice from the University of Michigan. As a toxics campaigner for Greenpeace, she assisted in the victories of key environmental justice battles in Louisiana against a uranium enrichment facility in north Louisiana and the nation's largest proposed plastic manufacturing complex in Convent, Louisiana. Dr. Ducre combines her experiences in the environmental justice movement and academic training for a unique perspective on the environment, linking race, space, and exploitation.

Thomas J. Durant: Thomas J. Durant, Jr., PhD, Sociology, University of Wisconsin, is Professor of Sociology and African and African-American Studies at Louisiana State University. His research and teaching interests include social stratification, minorities, Southern culture, gerontology, criminology, and global development. He has authored or co-authored three books: *Plantation Society and Race Relations*, *Our Roots Run Deep*, and *The Charity Hospitals of Louisiana, A Story of Poverty, Politics, Public Health and the Public Interest*.

Kevin Michael Foster: Kevin Michael Foster received his BA from The College of William and Mary (1991) and his MA and PhD from The University of Texas, Austin (1994, 2001). He is an educational anthropologist whose work looks at the social, cultural and structural factors affecting African Americans students' educational outcomes. His published work can be seen in publications including *Transforming Anthropology*, *International Journal of Qualitative Studies in Education*, and *Anthropology News*. He began serving as President of the Association of Black Anthropologists in 2005 and is Associate Editor of *Anthropology and Education Quarterly*.

Mindy Thompson Fullilove: Mindy Thompson Fullilove, MD, is a research psychiatrist at New York State Psychiatric Institute and a professor of clinical psychiatry and public health at Columbia University. She was educated at Bryn Mawr College (AB, 1971) and Columbia University (MS, 1971; MD 1978). She is a board-certified psychiatrist, having received her training at New York Hospital-Westchester Division (1978–81) and Montefiore Hospital (1981–82). She has conducted research on AIDS and other epidemics of poor communities, with a special interest in the relationship between the collapse of communities and decline in health. From her research, she has published *Root Shock: How Tearing Up City Neighborhoods Hurts America and What We Can Do About It*, as well as *The House of Joshua: Meditations on Family and Place*. She has also published numerous articles, book chapters, and monographs. She has received many awards, including inclusion in many "Best Doctors" lists and two honorary doctorates (Chatham College, 1999, and Bank Street College of Education, 2002). Her work in AIDS is featured in Jacob Levenson's *The*

Secret Epidemic: The Story of AIDS in Black America. Her current work focuses on the connection between urban function and mental health.

Stephanie Houston Grey: Stephanie Houston Grey is an assistant professor in the Department of Rhetoric and Public Address at Louisiana State University. She earned her PhD at Indiana University. She has received a number of awards for research and teaching excellence. Her published work can be seen in national rhetoric and media studies journals such as *The American Communication Journal, The Quarterly Journal of Speech, Critical Studies in Media Communication,* and *Text and Performance Quarterly.*

Melissa Harris-Lacewell: Melissa Harris-Lacewell is associate professor of politics and African American Studies at Princeton University. She is the author of *Barbershops, Bibles and BET: Everyday Talk and Black Political Thought.*

Chester Hartman: Chester Hartman (chartman2@aol.com) is Director of Research for the Washington, DC–based Poverty & Race Research Action Council and Adjunct Professor of Sociology at George Washington University. He has served on the faculty of Columbia, the University of North Carolina, Cornell, Yale, American University, and the University of California-Berkeley. He is founder and former chair of The Planners Network, a national organization of progressive urban planners, and he serves on the Long-Range Planning Task Force of the Louisiana Recovery Authority. His most recent books are *A Right to Housing: Foundation for a New Social Agenda* (Temple University Press, 2006); *Poverty & Race in America: The Emerging Agendas* (Lexington, 2006); *City for Sale: The Transformation of San Francisco* (University of California Press, 2002); and *Between Eminence & Notoriety: Four Decades of Radical Urban Planning* (Rutgers Center for Urban Policy Research, 2002).

Chelsey Louise Kivland: Chelsey Louise Kivland is a PhD candidate in anthropology at the University of Chicago. A native of Haiti, her interests include social theory, culture, public suffering, tragic narratives, and performance. She is currently conducting research on carnival bands in Haiti.

Julianne Malveaux: Recognized for her provocative, progressive and insightful observations, Dr. Julianne Malveaux, an economist, author, and commentator, is the President and CEO of Last Word Productions, Inc, a multimedia production company. As a writer and syndicated columnist, her work appears regularly in *USA Today, Black Issues in Higher Education, Ms.* magazine, *Essence* magazine, and *The Progressive.* Her weekly columns appear in numerous newspapers across the country, including the *Los Angeles Times, The Charlotte Observer, The New Orleans Tribune,* the *Detroit Free Press,* and the San Francisco *Examiner.* She appears regularly on CNN and BET, as well as on Howard University's television show, *Evening Exchange.* She has appeared on PBS's *To The Contrary,* ABC's *Politically Incorrect,* Fox News Channel's *O'Reilly Factor,* and stations such as C-SPAN, MSNBC and CNBC. She is the editor of *Voices of Vision: African American Women on the Issues* (1996) and the co-editor of *Slipping Through the Cracks: The Status of Black Women* (1986), and she recently co-edited *The Paradox of Loyalty: An African American Response to the War on Terrorism* (2002). She is the author of two column anthologies: *Sex, Lies, and*

Stereotypes: Perspectives of a Mad Economist (1994) and *Wall Street, Main Street, and the Side Street: A Mad Economist Takes a Stroll* (1999). A committed activist and civic leader, Dr. Malveaux serves on the boards of the Economic Policy Institute, The National Committee for Responsive Philanthropy, Women Building for the Future (The Future PAC), and The Recreation Wish List Committee of Washington, DC. Dr. Malveaux received her BA and MA degrees in economics from Boston College and earned a PhD in economics from MIT. A native San Franciscan, she currently resides in Washington, DC.

Suzette Malveaux: Suzette Malveaux is an associate professor at Catholic University of America's Columbus School of Law. Professor Malveaux earned her JD from New York University School of Law and graduated with an AB, magna cum laude, from Harvard University. Professor Malveaux joined Catholic University's faculty from the University of Alabama School of Law, where she had taught complex litigation, civil procedure, public international law and employment discrimination since 2003. Today, Professor Malveaux writes primarily in the areas of arbitration, class actions, and civil rights. Her interest in civil rights extends beyond the classroom. Malveaux's recent publications include co-authorship of *Class Actions and Other Multi-Party Litigation; Cases and Materials*, 2nd edition, (West Group, 2006); "Statutes of Limitations: A Policy Analysis in the Context of Reparations Litigation," 74 Geo. Wash. L. Rev. 68, (2005); and "Fighting to Keep Employment Discrimination Class Actions Alive: How *Allison v. Citgo*'s Predomination Requirement Threatens to Undermine Title VII Enforcement," 26 Berkeley J. Emp. & Labor L. 405 (2005).

Gene B. Preuss: Gene B. Preuss is Assistant Professor of History at the University of Houston-Downtown. He obtained his BA and MA from Texas State University-San Marcos, and his PhD from Texas Tech University. His academic areas of focus include oral history, ethnic and immigration history, and childhood and education.

Loren K. Redwood: Loren K. Redwood is a PhD candidate in American Studies at Washington State University at Pullman, Washington, where she is a graduate teaching assistant in the Department of Women's Studies. Her current research centers on issues of gender inequity, immigration and migration, labor exploitation, and tourism as a force of internal imperialism. She is currently writing her dissertation, which focuses on the civil and human rights abuses of immigrant women laborers in post-Katrina New Orleans.

D. Osei Robertson: D. Osei Robertson is Assistant Professor of Government in the Department of History and Government at Bowie State University (MD). He obtained his PhD in Political Science from Howard University and holds a Masters in International Affairs in Development from Clark Atlanta University, as well as a BA in History from Xavier University (LA). Professor Robertson's research interests include black politics, comparative politics, research methods, and political theory. He has presented papers in these areas at the annual meetings of the National Conference of Black Political Scientists, the American Political Science Association, and the Association for the Study of African American Life and History.

Gregory D. Squires: Gregory D. Squires (squires@gwu.edu) is a Professor of Sociology and Public Policy and Public Administration at George Washington University. Currently he is a member of the board of directors of the Woodstock Institute and the Social Science Advisory Board of the Poverty & Race Research Action Council in Washington DC. He has served as a consultant for civil rights organizations around the country and as a member of the Federal Reserve Board's Consumer Advisory Council. He has written for several academic journals and general interest publications including *Housing Policy Debate*, *Urban Studies*, *New York Times*, and *Washington Post*. Recent books include *Privileged Places* (with Charis E. Kubrin; Lynne Rienner, 2006) and *There Is No Such Thing as a Natural Disaster* (with Chester Hartman; Routledge, 2006).

Alan H. Stein: Oral Historian Alan H. Stein is Head of the Louisiana Division and City Archives at the New Orleans Public Library. Mr. Stein is a regular contributor to *Oral History Review* and editor of the *Northwest Oral History Association Newsletter*. He has conducted many oral history interviews, including interviews of Jewish veterans for the Library of Congress and the Institute of Southern Jewish life.

Dr. Dawood H. Sultan is Assistant Professor of Sociology in the Department of Sociology, Anthropology, Social Work and Criminal Justice at the University of Tennessee at Martin (UTM). He is also a UTM College of Education and Behavioral Sciences Distinguished Research Scholar and is a member of UTM's Graduate Faculty. He is published in a number of refereed journals. He received a BSc (Honors) in economics from the University of Gezira (Sudan), an MA in development studies from the University of East Anglia (England), and a PhD from Louisiana State University (LSU).

Lynell Thomas: Lynell Thomas is Assistant Professor of the Department of American Studies at the University of Massachusetts-Boston. She earned her PhD at Emory University, and her academic areas of focus are African American Studies, American Literature and Culture, and New Orleans Culture and History.

Ronald Walters: Dr. Ronald Walters is internationally known for his expertise on the issues of African American leadership and politics, his writing, and his media savvy. Walters carries three major titles. He is director of the African American Leadership Institute and Scholar Practitioner Program, Distinguished Leadership Scholar at the James MacGregor Burns Academy of Leadership, and professor in government and politics at the University of Maryland. Walters received his Bachelor of Arts degree in History and Government with Honors from Fisk University (1963) and both his MA in African Studies (1966) and PhD in International Studies (1971) from American University. In 1984, Walters served as deputy campaign manager for issues of the Jesse Jackson campaign for president, and in 1988, he was consultant for convention issues for the Jackson campaign directed by former Secretary of Commerce Ron Brown. Dr. Walters is the author of over 100 articles and eight books. His book, *Black Presidential*

Politics in America (SUNY Press, 1989), won the Ralph Bunche Prize, given by the American Political Science Association and the Best Book award from the National Conference of Black Political Scientist (NCOBPS). *Pan Africanism in the African Diaspora* (Wayne State University Press, 1993) also won the NCOBPS Best Book award. His most recent books are *White Nationalism, Black Interests: Conservative Public Policy and the Black Community* (Wayne State University Press, 2003) and *Freedom Is Not Enough: Black Voters, Black Candidates, and American Presidential Politics* (Rowman & Littlefield Publishers, Inc., 2005).

Michael White: Michael Gerard White, PhD holds the Keller Endowed Chair in the Humanities at Xavier University of Louisiana. Since 1980 he has also served as a Professor of Spanish and as Instructor of African American Music. The New Orleans native has been active for over thirty years as a jazz clarinetist, composer, bandleader, lecturer, and consultant. He has numerous publications and has appeared in various documentaries and feature films. White tours regularly and has performed on over three dozen recordings. He is the recipient of several honors and awards, including the French Chevalier of Arts and Letters.

Index